Library of Davidson College

THE PHILOSOPHICAL REFLECTION OF MAN IN LITERATURE

ANALECTA HUSSERLIANA

THE YEARBOOK OF PHENOMENOLOGICAL RESEARCH

VOLUME XII

Editor:

ANNA-TERESA TYMIENIECKA

The World Institute for Advanced Phenomenological Research and Learning
Belmont, Massachusetts

THE PHILOSOPHICAL REFLECTION OF MAN IN LITERATURE

Selected Papers from Several Conferences Held by the
International Society for Phenomenology and Literature in
Cambridge, Massachusetts

Edited by

ANNA-TERESA TYMIENIECKA

*The World Institute for Advanced Phenomenological Research and Learning
Belmont, Massachusetts*

D. REIDEL PUBLISHING COMPANY

DORDRECHT : HOLLAND / BOSTON : U.S.A.

LONDON : ENGLAND

Library of Congress Cataloging in Publication Data

Main entry under title:

The Philosophical reflection of man in literature.

(Analecta Husserliana ; v. 12)
Includes index.
1. Philosophy in literature—Addresses, essays, lectures.
2. Phenomenology and literature—Addresses, essays, lectures.
I. Tymieniecka, Anna Teresa. II. International Society for Phenomenology and Literature. III. Series.
B3279.H94A129 vol. 12 [PN49]   142'.7s   81–12010
ISBN 90–277–1312–X   [809'.9338]   AACR2

Published by D. Reidel Publishing Company,
P.O. Box 17, 3300 AA Dordrecht, Holland.

Sold and distributed in the U.S.A. and Canada
by Kluwer Boston Inc.,
190 Old Derby Street, Hingham, MA 02043, U.S.A.

In all other countries, sold and distributed
by Kluwer Academic Publishers Group,
P.O. Box 322, 3300 AH Dordrecht, Holland.

D. Reidel Publishing Company is a member of the Kluwer Group.

All Rights Reserved
Copyright © 1982 by D. Reidel Publishing Company, Dordrecht, Holland
No part of the material protected by this copyright notice may be reproduced or
utilized in any form or by any means, electronic or mechanical,
including photocopying, recording or by any informational storage and
retrieval system, without written permission from the copyright owner
Printed in The Netherlands

TABLE OF CONTENTS

ANNA-TERESA TYMIENIECKA / The Theme   ix

ACKNOWLEDGMENTS   xi

ANNA-TERESA TYMIENIECKA / Introductory Essay: *Poetica Nova*   1

PART I

PESSIMISM AND OPTIMISM IN THE HUMAN CONDITION: THE LIMIT SITUATIONS OF EXISTENCE

A. M. VÁZQUEZ-BIGI / The Present Tide of Pessimism in Philosophy and Letters: Agonistic Literature and Cervantes' Message   97
DAN VAILLANCOURT / Is the Battle with Alienation the *Raison d'Être* of Twentieth-Century Protagonists?   125
STEPHEN NATHANSON / Nihilism, Reason, and Death: Reflections on John Barth's *Floating Opera*   137
HUGH J. SILVERMAN / Beckett, Philosophy, and the Self   153
GILA RAMRAS-RAUCH / Protagonist, Reader, and the Author's Commitment   161

PART II

THE HUMAN SPIRIT ON THE REBOUND

DAVID L. NORTON / Nature and Personal Destiny: A Turning Point in the Enterprise of Human Self-Responsibility   173
VEDA COBB-STEVENS / *Amor Fati* and the Will to Power in Nietzsche   185
STEPHEN L. WEBER / Jorge Luis Borges – Lover of Labyrinths: A Heideggerian Critique   203
JAMES B. SIPPLE / Laughter in the Cathedral: Religious Affirmation and Ridicule in the Writings of D. H. Lawrence   213
REINHARD KUHN / The Enigmatic Child in Literature   245

PART III
THE GIFT OF NATURE: MAN AND THE LITERARY WORK OF ART

JEFFNER ALLEN / Homecoming in Heidegger and Hebel	267
MARGARET COLLINS WEITZ / Pastoral Paradoxes	277
JEAN BRUNEAU / Reality and Truth in *La Comédie Humaine*	291
MARIA DA PENHA PETIT VILLELA DE CARVALHO / Man and Nature: Does the Husserlian Analysis of Pre-Predicative Experience Shed Light on the Emergence of Nature in the Work of Art?	301
ALPHONSO LINGIS / The Language of *The Gay Science*	313
HENNY WENKART / Santayana on Beauty	321

PART IV
GENESIS OF THE AESTHETIC REALITY: WAYS AND MEANS

BEVERLY ANN SCHLACK / Heroism and Creativity in Literature: Some Ethical and Aesthetic Aspects	329
ROBERT MAGLIOLA / Permutation and Meaning: A Heideggerian *Troisième Voie*	353
EUGENE F. KAELIN / Criticism of Robert Magliola's Paper	385
VEDA COBB-STEVENS / *Mythos* and *Logos* in Plato's *Phaedo*	391
RAMONA CORMIER / Sartre's Conception of the Reader-Writer Relationship	407
CHRISTOPH EYKMAN / "Souvenir" and "Imagination" in the Works of Rousseau and Nerval	415
RICHARD T. WEBSTER / Intuitions	429
CHRISTOPH EYKMAN / Eidetic Conception and the Analysis of Meaning in Literature	443
ANNEX: Programs of the Conferences from Which the Papers Were Selected	455
INDEX OF NAMES	481

For Ludwik, Izabela and Jan Tymieniecki Houthakker
to whom this existential inquiry owes its impetus and inspiration
and for the pessimistic thinkers of future generations . . .

Professors Eugene Kaelin, Marlis Kronegger, Anna-Teresa Tymieniecka, and Stephan Erickson

ANNA-TERESA TYMIENIECKA

THE THEME

POETICS AS THE TWO-FACED MIRROR OF THE HUMAN BEING IN PHENOMENOLOGY AND LITERATURE

It seems that in our time and age philosophy and literature manifest themselves in an almost interchangeable way. Not only do they encroach mutually upon each other's seemingly unique territory, but they intermingle to the point of appearing indistinct. Existential thinkers like Gabriel Marcel, Ortega y Gasset, Unamuno, and Buber, as well as the most recent writings of the French philosophers recur to the literary type of experience as reflected in life on the one hand; whereas, on the other hand, the contemporary novel's experiments with philosophical issues and the contemporary drama of Ionesco, Beckett, Brecht, Sartre, Camus are philosophical to their very core. The theories of literary criticism are themselves either sectors of philosophies or philosophical theories themselves.

Why then the need to investigate the relationship between philosophy and literature, a need so imperative that, to respond to it, I have founded the first international society (ISPL) devoted to such an investigation?

No doubt the relationship between these two creative manifestations of the human being is, by its nature, most intimate and complex. The closer they come together in practice, the more confused, due to this closeness, become their respective aims, meaning and significance, the more acute demands are put by them upon the reader or spectator for their understanding and interpretation. Paradoxically, in order to be understood to their profound common core, they have to be severed from each other, kept clearly apart, and envisaged each in its own right. It is only within a clear distinction of their respective aims and means that their profound union may be understood. It will shed light upon both.

But first of all, by the simultaneous mapping of their respective fields and by disentangling the knots of their ultimately shared concern, we might reach the very soil from which their common roots, and for that matter the roots of *what makes man a human being*, spring. *To be* human and *to become* human through man's own *heroic efforts* are one. Poetics is the study of this generative progress through man's own modeling of himself.

Where else can this self-modeling progress be most directly exposed to light than in the creative unfolding of man's existential endeavor?

Phenomenology is particularly apt to investigate this relationship. There is a recurrent need in literary appreciation to come to terms with the philosophical reflection present in the literary text and unmistakably constituting its deepest sense. There is also a wealth of insights that literature could yield for the philosophical reflection proper. The bridge of "interdisciplinary communication" is then to be carefully built between the two.

To throw such a bridge and to elaborate it carefully is the aim of the International Society for Phenomenology and Literature whose selective work is presented in this volume. It opens with a new *poetics* conceived at the heart of the phenomenological investigation. Indeed, not only is phenomenology uniquely apt to interrogate the springs of the *Human Condition* in its philosophical perspective, but being attentive and receptive to all types of human experience it is also uniquely apt to investigate the Human Condition as it appears within literary creativity.

So far the influence of phenomenology upon literary criticism has been incontestably strong and extensive. It has made the critic attentive to beware of twisting literature to psychologistic, ideological or metaphysical biases assumed as a preconceived world-view. It has, however, by the same stroke overemphasized the autonomy of the literary text and chained its interpretation to its static form.

It is again up to phenomenology to break also this last dam which hinders the flux of the powerful "sacred river" of literary inspiration as it infuses its significance into the text through the creative process.

This "sacred river," to use Coleridge's expression, of higher inspiration for both literary creativity and specifically human existence will be brought to its own in the new *Poetics*. We will find in it all the already validly established approaches to literature falling into their place like streamlets fall into the main stream.

ACKNOWLEDGMENTS

Three conferences of our International Society for Phenomenology and Literature were needed to bring forth this spectrum of issues as a basis for the meeting of philosophy and literature. The idea of an intrinsic dialogue between the two was, and still remains, so radically new within American culture and scholarship that it was necessary to mobilize latent forces to initiate it.

Our Society, which has set for itself this goal, was organized in Quebec in May 1975 as the International Association for Philosophy and Literature. The founders were Anna-Teresa Tymieniecka in cooperation with Professors Claude Levesque (University of Montreal) and Waldo Ross (McGill University); and they were later joined by Jean-Jacques Demorest (formerly of Harvard). The Society's present name better corresponds to its orientation and to the work of the World Phenomenology Institute (of which it is an external branch).

In recognizing the support we have received, we owe thanks, first, to Mrs. Brita Stendahl, a fine Kierkegaard scholar, for her valuable help in organizing the first conference, and to Dean Krister Stendahl for the hospitality shown by the Harvard Divinity School. We must also thank Professor Christoph Eykman, from the Department of Germanic Studies of Boston College, for giving an abode for our second conference. Since then our research seminars and conventions have taken place at the Institute in Belmont, Massachusetts, or at our location at 4 Ash Street in Cambridge.

We owe many thanks also to the members of the Society's board: to Professor Veda Cobb-Stevens (Lowell University), Michel Masson (The Chinese University of Hong Kong), Cécile Cloutier (University of Toronto), John Hoberman (Harvard University), Eugene Kaelin (University of Florida, Tallahassee), and Robert Magliola (Purdue University). Special recognition is due to our esteemed collaborators, Professor Benjamin Schwartz of Harvard University and Professor Beverly Ann Schlack of Manhattanville College, for their constant encouragement and profound understanding of our cause. Professor James Iffland of Boston University deserves our thanks for his invaluable editorial assistance. Miss Rebecca Ramsay deserves my personal thanks for her expert typing.

THE EDITOR

ANNA-TERESA TYMIENIECKA

Poetica Nova

THE CREATIVE CRUCIBLES OF HUMAN EXISTENCE AND
OF ART

a treatise

in

THE METAPHYSICS OF THE HUMAN CONDITION AND
OF ART

PART I

The Poetics of Literature

To Philip Suter, the noble friend
Belmont, 1978/1981

A.-T. Tymieniecka (ed.), The Philosophical Reflection of Man in Literature, 1–93.
Copyright © *1982 by D. Reidel Publishing Company.*

TABLE OF CONTENTS

I. THE CREATIVE QUEST: DESTINY AND "HUMAN REALITY"
   1. Destiny and "Human Reality" as the Common Focus of Literature and Philosophy
   2. The Creative Context of Literary Work and the Privileged Vantage Point of the Creator at the Apex of His Existential Situation
   3. The "Creative Context", Its Coordinates in the Human Condition, and the "Locus" of the Work of Art
   4. Literary Creativity and the Philosophical Coordinates of Human Existence: Human Reality and Destiny
   5. Human Reality and Creative "Ciphering"
   6. The Human Destiny
   7. Poetics as the Exceptional Access to the Situation of Art in the Human Condition

II. THE SUBTERRANEAN QUEST OF THE "REAL" AND THE "LASTING" IN THE HUMAN CONDITION: THE EXAMINATION OF THE CREATIVE GROUNDWORK OF MAN'S EXISTENTIAL MEANINGFULNESS AND HISTORICITY
   1. The Real and the Lasting: A Subterranean Quest
   2. The Radical Self-Examination and the Current of Man's Life
   3. Two Modes of Systematic Examination in the Natural Current of Man's Life
   4. The Common "Coordinates" of These Two Methods of Examination: The Irreversibility of Formative Advancement
   5. The Countercurrent of Reflection on the Past, and the Perspectives of Interpretation
   6. The Identity of the Interrogating Self

III. ON THE SEARCH FOR THE CRUCIBLES OF POETICS: THE BASIS OF LITERARY "SCHOLARSHIP" IN THE PHILOSOPHICAL INTERPRETATION OF THE CREATIVE ORIGINATION OF LITERARY ART
   1. The Investigations of Literary Scholarship and Their Philosophical Involvements
   2. The Epic Coordinates of Man's Self-Interpretation in Existence
   3. Drama as Orchestrating the "Lived Engagement" between the Created World of Fiction and the Individual within the Real Life-World
   4. The "Lyrical Cadences" of Pure Poetry and of Human Existence

IV. THE RADIUS OF POETICS IN PROBING INTO THE EXISTENTIAL SIGNIFICANCE OF LITERATURE AND HUMAN EXISTENCE
  1. Poetics within the Entire Spread of Literary Investigations
  2. The Poetics of Literary Creativity and Its Proper Place in Literary Studies
  3. *Poetica Nova* as Metaphysics of the Human Condition

V. THE EXISTENCE-SIGNIFICANCE OF ART AND THE "DECIPHERING" OF LITERARY WORK
  1. The *Existence-Significance* of Art
  2. Interpretation: The Creative Coordinates for the *Existence-Significant* Exfoliation and "Embodiment" of the Literary Text
  3. The Threefold Pole of the *Creative Context*

NOTES

CHAPTER I

THE CREATIVE QUEST: DESTINY AND "HUMAN REALITY"

1. DESTINY AND "HUMAN REALITY" AS THE COMMON FOCUS OF LITERATURE AND PHILOSOPHY

Philosophy and literature share their innermost concerns. Throughout history poets have thrown the searchlight of intuition upon the deepest feelings agitating the human heart, and writers have attempted to pursue, and to reveal, the most subtle threads of human destiny. On their side, philosophers have attempted to explain and clarify the balance between feeling and reason, that is, they have tried to reveal the very condition upon which the mysterious and puzzling meanders of this destiny depend. Beginning in antiquity with Hesiod and Homer, Parmenides and Heraclitus, we witness a complete intermingling of both, in which either the philosophical reflection is made the heart of the poem, or, as in our time, philosophers like Nietzsche, Jean Wahl, Unamuno, and Croce, make poetic sensibility and intuition the heart of their philosophical quest.

In fact, it is the human being, the enigmatic situation of his existence and his destiny, that is the common ground of both endeavors; to evoke, reveal, and interpret it is the aim of the one, while to grasp it in its causes and to explain it is that of the other; yet not all that this common groundwork embraces enters into the concern of one or the other or both.

It is the attitude toward man and his humanity that differentiates them radically. In a revolt against classical philosophy — which he denounced for its approach to the human being — Ortega y Gasset writes that "Because the adjective *humanus* is as suspect as the abstract substantive *humanitas*, humanity," it is neither humanity nor the human element distilled by philosophy which may pertain significantly to the human being. Neither a simple or complex adjectival form nor that of an abstract substantive, it is the *concrete* human being which matters: "Man in the flesh, the one who is born, suffers and dies — first of all dies — the one who eats, drinks, plays, sleeps, thinks, and loves; man whom we see and whom we hear, a brother, the true brother."[1] Whereas there can be no doubt that literary effort aims at evoking the living man in the full spectrum of his vibrant unique concreteness.

To clarify ever anew their respective aims, ways of procedure and role which their products play within the human existence, leads back, first of all, to this common root in order to bring back to light the lost dimensions; but, ultimately, *it means to elucidate the hidden springs of the human being, of his existence and of his actual, as well as "possible," worlds.*

(1) The writer does not need philosophy to perform this task, but he can hardly avoid wondering and meditating about the way he goes about his work, about the relation of his work to his own existence and, further, to human existence in general, its significance for the meaningfulness of his life and for the reader whom he addresses. Thus, he is infringing upon the territory of the philosopher's reflection. The philosopher, in turn, does not need — as often in fact happens — to be a literary man nor to seek material for thought in literature; and yet in order to grasp the essence of the human world, of the human being and of Nature and to elucidate their role in human existence, he must seek to obtain not only a vast spectrum of observations about them — first in the natural, naive way, then by interpreting all the scholarly insights — but also to gain access into their experiential genesis and their role in the origin of the human life-world — this in the same way the poet does.

The philosopher and writer part with respect to their direct aims and means of presentation: the one seeks to transmit information by establishing a reliable body of knowledge, the other seeks to conjure forms, such that, by means of his work, his deepest convictions may be brought to life for other human beings.

(2) Again, both the philosopher and the writer, especially the poet — like T. S. Elliot or Paul Valéry — are tormented by the question of "germs of reality," the ultimate source of human endeavor and experience. Both the poet and the philosopher, whether in a poetic or a scholarly way, attempt to penetrate to the hidden springs of experience — what the phenomenologist calls "pre-given," "pre-objective," or "pre-predicative"; to designate it, poets use an infinite variety of expressions: it is enough to mention "the sacred river" of Coleridge, "the Real" of T. S. Elliot. They both share a similar metaphysical inquisitiveness.

(3) The very sense and validity of their respective enterprises raise a question about the reality of life, human existence, and life-experience as such and becomes an object of wonderment concerning their relation to the philosophical investigation, on the one hand, and to the literary transmutation of life, on the other hand. Whether we consider it as an intellectual reconstruction, or a mediation between the bubbling sap of life and the uniform structures of the human mind, or even if we assume the literary

work, as well as the philosophical theory, as an independent autonomous reality, as is the fashion today, we hit at the main issue, the core of the matter, where the knot lies to be disentangled – *the knot in which human sensibility, emotional powers, cognition, and the swing of the inventive and imaginative virtualities tie together: the creative faculty*.

(4) Hence, on the one hand, philosophy and literature share not only the question concerning (a) the proportion in which they stand to reality and (b) the ways in which the ultimate Real, which the most profound of them seek, may be captured by literary or theoretical means, but also with the ways in which it may be communicated to the recipient. That is the concern with expression, on the one hand, and with interpretation of the work, on the other – its modalities and the postulates, which an ever-changing cultural progress sets forth. Finally, there remains the unchangeable framework of the human faculties on one side, and the correlative system of possible forms of objectivity on the other side.[2]

(5) This brief outline discloses the points at which philosophy and literature intersect or interweave in the vast intellectual enterprise, or at which they seek to accomplish the same task, culminating in the inward realization of the creator/recipient's being, by exploring the human virtualities within the universal framework of existence, albeit in different ways.

First of all, however, it has to be pointed out that *the human significance* of literature lies in *enhancing the life experience toward its reaching the higher level of significance of HUMAN EXISTENCE*. Whether it be by way of raising aesthetic experience from life-subservient functions, or by permeating with lyrical sap the dry commonplaces of life, or by disentangling the knots of the common doom of mankind, or by spurning our courage to resist the pressures to submit to or to compromise with passions menacing our lifehood and rise to heroic attitudes and bold deeds, literature, maybe all art, for that matter, is essentially the *self-interpretation in existence* of the creator and *self-illumination in existence* by the recipient.

In this summary of the five major concerns shared by literature and philosophy, I have sketched the four major lines for the adequate understanding and enjoyment of literary works: (i) psychology, sociology, and history of literature; (ii) metaphysics and ontology of the literary works; (iii) theory of literature with the study of the creative process at its center; (iv) literary criticism focusing upon the modalities of the creative-context, and its inerpretative "deciphering," interchangeable with philosophy of literature, and the appreciation of the way in which literature performs its *role of life-significance in man's self-interpretation in existence*.

(6) The key to the unfolding of this understanding of the radius of *existential* spread which literature projects on the wings of philosophy is well hidden in the *crucibles of the human being and his existential condition*. Hence, both the philosophical and the literary, or for that matter, all artistic endeavor, spring from the *creative function* of man.

Only by asking, "What makes literature an artistic literary work, and philosophy, philosophy?" can we penetrate into the meaningfulness and root-linkage of the various aspects of questions concerning the interpretation of the work of art. This root-clarification in which philosophy and literature are interchangeable is *Poetics*. Nevertheless, a doubt arises: "Can philosophy, opposing the imaginative molding of thought, and hindering with its conceptualism, its technical language, the evocative mode of expression and conjuring up of the 'bodily presence' of its objects, ever come to a point of complete convergence with literature?"

In countering this doubt, I ask: "Is there any intrinsic reason for philosophy to ape science by squeezing its vibrant reflection into a pseudo-scientific language and mostly irrelevant scholarship, thereby hiding by its volume the scant truly philosophical import?"

Cannot and should not philosophy, like literature and science, have a language of its own? Was it not philosophy that in the Occidental culture generated science? The great original thinkers of Occidental antiquity until Aristotle, as well as those of other cultures, had each invented his own language to suit his reflection. Is it correct to conclude that, since Aristotle's mingling of scientific and philosophical inquiry, a strictly rational standard of scholarship entered into philosophy, dominating it to this day, and that therefore this mingling is perfectly justifiable?

With the change in the procedure of science itself – which has moved away from conceptual rationality to "intuitive operationalism of the workings of Nature" – this ideal of pseudo-scientificity of philosophy appears not only obsolete, but arbitrary and a distortion of its ideal. Moreover, since our phenomenological investigation attempts to give to each type of experience a forum to voice its case in its own way, we will for this crucial reason forego the singling out of one language for articulating experience; we will find room and proper voice for all of them instead.[3] In this we also intend to avoid the trap of another type of prejudice stemming from the theories of language themselves.

In what follows, let philosophy speak in its own language!

2. THE CREATIVE CONTEXT OF LITERARY WORK AND THE PRIVILEGED VANTAGE POINT OF THE CREATOR AT THE APEX OF HIS EXISTENTIAL SITUATION

The Literary Paradox and the Writer's Dilemma[4]

The literary work — unlike a historical account of facts or a scientific theory, which attempts to explain them — is in a peculiar position with respect to reality, the world, and the living man: a position which is rooted in at least five paradoxes:

First, although it does not primarily present human reality, yet even when it seems to do so, it does not represent the reality of the life-world, of the *real* world. Nothing concrete may be properly learned from it about reality.

Second, everyone agrees that in spite of the fact that the literary work is a work "of fiction," that is, its content offers the reader a "fictitious reality," even if conceived in its furthest remoteness of objective reality of life, some point of reference to objectivity is unavoidable and indispensable. Without any link to it, there would be no "work of art," but an absurd, insignificant concoction of elements.

Third, the presentation of a fictional reality, of a unique kind of "world" and "life," carries an innermost concern with the "truth" of the objects represented, things, beings, life, experiences, images, sounds, etc.; and yet the views, ideas, convictions which the work of fiction entails cannot have any claim to truth, not only concerning their adequacy with corresponding objects, events, experiences to be referred to within the real world, but to anything outside the fictional reality they incarnate. It may then appear that the unquestionable function of the literary work of art is exclusively aesthetic: to bring forth "beauty" in whatever shape it be — from the radiant to the gloomy and from the beautiful to the ugly — and to generate the aesthetic experience in its entire spectrum — from the mere aesthetic pulsation to the Aristotelian completion and total consuming of it in a catharsis.

And yet the greatness of a literary work, or of any work of art, for that matter, is ultimately crowned by its profound commitment to bring forth and to *communicate a message about life and human destiny*. What can it mean?

Fourth, although, as we know, the work of art originates within a personal human psyche and is created by the deep personal efforts, struggles, sacrifices, and joys of the poet, writer, painter, sculptor, etc., yet once it is produced, emerging from the recesses of his flesh and spirit and carried in its long meandering of progress by the sustained power of his will, it detaches itself from its maker. Severed from the innermost ties that sustained it in its making, it stands out independently on its own — and beyond the reach of its creator.

The question then arises, "What are the reasons for this paradoxical situation?" Further, "What is the *locus* of the literary work — or of a work of art as such — with respect to the real world, the human life-world and individual lives of spectators or readers for whom it is intended and of the creator as a human being?"

Finally, through its creator, who is an integral member of the life-world of the given period — with its own distinct social face, social style of life, cultural spirit laden with viewpoints, specific biases, dominating trends — the work of art is not only forged at the cross currents of these prevalent tendencies and fashion, but as the inheritor of traditions which have prepared it. And yet, it is recognized as a work of art inasmuch as it challenges the current fashion, the "already seen." To be recognized as a work of art it has to stand up against all which would mean a "reproduction" of the already established. In short, it is original and new with respect to what it originated from and what it enters into as a "new" element of the life-world.

As the five paradoxes announced above indicate, with respect to its meaningfulness, significance, purposefulness, generation, and type of existence, the literary work is situated at the junction of several *antennae*, knotted by its maker in a novel way; it is taken from the natural threads by means of which its maker expands within the human world as a person, as well as a member of a cultural and social world with reference to other human beings, and, in this unique relationship of their possible sharing in his creative undertakings. The literary work — like any work of art — can be explicated as such only with reference to these *antennae*. They circumscribe a constant, although fluctuating, *Creative Context* in which the work of art remains as a chrysalis hidden within her cocoon and may be transformed into a butterfly opening its wings each time the entire network of interrelations of this context opens to the light of appreciative attention.

When we have to face the crucial question, "What makes a work of art a work of art?" we do not plunge into a mental labyrinth in which we would find ourselves inevitably "encircled," attempting, as Heidegger does, first, to explain art by the work of art and simultaneously attempting to explain what a work of art is by the essence of art itself. In contrast, in order to answer by way of a *truly foundational* inquiry the question, "What makes the literary work a work of art?" and "What makes a literary work a *literary* work?" (and not that of history, scientific theory, journalism, or popular writing), we *have to establish its bearings within the creative context.*

We will come to it later in focusing upon the first question we raised, that of the origin of the work of art, which will bring us naturally to the core of our intent.

The Generative Locus of the Work of Art

The work of art is in course of being established in its own *existential circumference* progressively throughout the entire progress of the *creative process*. Thus, it is, as indicated above, initiated and carried out within the *existential circumference* of its maker. In order to investigate its bearings which outline the radius of its *existential circumference*, we have to inquire into the *conditions* of its origin. That is, we must start by mapping the ground of the *existential locus of its origin*.

In fact, the literary work — the work of art as such — is generated through decisions taken by its creator upon controversial points which he receives as challenges and which occur in his life-world. First, the writer, as every creator, engages in the creative endeavor and carries it on, from and within his own existential situation in the real world, from and within his specific personal position in and toward the life-world, that is, the cultural and social world in which he lives. His "creative urge" stems, as I have emphasized elsewhere, from a "dissatisfaction" with the form, shape, and fabric of the given phase of the life-world. Naturally, he is caught in the life-world routine, which is distinguished by its given forms, schemata and principles of validity established by the society for the sake of everyday survival. Yet this "creative urge," that is this need which surges from his innermost being to manifest, express, bring forth something which he feels to be the pervading element of existence — which gives it unity and significance, but which is not only obscured by the life currently lived but even lost out of sight in actual life — prompts him to discover or invent shapes and forms, novel and unprecedented, reaching to the very quintessence of understanding the essentials of "human reality,"[5] that is, to forge modes of expression, build ways of bringing them together in a strategy that enables this quintessence to come forth and show itself. This strategy aims at bringing the final expression of this quintessence of human "truth" to the life-world; an expression more adequate than those within which the artist "naturally" stays intertwined and ingrown. He has to establish this new "universe of man" with his own means and according to his own blueprint.

To follow his urge for discovery and invention, the artist or writer is carried by the nostalgia of the yet unknown, hidden, unprecedented, that is, of the "new." To follow it, he has to suspend the validity of the established shapes, forms, matter — setting them aside as the "old." Being one flesh with it, still he has to extricate himself from the dominion of the established universal life-current he passively flows with, from its style, conventions, prejudices, codes of values, etc. He has also to invert within his own life flux the current of existence; while continuing to function as "everyone else," he has nevertheless to transform his hitherto single-line oriented conscious soul-body-spirit system of meaningful purposefulness, into a *double-line* complexity. That is, his naturally unfolded life rhythms, cadences of moods, spectra of qualities of experience — with their intrinsic principles of light at the one and darkness at the other extreme — values of instances and of duration, and, ultimately, evaluation of life's objectives and means without losing their now sharply evident survival status, have to be as if "doubled" by other rhythms, cadences of moods, qualitative complexes of experience, criteria for evaluation, etc. The second line with which the specific process of discovery will proceed takes precedence over the first within the creative life of the artist.

He does not then cease to participate in the mainstream of life. It means only that, on the one hand, he stays and maintains his existential roots solidly in the ground of the life-world — and of the natural world — and, on the other, simultaneously inwardly inverts the course of its mainstream: he flows with the mainstream of life linking it and trimming, advancing, and going backward as befits the creative process — that is, he incessantly counters the flux of the mainstream he flows with by his own line of creative endeavor; and he has constantly to make a pact with the exigencies of current life.

This situation in and out of the life flux, within and above the current life-world with its horizons of past and future, gives the artist — the writer — an *extraordinary vantage point upon human existence*. It also situates the work of creation emerging from it; it delineates the arteries through which its *locus* within the radius of human existence and, primordially within the *Human Condition*, is set.

This vantage point, in which all the strings of intellect and imagination shaping the work of art, and all the forces and dynamic resources of the human being prompted by the subterranean stream of spontaneity, flow together, shows how the work of art is generated and formed at the cross currents of human dynamic and constructive virtualities: those which bring man into existence among other living beings and things; those which allow him to unfold a specifically gregarious existence projecting a human life-world; and last — but not least — those which prompt him to launch this existence toward a "transnatural telos."[6] The work of art approached within this context, which its origin circumscribes, reveals the hidden springs of the Human Condition. And it is only from the perspective of this condition that we can ask, "What is art?" But it is also through the investigation of its unique nature which crowns the hierarchy of human virtualities that we may dig into the roots of what makes a man "human"! The *creative context* in which both meet is the ground from which to investigate them both simultaneously.

The recognition of the extraordinary vantage point of the artist, of his condition as creator, which is expressed by his work, may well be the reason that, although the literary work, because of its fictional nature cannot claim truth validity, yet we naturally seek for, and naively see, in the literary work a "testimony" about life, man, personal experience, and the meaning of existence. It is the proper understanding of its role that is here at stake.

3. THE "CREATIVE CONTEXT," ITS COORDINATES IN THE HUMAN CONDITION, AND THE "LOCUS" OF THE WORK OF ART[7]

Let us now briefly sketch the arteries of reality, through which the forces in the generative womb of the work of art in progress flow and the antennae upon which its generative process is suspended, as they all converge in the extraordinary position of the creator.

How is this vantage point attained? How is it sustained? What does it encompass?

The situation of man as artist is, to start with, supremely precarious, as I have shown in the first part of my treatise. His entering upon a creative endeavor means a sustained effort to raise himself above his "spontaneous passivity" — as Husserl would say — of everyday existence, into which, however, he is constantly and repeatedly drawn by this co-natural "passivity" of the flux of natural existence. His efforts have to be constantly repeated, so that he may keep his accounts with everydayness quickly settled, and so that he may maintain himself in the *other* course which must incessantly remain the focus of attention.

As the *crucibles of the creative activity*, we have then, first, the artist's *natural* subjectivity; his natural subjectivity, not self-enclosed, but with its entire endowment of the sensitive, emotional, passionate background of the soul, and which has as its axis the intellectual apparatus of reason, mind, and will. It encircles and remains, on the one hand, embedded in the individual's bio-psychical arteries of functioning, which along with the organs of the senses, attentiveness, and absorbtion of the exterior and interior fluctuation of life we call 'body'. And on the other hand, it is expanding in all its functions toward other beings and things, and retains its own functioning in an essential interplay with elements of this existential radius.

The interplay through which the very existence of the natural human being delineates itself attains at the conscious level, at which the human "subjectivity" emerges, an "objectifying" tendency. What we call "subjectivity" and "objectivity" are conceptions pertinent to the phase in which man is working out his *existential* course within his *existential* network. They do not oppose each other, as is often naively asserted. On the contrary, they mutually imply each other — or are correlative — within the common network of the *meaning-bestowing existence of man*, who lives both as an experiencing "subject" and as an acting "objective" being. Reworking the material brought forth in vital interplay, through the conscious apparatus, the living individual establishes his *existential radius* in the form of an "objective" life-world. In this distinctly organized "objectivity" emerge other subjects, not as separate beings, but as *vital counterparts* of the human psychic universe. They form together an intersubjective-objective world. (The emergence of the human being as a subject together with other subjects is a "natural," passively accomplished course of Nature).

But as I have mentioned above and shown in my previous analysis of creativity the "creative urge" surging up within the natural human subject within the intersubjective-objective world, runs counter to the passive flux of natural existence. Natural "subjectivity," as the meeting place of the forces of Nature — interwoven into the design and the intersubjective life-world and expanding the subjective self, while embroidering its own pattern with the self as a center within it — is merely a mighty root from which the creative self can grow. Indeed, as I have already pointed out, the *creative urge*, which springs from this *Elemental* soil,[8] has to be taken up by the conversion of a self from a "passive" position in the natural flux to that of a *creative agent*, who will take the lead and bring it into his very own mold of vital force; and this not by the force of the passive flux of the *natural* subjective unfolding, but through *his very own Spontaneity* and upon *his very own decision and responsibility*.[9] Always gathering the fuel and receiving the prompting from the very pulp of his natural life-world and subjective endowment, he differentiates himself from its pattern, from its laws and entanglements. In his quest for raw material and guidelines to "cipher" reality in its "true" face, and to give it expression in a work of art, the creative agent works in the midst of the varying aspects of the life-world. He investigates them in a closely scrutinizing process. He seeks and discriminates, selects and rejects, tries for fit, in the last instance, he *invents*. There is nothing concrete at stake, but it belongs to the significance of this fulfillment that from his seemingly self-encircled territory he is aiming at entering the intersubjective life-world with a new form.

From the inside of the inherited culture which forms the core of his personal self, he wants to effect its transformation, and in this effort he joins the past both to the present phase of the life-world and to the future. He thereby encroaches upon the otherwise discreet phases of history, fusing them together in his own work; he encompasses the present, which he denies but in which he feeds himself; the past, which he attempts to leave behind but of which he partakes; and the future, which he wants to shape, which is not yet there, and which in fact might never be such as he envisages it. In the uniqueness of his vantage point, in the midst of the *Elemental* forces with or against which he struggles — and which he puts simultaneously at a distance — not only all the elements of his condition flow together into his endeavor, but in his strenuous exertion to hold together all their strings — none of which, alone, can support him — he generates his work as the quintessence of the historico-cultural unfolding of the *human endeavor* as such.

The return upon itself of the *creative urge*, the return upon the subject has, as hinted at above, a unique significance. Not only does the creative effort stem from the very center of the person's subjective-objective intent: it also carries with itself a *crucial significance* for its bearer. In fact, as pointed out above, the creative urge crystallizes into a pursuit of its realization at the same time that it is the most significant fulfillment of the person carrying it out. In its return upon the subject it is directed above its immediate objective toward the subject's highest fulfillment: the ultimate human *telos*. By this orientation the creative endeavor is suspended also upon the *coordinate* of the "transnatural destiny"[10] of the human being. It then reaches beyond the immediate course of the cultural evolution of mankind, toward the highest vocations which human beings and humanity can propose to themselves.

It is at the cross currents of these lines of human subjective-objective condition, those of the intimately personal and interpersonal (social), and finally those of the individual consciousness and the spirit of the human person and of humanity, that the work of art finds its place.

Emerging at the cross currents of vital forces with spiritual ones, currents and countercurrents of the present tide with the remnants of the distant flux of the past, nurturing the onset of the future of the singular being crisscrossing the network of the infinite variety of beings around him, (along with givenness in its transient essence and nostalgia and striving for lasting fulfillment) the work of creation surges, drawing all their juices into its crystallized unique entity. Thus it processes them through its molds in progressing toward a novel form. To understand it, we have to reach deep into its coordinates which project the *creative context*.[11]

4. LITERARY CREATIVITY AND THE PHILOSOPHICAL COORDINATES OF HUMAN EXISTENCE: HUMAN REALITY AND DESTINY

Originating from a common source and also unfolding on the wings of the *creative function* makes philosophy and art (literature in occurrence) progress — albeit along two different lines — along two contrastingly formulated yet united, equally essential, quests. The one is the query about the significance of human life plunging to its deepest in the quest after human destiny. The other is the simultaneous search for the bearings of this significance with respect to the *life-individual-world* framework at large — that is, within the *human reality* — as well as the means and ways to retrieve this significance from the abysses of the Human Condition and *give it expression*.

This double interrogation runs through the entire span of human existential origination down to the threshold at which the *Elemental virtualities* of the human individual are in transaction with the circumambient elements of his situation. The attempt at understanding the *sense of human existence*, which we may identify with its current understanding of the "meaning of life," interlaces and intertwines within the radius of this interrogation, which comprises all that is human, along with the equally extensive deep-seated field of man's manifestations in its origins, that is, means of expression, sources, rules, and principles of what has been currently called the "richness" of the human meaning projecting into the creative constraints leading to "human reality."

Great works of art stand out in their mastery in presenting original and exemplary interpretations of "human reality" by bringing to light the otherwise unobservable elements and forms of human existence that carry its sense. We need only mention Dante's *Divine Comedy*, in which the great drama of existence is played in its full panorama of expressive modes in accordance with the full expanse of the sense of life carried by the blind passions of greed, merciless enjoyment of power, and instinct, from the "infernal" darkness up to the "celestial" purity of heart, worldly innocence, and selfless devotion.

We will introduce the study of poetics by this two-fold entrance and begin by delineating the respective lines of the quest of literature and philosophy as they may be discerned in their otherwise joined endeavors by focusing upon their originary points of junction.

5. HUMAN REALITY AND CREATIVE "CIPHERING"

We may say that in the final analysis, both philosophy and literature are concerned with reality. First, I use this term to denote everything which confronts the thinking man: Life, world, Nature, cosmos, his own being, and his own contingency — which he cannot accept. Yet it is not in a vague but rather in a quite precise way, so precise that it is almost impossible to grasp, that the two different channels of reflection — literary creativity and philosophical discovery — approach reality. Tentatively I suggest that both of them at the very threshold between a quest and its terminal, creative endeavor, pry into the maze of infinitely vast and ever-expanding field of *man-and-the-human-condition* with two intentions: (1) to synthesize provisionally this infinitely vast field and (2) to discover its first and last principles. Its synthesizing endeavor in inventing the SIGNIFICANT KNOTS OF THE SPECIFICALLY HUMAN REALITY and its principles leads to the discovery of relevant units of SENSE and to express it with the CIPHER. We recognize in the first the concern of man undertaken in both literature and philosophy with the HUMAN DESTINY. In the second we find the concern of literature with "MIMESIS" in its full-fledged meaning, and in philosophy with the MODES OF RECONSTRUCTION in a peculiar apparatus of these principles in order to account for the givenness.

Reality and the coordinates of the Human Condition: the "ciphering"

In fact, the pivotal point of wonderment for philosophy in all cultures has been, is, and remains the question of "reality."

Whether it is the metaphysical marveling of the Pre-Socratics about the composition of Being and origin of beingness or Eastern thinking about the essential and "ultimate" reality of man's "real" existence, which in innumerable variations pervades the history of philosophical thought, the philosopher seeks criteria to distinguish the various dimensions of reality revealed to man. The quest of the arts expresses concurrently the deepest concern of the human being to find the lasting element within the fleetingness of life and to reveal it in a uniquely apt form.

Although wonder about "reality" did not emerge before man asked this question in self-reflection, that is, before he reached the stage of lucid awareness in his self-interpretative existential progress and sought for "certitude" within the whirl of life-options competing for his attention, yet he has been "dealing" with the problem of reality and certitude during the whole of his organic and vital *self-interpretation in existence.*

He first had to discriminate within his vital situation elements of constant intercourse and to identify a relatively stable and ordered life-world. In order to accomplish this, he had to differentiate, evaluate, and select from among the options confronting him during his mute but concrete, operative commerce with life-conditions, that is, he had to absorb and reject, drawing freely upon or defending himself against the outward elements of his *life-situation* which intrude upon him from the "outside" or emerge from his "inward" realm. Running through man's perception, appreciation, willing, and thinking — indeed, through the entire span of his vital-existential route — is the need for certitude — a need which leads to a lucid discrimination of the various types of the "real" into the imaginary, fictitious, hallucinatory, ephemeral, and the like (all designated as 'unreal') and the 'real,' regarded as the basic characteristic of 'facts,' as the core-confrontation with 'reality.' This need for certitude in assessing the degrees of reliability in 'real life' becomes more urgent with the unfolding of progress in man's lucid awareness, and eventually leads to the question, "Am I myself real?"

The puzzlement about the specifications of the criteria ranging from the cognitive determination, originary evidence, intrinsic support to its extrinsic factors of the 'real' as opposed to the 'nonreal,' leads into the fundamental philosophical issues of man's commerce with Nature, of the nature of his life-world, of the *ultimate telos* of his fragile existence as well as of the *elementary* that is pre-objectified, pre-thematized, pre-established existential conditions of his beingness. It presides over the profoundly searching questions concerning the functioning of the human faculties in their three-fold commerce with: (1) the elements of the vital ground from which it stems and unfolds; (2) the conditions of the self-initiated human agent to accomplish his highest specifically human aspirations; and (3) man's intersubjective situation within the life-world. In short, it opens up the basic issues that emerge with respect to the Human Condition, with all their intricacies.

Art, scientific invention, outstanding technical creations — these are all products of human creative activity and confront us in a face-to-face encounter with reality in its various dimensions, enigmatic appearances, and presentations. No wonder that this encounter brings philosophical reflection into the crux of his concern with reality and the "Real."

It is precisely at the heart of the artistic endeavor that the struggle with the issue of reality takes place.

Does not all artistic creativity focus on the issue of "mimesis" and its right proportion with respect to the "Real"? All art refuses the 'brute' reality of everydayness — even when it is taken as the 'object' of its 'presentification' — the rightful mode of "presentifying." Aristotle binds the issue of "mimesis" as its right proportion to "reality" by insisting, first, that the personages "presentified" in a tragedy should be "either greater" or "smaller" than the people in real life-drama, but cannot be presentified as just so.[12] Second, he considers it an aim of the artistic accomplishment to give the human being a reprieve from the drudgery of his everydayness and his real life, which hold him captive, by enabling him to shake off their overwhelming absorbing power, be it of the time of the dramatic sequence on the stage. This shaking off of the factual reality gathers momentum in the full range of aesthetic experience until it culminates in the moment of catharsis in which the real and the fictitious are suspended, and the human being reaches deep down into the crucibles of his experience and its conditions.

It lies in the existential efficaciousness of art that it neither produces nor reproduces factual reality; when it "presentifies" it, it accomplishes it in its own peculiar way. Each of them brings forth for the reader, spectator — performs virtually one or several of its various dimensions in a befitting way. Here lies the crucial issue for art — for literature, in particular — concerning "mimesis" and "reality". If art can accomplish this mediating role with respect to reality, by keeping it at a distance, maintaining a mid-air position, while at the same time plunging deep down into it, it is because of its intentional existential status. The intentional existence of the work of art has been universally recognized.

Yet it is one thing to account for the "presentificational" neutrality and another to explain how such a "presentification" may be accomplished.

The intentional network of human consciousness spans through the entire conscious spread of human functioning and is the factor to which we undoubtedly owe the "objective" and "intersubjective" meaningful ordering of the factual reality of human life. And yet when we approach the work of art, and especially the work of literary art, its specific art-significance consists in the fact that, as pointed out above, it dissects this intentional true-life presentation of reality; it makes it disintegrate and from its dismembered forms transforms their elements, transmutes their signals, transfuses their sap into new channels, so that the presentified "reality" is a *recreation* of the actual one; and considering the radical originality of the form in which it is presentified, it is a new "creation." From beneath the factual reality of the human life-course, the fruit of the intentional constitutive function, lurks the fruit of man's *creative function* which frees itself from the constraints of life-subservient rules, and imagines and invents forms to respond to its aspirations which are higher than mere survival: forms of the specifically "human reality."

In fact, how could the same functional system account for the stereotypic and automatized repetitive order and for its transformation from the break-in of a novel spontaneity?

How could the same function account for repetitiveness and radical renovation of the vital world into the form of a cultural world?

Of all the arts, it is literature which offers us the most acute case in point. Whereas painting presentifies the "reality" of colors and shapes, surfaces, designs, perspectives, etc. in colors and shapes, and sculpture presentifies the "reality" of shapes, surfaces, volumes, in volumes, surfaces, shapes in "the same" modalities, literature, by contrast, presentifies factual, imaginary, etc., dimensions of events, feelings, passions, plots, inward developments, aspirations and strivings of the human beings on the one hand, and all the aspects of life, world, society, Nature, on the other and yet not in their actual modalities suspended on the axis of actual space and motion. Literature does it by the medium of "meanings" with which this fleeting, irretrievably passing flux of the actual present is being endowed by the human being establishing: (1) its life-significance and (2) its intrinsic order.

It is through this system of meanings that the otherwise insignificant and anonymous flux is lifted from an unavoidably menacing oblivion into a relative stability, and simultaneously retrieved in its salient life-significant points as a network shared in principle by all human beings. Signs, symbols, meanings and concepts, in their intrinsically founded concatenations as the basis for expression in words and carriers of messages within words syntactically phrased, form the human language with which literature presentifies the inherently alien stream of life-world-Nature. On the archeology of language and its lengthy genealogy from the mute and concrete vital process of the human individual till its literary expression enough has been written. Our present arguments begin where this genealogical searth gets stuck by stressing the seemingly autonomous status of language or by showing its complete arbitrariness with respect to the human functioning at large, before it can give an account of the crucial factors of its origin. This origin, as I have shown elsewhere, must indeed be sought at the pre-intentional level, but it needs a hitch by means of which we can salvage from the helplessly dispersed elements of a disintegrated intentional consciousness the crucial instructions concerning the generative forces, factors, and intermediary links between the signals of the vegetative and organic order (as well as the human *subliminal virtualities*) and the SENSE which man forges from them in order to establish the meaningfulness of his own life-course.

It is at this threshold of human functioning, which already marks man's unique creative faculty as well as initiates his *self-interpretation in existence*, that literary art and philosophy most intimately meet each other in the issue of "mimesis." This issue has received a most extensive treatment in literary studies as well as in the theory of literature at the borderline of philosophy.

Leaving aside the concept of "mimesis" as it has been and is extensively discussed in the theory of literature, I propose to understand it as the *transposition of the "signals" received by our Elemental virtualities from the preconscious givenness into the intersubjectively conscious network of the human by endowing them with a specifically human SENSE*.

Philosophy (and phenomenology for that matter), has vainly pursued the intentional spread of the human faculties constitutive of experience in the search for its origin. The thread of the constitutive-intentional faculty of man breaks down precisely at the junction between the furthest outposts of its reach and the pre-constitutive, pre-conscious, pre-intentional. Yet it is precisely at this junction that we may expect the human virtualities instrumental in objectivizing to set work. At this level it is decided by their differentiation whether the *Elemental* or *Subliminal* virtualities, as I call them, will orient the human faculties to being stereotypic life-world, or whether the human being might at the same time unfold a line of *his very own*. This is where the decision falls whether man will be confined to establishing a pre-traced meaningful system — "meaningful" by its relevance to the survival values — or whether he might also simultaneously attempt to invent — free from submission to life — *his own* meaningful system according to the SENSE he projects HIMSELF.

It is on this point that art, or all creativity for that matter, challenges the common ground-reality of factual existence which is the object of the philosophical reflection in the intentional constitutive investigation; taking as its foundation the perception of factual reality in the common sense experience and thereby taking the line of the stereotypic life-subservient function, it encounters the philosophical quest at the point of junction between the furthest outposts of the constitutive faculties and the pre-constitutive, pre-conscious operations. Where philosophical analysis of intentionality stops, the *creative analysis* of the double, literary creative reflection and philosophical interpretation of its findings, begins.

In his vital progress the human being individualizes himself and grows by means of the constant confrontation with factual reality, following the preestablished route of life — anonymous and stereotypic. But in the work of creation, each confrontation with reality, the reality that the literary work presentifies, is unprecedented and unexpected for the reader or spectator. It is not one more mesh of a continuing chain of encounters — like those of his current life business — but a shock breaking the current life chain.

It is a shock of a different "world," or "human reality" — "different" because it is established through a meaningful system based on principles of significance which are different from those of the life-subservient intentional meaningfulness.

Our double-faced plunge into the creative forge wherein literary creativity, the "ciphering" of the new "sense," occurs will show us, first, the *creative coordinates of the human condition* as tied into the *creative context*, in such a way that the new *sense* may be triggered. Second, we will see what SIGNIFICANT KNOTS for a new *meaningfulness* the literary creativity offers to the human being.

6. THE HUMAN DESTINY

Looking anew and from a distance at the essential objective of literature as a human enterprise, we cannot fail to see that it is the *elucidation of human destiny*. From beneath a vast range of interests, most of which refer more or less directly to insatiable wonderment about the enigma of the human struggle to survive, the writer is prompted to create a seemingly fictitious world of an epic, drama, novel, or poetry on the one hand, and the reader or spectator is urged to interrupt for a while the current of his ceaseless life-concerns in order to plunge into the "reality" of art as into another universe, on the other.

Whether it be curiosity about the reasons and causes underlying events in human life and culture in general or the personal need to understand the entanglements of human emotions, feelings, and passions in order to draw from them a parallel with our own, it is first of all interest in *human destiny* which reveals itself in the lives of outstanding personalities, social groups, or nations as well as in the life-course of the "common man," embedded in the passing stream of the progress of humanity and forgotten there, having failed to inscribe himself in its course.

Whether Homeric poems, Scandinavian sagas, Romance or Slavic epics and tales, or the epics of the modern fiction like Manzoni's *The Betrothed*, Galsworthy's *Forsyths*, Thomas Mann's *Buddenbrocks*, Sienkiewicz's *Potop* and *Ogniem i Mieczem*, Grabski's *Saga o Jarlu Broniszu*, etc., it is the pursuit of the consistent course followed by a person, family, nation, or social group, the ideals which it has maintained, the aspirations which it has nurtured and sought to actualize within the topsy-turvy course of events, the haphazard turn of affairs of the social and political order, long, devastating wars and short respites of peace, which underlie the motivation of the writer and the reader to awaken and sustain their interest. It is as if, relishing the melancholy feeling of the irrevocable passing of things and the uncontrollable disasters and defeats in which the individual projects, nostalgia and strivings lie (the profile of the very consistency of the life course terminate) living through them in the disguise of another person — all of them, fictitious but revived through our experience — gives man an appeasing satisfaction not to live in vain. Indeed, the writer, reader, or spectator penetrates with his own feelings into the fierce struggles which human beings fight within their inner self as well as among themselves, and into their heroic deeds (following them with fascination as they unfold in the orbit of human affairs in reaching or not reaching fulfillment), and transmutes with the creative imagination the general into the uniquely personal, the repetitive into the singularly meaningful, lifting the human existence from the engulfing morass of a trivial and almost anonymous survival to that of the *human destiny*. In witnessing the human existential plight in its transfigured sense — human life endowed with a *new significance* — man exults in his own victory over the cyclic, ephemeral existence of other types of organic beings by giving that victory a *sense* which, in retrospect, comforts him for what he might otherwise consider "failure," and what, by playing its part in this significant streamlike effort, turns into a "noble defeat." It allows him to see retreats as "elevated sacrifice," and otherwise trivial successes as "victories" over blind, overbearing forces. This new significance sheds a light illuminating the entire unfolding of human life.

But the roads for accomplishing this conversion divide. The writer attempts to wring from the ever-recurring trivia of life this new significant course and make it plausible to the universal reader through literary form. Since as creator he will remain an experiencing person through the prism of whose experience life is seen, his literary quest is ultimately identical with his most profound personal quest: he seeks, through the form of existence he thus forges, to salvage *his own* existence from the oblivion of the meaningless current of events. The literary autobiography of a writer is experienced by him undoubtedly as his *existence*, no matter what the *life* circumstances were. Some classic examples of this identity of the writer's ultimate intentions with his own quest after fulfillment as creator are Goethe's trilogy of *Wilhelm Meister*, Thomas Mann's *Doktor Faustus, Lotte in Weimar, Tod in Wenedig* and Virginia Woolf's *Orlando*. In his major work, *The Undivine Comedy*, the celebrated Polish Romantic poet Krasiński relates the self-examination of his hero, a poet, and as a poet makes him assert: "You invent a drama."

The reader seeks to interpret the fictional life events, personages, conflicts, joys, etc. with reference to *his* feelings, observations, views, and the experience of *his own* life, *his own* aspirations and disappointments, and to understand and interpret them by confronting *his* situation, measuring *himself* with its challenges as well as with personages who emerge as the exemplary models, illuminating his own existence with their light. Indeed, all literature, insofar as it centers upon one or other fragmentary element of life experience, and takes sides upon some singular issue concerning it, contributes to the great controversy lying at the heart of the literary, as well as the philosophical, enterprise — a circumstance which can be formulated as the question, "What is the lasting element in the fleeting essence of life?" or "What is the human destiny?"

Philosophy and literature meet upon this crucial ground of shared interest, but their respective tasks separate. Although issues directly confronting human destiny belong to certain sectors of philosophy, yet it may be argued that at the origin of Western philosophy this question remained in the shadow and has only marginally entered the core of philosophical questioning. Whereas the major philosophical question which, differentiating the endeavors of both yet bringing them also together — precisely upon this basic ground — had, from the very origin of Western thought throughout its history, their center in the search for the *ultimate condition of the human existence within Nature, the universal order of being and the specifically human virtualities*.

In the literary search there comes into focus the struggle against the forces of this condition, which appears as in principle congenitally threatening the human being's vital progress and adverse to the human being's higher aspiration and desire. Literature tries to elucidate the tactics, strategies, efforts in man's taming of the hostile elements toward establishing with them a kind of truce in order to attempt the fulfillment of his strivings during this period of harmony with Nature's rules; whereas the philosopher attempts to dig deeper into the nature of the human being with its actual and virtual potentialities, dynamics, forces, powers of reason and reflection in their interplay with the forces and rules of the circumambiant organic and social world.

The philosopher seeks to discover and establish the capacities of man which allow him to occupy his place among the other living beings and forces of the universal order and cosmos and the point at which they determine his own. Finally, the philosopher seeks to find the reasons for and to determine the limitations of man's nature itself, balancing his irremediable contingency and transiency with his innermost desire for self-perpetuation. He ponders man's extraordinary perdurance through time, while he cannot preserve even a precious instant. He marvels about the consistency of man's being "himself" through temporal phases in spite of the fact that he can neither control the outside world— to steer an even course as he projects it — nor maintain an inward course of personal attitudes and conduct along a line which he proposes to himself throughout the entire course of his life until it is mercilessly broken in a definitive and final way. For the philosopher, man is a paradox which he endlessly toils to understand: everything seems to indicate that he is merely a play of circumstance, and yet he is the only living being who is essentially capable of self-orientation and self-direction in shaping his existence and devising its significance!

No wonder that, as I have discussed elsewhere, this paradoxical status of man makes the probing into the sense of his effort crucial and simultaneously controversial for the writer, philosopher, and reader/spectator alike. It projects an aporetic that is in its very nature philosophical, and pertains directly to the stand taken upon principles of life and of conduct. It centers in the dilemma about the *significance of human* existence with the perennial controversy, which I have formulated as the controversy about PESSIMISM VERSUS OPTIMISM IN THE HUMAN CONDITION at its heart.[13] Emerging from considerations of the human paradox, it centers upon *human destiny* and the direction of the propensity to approach it, a direction which is "optimistic," if it stresses the human power of self-determination reaching beyond the constraints of the circumambient world as well as his vital-condition with respect to elementary nature or its opposite; while it is "pessimistic" if it inclines to view the human being entirely as the fruit of his vital situation and as conditioned by them in all his manifestations.

In this issue come together all the questions which man, an experiencing and thinking being dissatisfied with simply observing and stating the facts and states of affairs, and prompted by the trickiness of his survival-situation to ponder, to deliberate, to decide — which, presupposes weighing the forces of circumstance over against his own and measuring his by the yardstick of the resistance of the matter with which he proposes to deal and of the task at which he aims — has been asking himself and will continue to ask, albeit in changing formulations, throughout the course of humanity. In the last instance they all put at stake the human individual's self-directing potential, his self-identity, and his autonomy with respect to the current of anonymous organic development and within the neutral current of the circumambiant world. At every turn of the human life-course, the decisive question is whether man has been the controlling factor in the situation and play of circumstance; whether it has been his very own, lucid decision that has been resolved — even whether he has been in a condition to deliberate and choose — or whether he has been confused by his own unclarified tendencies, blinded by the persuasion of others, or too weak to reach and take a stand.

At every turn of human existence the unavoidable question is that of the scope and limitation of human discernment and penetration. Each culminating point of man's life-course that may decide whether man sinks into the anonymity of the common doom or rises to the level of *inscribing his existence within the universal human history*, carries within itself the challenge to a heroic undertaking, the decision to sacrifice his narrow self-interest, to bold ventures, as well as an aspiration or conviction that goes against all the already established canons, that nothing could stop being put into realization. In every significant stream of existence, we witness extraordinary intuitions, instances of indomitable determination — unforeseeable and unforeseen — that go counter to all rational law and petty calculations and scheming that enter into ordinary life strategies.

As I have emphasized in the above-mentioned study, this controversy lying at the heart of the literary endeavor and constituting its most significant factor cannot be decided legitimately by means of the reflection carrying the literary work itself; its nature and its roots lie in the line of philosophical reflection.

Thus, the investigation of the paradoxical and seemingly dilemmatic existential situation of man comes down to the puzzling question: "Is there within man's entire spread of functioning (including the biological functions promoting the natural life-course of the living individual) a function or a faculty, which could effect a break from the bonds of the natural life's functional system, allowing him to exercise a 'free' choice and decision; or does he exclusively follow with a 'spontaneous,' to use Husserl's term, passive/active enacting of his life-course entirely relative to his vital progress, the gregarious life conditions?" Husserl himself had a decisively "optimistic" bent: to account for the human freedom expressed in the cultural and spiritual manifestation of man, within the life-world, he differentiated within man, first, body, flesh and soul. Then, while acknowledging the dominion of Nature reaching up to the soul, he brought out the ethical decision surging at its borderline as operating the turning point at which the freedom of the spirit enters.

And yet, Husserl was not in a position to explain how the ethical decision that was supposed to revert the course of nature may occur. We have to question the free decision and pursue it all over again.

It is, then, in this crucial issue of human destiny that philosophy and literature fundamentally meet. Each of them emphasizes a different line of reflection, quest, aim; yet even when they lose track of each other, they both remain, ultimately, interdependent. Each of them, in emphasizing its own special interrogation or scrutiny and in using its own means to its task, remains implicitly concerned with the other. We have to find their common territory or the point of their origination at which also the condition of human freedom resides. But there is a difference between the concern of the literary writer, theater director, actor, or spectator and the philosopher with his concrete data of experience in the concrete "expression" or "presentation" of life, and the quest for its significance. In the search for this *thread* running through the disparate and transient instances that would constitute the lasting direction — which, whether they result in disaster or in triumph, make man greater than himself — we seek to "invent" or "discover" meanings fit to express their "sense," that is, to "interpret" them.

The interpretive quest of the writer, as well as of the critic or of the perceiver, is not literary inspiration alone; it is the philosophical reflection *par excellence*. It is the reflection of the philosophical nature, drawing upon all the resources of the philosophical mind, that accounts for the significant reflection which transmutes animal vegetation into human destiny — the destiny of both the individual and of the nation or humanity.

Literature depicts man caught in the human drama, whereas philosophy — taken for itself — studies its "mechanism"; to see it and to understand it, we need literature; to find its significance for its universally human validity, that is, in reference to the basic factor of life and existence, we need philosophy.

7. POETICS AS THE EXCEPTIONAL ACCESS TO THE SITUATION OF ART IN THE HUMAN CONDITION

The major question which imposes itself when we confront the literary work of art in its interpretation is, "How do we distinguish the different points of view present in the direct study of literature so that we may become aware of the biases with which the work itself has been approached?" Each distinct point of view might have done some justice to it and at the same time have distorted it in some respect. "How can we grasp all the wealth of ideas, experiences, intuitions in the overall aesthetic experience which comprises them all?" Willi Fleming tells us that the *Literaturwissenschaft*[14] approaches the literary work in its basic material from two radically opposed viewpoints: (1) as a work it treats it from the perspective of historical inter-influences, connections, and development, (2) then, in its composition, as poetry (*Dichtung*). He claims the necessity of a systematic literary science (*Literaturwissenschaft*), that is, of *Dichtung*.

Undoubtedly the literary work imposes itself first in presenting a self-enclosed universe that challenges us by its strangeness to our everyday life; which invites us to enter it, and which awakens our attention. If, receptive to its alluring emergence within our own streamlike everydayness, we respond to it, seeking reluctantly at first for a point of familiarity with our own experience, and then progressively blending the personally known with the new possibilities offered, we will be drawn into this universe, first, just enjoying it, then attempting to interpret it by seeking a universal significance. But even this strictly subjective personal encounter with the literary work in enjoyment does not occur in a cultural void. It is true that the reader discovers by himself the work in the modalities in which it presents itself. Seeking to relate to it, to find its relevance to his own life-world, to understand its general meaningfulness, he becomes a "critic" of the work. But as spontaneously personal and unintentional as his interpretation of experience, forms, and meaning of the work may be, it partakes of the prevailing scholarly or "professional" trends in literary criticism, and of those who claim the authority in their milieu and period. Literary criticism involves consideration of innumerable questions which in their treatment are related to many issues and approaches of other types of investigations concerning the literary work. If we follow Northrop Frye's approach in his *Anatomy of Criticism*[15] to the study of literature as a whole we would differentiate what he calls "an incomplete attempt, on the possibility of a synoptic view of the scope, theory, principles, and techniques of literary criticism," and the criticism itself as lying at the center of the organic — although discrete — unified picture reflecting the essential aspects of literature. There are, beyond those reasons we have mentioned, others with a stronger impact on this position. As Frye emphasizes, poetry, drama, etc. are "speaking" to the public through the critic in various powerful ways, and to a certain degree the literary work — and for that matter any work of art — is enhanced by this "criticism" and comes alive because of it.

The last question which the encounter with the literary work stimulates is: "What, if any, is its significance with respect to the great concerns of the human being and his existence?" With respect to these three questions several approaches have been identified, each emphasizing its own specific role. Yet what is crucial in the enterprise of appreciation seems to me to be the philosophical significance of the literary work with respect to the predicament of *human existence*.

CHAPTER II

THE SUBTERRANEAN QUEST OF THE "REAL" AND THE "LASTING" IN THE HUMAN CONDITION:

The Examination of the Creative Groundwork of Man's Existential Meaningfulness and Historicity

1. THE REAL AND THE LASTING: A SUBTERRANEAN QUEST

I have already introduced the notion of the 'cipher'[16] as the means by which the human being, defying the rigidity of automatized processes by which he establishes the vital meaningfulness of his circumambient life condition in terms of objects and processes of the circumambient world, reaches deep down into the fountain of life itself and, by his own means, endows the *elemental* stuff with a significance uniquely his own. I consider the "ciphering" as the passage between the organic functions in submission to Nature's demands forging his vital existence and the specifically human *creative function* meant to wring from the pre-predicative, pre-thematized, pre-structured ground a unique response to the human nostalgia for the *Lasting* and the *Real*. In fact, metaphor and symbol, which are a further belaboring of the *meanings* of the cipher, have to be provided first with its significant material. It is the cipher which at the point of germinal meaningfulness of the *human* reality, where man's vital function in its condition encounters his specific quest for the *Lasting* and the *Real*, expresses the generation of the thus created *original sense*. Yet it is not at a specific point of man's vital course that there occurs this extraordinary break in man's innermost desire for *his own existential meaning* in the anonymous process of his life.

It originates as the result of a long *subterranean quest* to which man subjects himself. This grab bag origin of man's creative endeavor, which finds expression in works of creation as well as of his *self-interpretation in existence* to which this first offers "knots of significance," opens a unique access to the *human condition*.

I will first discuss the *subterranean quest* in its basic form. Then, we will see how literary creativity may offer "significant knots" for *man's self-interpretation in existence*.[17]

2. THE RADICAL SELF-EXAMINATION AND THE CURRENT OF MAN'S LIFE

It has been said that man can be defined as the being who interrogates himself about himself. But how is it possible that in the flow of his occupations (so strictly and faultlessly bound one to the other by concrete logic), even though everything appears "natural" in the established order of things, one could begin to doubt everything and to wonder — "and afterwards, what is there ... ?" From this question arises *sua sponte* the irresistible desire to examine the individual and universal condition of our personal life, a life apparently so well-organized, so sure and untroubled. This desire moves manifestly in the inverse direction, against the ongoing structures of life, springing as they do from a natural spontaneity. The inverse desire sheds doubt on laws and forms. We used to adhere unreservedly to our actions; they carried us with them along the stream of life. Now a self becomes detached and departs from the current. This self no longer identifies with its own acts; instead, it endures them. The self subjects all our paths and customary involvements to critical examination, and they are all found to be narrow, futile, and banal.

This impulse to question ourselves about life, the world, and ourselves is manifestly one of the most profound acts, because it brings into play all our faculties, and our entire being concentrates there. Is this examination imposed on each of us? Does it belong to universal human nature, or is it indeed the result of particular personal developments?

It could be said that we all come into the world with a germinal restlessness, a need to better understand our existence. "This mystery at the bottom of our being, that we have no understanding of our fate," as Valéry puts it, is in fact our greatest treasure. One can see it emerging from the background of our life. We nourish and cultivate it in different ways. That nature is generous in this respect becomes clear. But with most people, the disquietude which asks, for example, "And what good is all that?" — becomes dormant and remains hidden in the familiarity of natural life, in the joy of living, under the exuberant waves of emotive life. With other people the disquietude allows itself to be exorcised by means of fearless initiatives which keep proposing new tasks, new aims. With still other people, the same restlessness is placated by the consolation of religion, or of an ethic or a philosophy of life.

In the case of certain personalities (Kierkegaard's, for instance), this restlessness arises under the appearance of an agonizing necessity to find "the guiding principle of life." By discovering this principle, one finds himself in its *truth*. "Life is like a rented, furnished house, which remains empty without this guiding principle." Without it, there is no life. It is situated outside of ordinary life, outside of the exterior conditions of existence, outside of the categories of reason and intelligence.

To find the radical bond of truth in which life as such could begin is the necessity which pushes us to the examination of familiar paths. By rejecting a life limited on all sides, dormant and anonymous, this necessity sets in motion the pressing impetus of an "exalted" existence, consecrated to the infinite and pursuit of absolute beauty, truth, the noble. Has not Socrates already affirmed that life is not worth the pain of being lived if we do not subject it to criticism? In this case it is not a question of a critical attitude nor of a method of thinking. We can indeed analyze concrete deeds, report them in universal terms, and thus raise ourselves toward the intuition of their intelligible essence and ideal structures. This can be done, however, without ever touching the pulsating depths of our existential preoccupation. The Socratic method, just as the eidetic analysis of phenomenology which has resumed it, has the great merit of unmasking our erroneous conjectures, of unveiling obstinate opinions that no facts can justify, of clarifying an underlying order which remains concealed in the confused density of facts, impressions, feelings, and judgments. In short, these methods bring about a clarification of knowledge (*savoir*). However, even if this clarification can contribute, as in the case of Socrates, toward the rectification of our judgments and conduct, the achievement remains on a strictly intellectual plane.

Indeed, the penetrating studies of the phenomenology of the lived world merely bring us the universal structures of the phenomena of life. These phenomena themselves remain unknown in their own, unique, pulsations precisely where the intimate meaning of man's authentic existence is decided. Though we are unable to deny the relations that our rational and discursive intelligence maintains with our moral attitudes, and also those relations on which the intuition of essences depends, the following observation, for better or for worse, is also true. Our moral life is not summed up in the attitudes, judgments, and states of spirit that can be justified by rational persuasion and judgment, according to an objective intellectual reflection on the "*données* of experience" held universally valid for humanity. Rational reflection does not penetrate to the deep dimension of being where our irrepressible tendencies are born, and where – amid obscure but powerful stirrings – the quality of our values is decided. These "true" and felt values, which often remain buried, can break a subterranean path and finish by prevailing over the results of intellectual reflection. The intellectual exploration of the meaning of life should only explore the intelligible universe of man. The intellectual exploration could never be anything but the first step of an inquiry really guided by our pulsed rhythms and deepest tendencies. The latter, emerging together with irresistible violence, and constituting themselves sole judges of life and death, break the bulwarks of intelligence and rational systems.

We wander about while casting our eyes sometimes on the past and sometimes on the future; the stages of childhood, adolescence, and maturity appear to us as just so many natural phenomena common to all and without particular significance for anyone. Whether a child turns out good or bad, whether he is studious or not, will doubtlessly change his possibilities for options in life. But whether it be one profession or another, one kind of success or defeat or another, we are only taking up and following in the tracks of those who have gone before, and we are only going on ahead of those who will follow after. This course, however, the way in which established theories of religion and of metaphysics may conceive of it, remains insignificant and is fulfilled in death. Everything could have been different without having changed anything. Apparently, beings are completely consumed in this anonymous wandering and, having reached their term, wonder sometimes what the meaning of it was. What do these little successes, these results which are the objects of so much effort, definitively *signify*? They are directly incorporated into the vast reservoir of collective efforts and before long are surpassed by other people. And what can be said about our failures? We believed that they had upset our lives, but they have left no trace other than the smarting memory we have kept of them. No essential link, no necessary relation appears between our desires and the objects of our desires.

Our natural impulses become reconciled somehow or other, without hoping for more than what presents itself as best at the time. Iris Murdoch makes a point of showing with insight and clarity that what we are supposed to believe, the most intimate impulses in us, our impulses toward love, are embodied naturally in a complete emotive complex which is proper to us as a species. Impulses of love sometime fasten on one human being, sometimes on another, according to the circumstances, without anything essential having been changed. Discouraged, men no longer try to understand, because these annoymous streams of individual existence do not have any meaning.

Everything could have been otherwise without making any essential difference. But in our concern to explain everything, we cannot be satisfied with a compromise. Our mentality requires examination not only of all the paths which are offered to us, but also of all the commitments, pursuits, and careers which ongoing life could suggest. And our mentality persists in asking, "And after?" This question is postponed again and again and finally remains unanswered.

Thus man questions himself about the meaning of human existence. The well-beaten paths of life, established values, spontaneous attitudes do not help us to discover this meaning. This quest for the meaning of existence — a quest in which intellectual reflection is subservient to the intuition of lived experience — shows that everything could still be different without producing any radical change whatever. No choice ever achieves a decisive and final result. Beckett's man, brooding over the workings of his past existence, only finds a bundle of functions parodied as value. Objects of desire, tasks, preoccupations are only disguises. They are as arbitrary and interchangeable as pebbles gathered in a haphazard manner to play a game of chance right on the spot. Each of these pebbles has particular traits, but they all suffice equally, and others could have been chosen in their place with very little essential difference. The chart of life indicates to us a variety of recognized routes, of "posted" roads. But this variety does not suppress the fundamental fact: it is the same anonymous route for anonymous travelers, the same rules, the same nature, and the same cyclic course of life and death.

Is organic and social existence, under the form of one civilization or another, anything but a collective work of man oriented toward the survival of the species? On the contrary, this urgent desire to penetrate the depths of our natural existence indicates a need to find a mode of living in which something *else* would be expressed. This would be something which would no longer be interchangeable, and at the mercy of circumstances, but which would be discovered and chosen in the most profound adhesion of our being. It would realize our spontaneity, our most intimate fits of enthusiasm. This path would not be defined by the practical rules of the lifestream, but by our discovery of the superior values which would illuminate our existence and open up for it a unique perspective. It calls for a *radical* examination.[18]

This radical examination needs to be investigated in its origin, that is, in its preliminary currents of interrogation to which man submits himself, in their initiation within the Human Condition and in the human faculties which perform it. Last, we will seek to elucidate its specific existential significance.

To begin, this formulation of questions appears unfounded. Why should we assume that several human faculties play a role in its process? Has the use of reason not been made for us to examine ourselves about our actions, our thoughts, our inclinations? Reason is at the service of life. But our self-interrogation is not solely the work of reason. We reflect upon ourselves on various occasions, for various reasons, and toward the various ends that life creates. Finally, self-examination belongs to the dealings of life itself: it arises at one of these moments, follows the ongoing current of life, and returns to this same current. We reflect on the interior of this current. However, among the various kinds of self-examination that can be distinguished, there seems to be one which — although appearing at one precise point in our life — does not follow it. On the contrary, this type rises up against life. It happens at the precise moment when we distance ourselves from familiarity with world and self. It happens at the sudden moment when our total adhesion to our own acts weakens; that is, when we sense ourselves no longer absorbed by our current life-course, and solidarity with this life-course breaks. Against the on-rushing current, we delineate a "radical examination."

In order to establish what a "radical examination" should be, and what its specific and privileged scope should be, we are going to sketch three other self-examinations at the interior of life's current.

3. TWO MODES OF SYSTEMATIC EXAMINATION IN THE NATURAL CURRENT OF MAN'S LIFE

Without a doubt, throughout our existence we ask ourselves questions concerning the consequences of our actions. We try to analyze our schemes and even our desires in order to reveal hidden tendencies which, when brought to light, will clarify the personal meaning of our enterprises. Moreover, we submit even our involuntary impulses and tendencies to critical examination. We are not naturally conscious of these involuntary impulses and tendencies, but such analysis will aid us in discerning the most intimate motives which animate them. This concern for understanding, which is at the base of these soundings of our ongoing existence, comes from a doubt: the doubt which appears in concrete circumstances apropos the personal validity of the course we have engaged. Thus, in diverse circumstances, we naturally have doubts about our acts, our tendencies, and the values which preside over our existential choices. In general, we doubt ourselves. It is enough that our long-sought enterprise fails; and that we clash with an insurmountable obstacle, or an obstacle obliging us to too much effort. Indeed, it suffices that our trust, placed in the beings around us, suffers deception, and we at once place in doubt the validity of everything.

Therefore, a critical examination and a suspicion accompany the life current, a current forever advancing from one event to another, from one project to another. Our life-course always advances by means of an infinite series, and through successive exploits which provoke new critical moments. Our existence refines the moments, specifies them, and renders them more universal. The doubt which accompanies our life-course, becomes as it were, an integral part of this current.

Our existence, promising but still uncertain, and open to yet indeterminate perspectives, advances ceaselessly toward ever new developments. And doubt flows in the vital bed of our life as a means of suspending, for an instant, intimacy, cohesion, the thoughtless familiarity we have with our life's course. Doubt causes us to reflect and to become conscious of our life's course, so as to better resume or even reorient it. The hidden aspects of our acts and of their possibility of succeeding in the real world (which the examination they provoke must clarify) are investigated as positive assumptions. These acts define us as real beings, taking part in the real world and pursuing a concrete existence which paves the way.

We are men endowed with certain capacities, desires, tendencies, and affective sensibilities. It is only a question of taking hold of what we are doing during "crisis" occasions. During such occasions the complications of the complete situation and our various engagements in the situation seem to reveal unknown capacities. Yet in reality our concrete, empirical, and real being — as well as the *données* in which the current of fragmented life flows — always escapes from being grasped with some certitude. We must satisfy ourselves with conjectures while admitting that in the final tally concrete man remains unfathomable, and concrete reality shuns conclusive analysis. But the fact that we do not look for a definitive truth belongs to the very essence of this doubt, and of the criticism which follows it — and finally of the purpose which animates it.

So as to pick up the lost thread of our vital dynamism and to orient it anew in a more propitious way than the previous one (which has proved to be closed, discouraging, or even impossible), it suffices on the one hand to grasp the essential, concrete *données* which connect with our felt tendencies and desires. These latter are manifested in our reactions and impulses, pushing us to assume either satisfaction, disappointment, distress, or even joy. On the other hand, it is enough to grasp some positive *capacities for* the realization of these desires in the world of which we are a part. In order that this new path may be attained in one form or another, we are seeking to determine it by the examination of the mundane "given" (whether the "given" is more or less favorable).

However, pushed to the limit, we may ask ourselves the following question when obstacles to our plans come from this world and from other beings: "Why is the world not different from what it is?" Yet even if we do ask such a question, it still remains true that we must face certain concrete features of the world and of life (which could include, after all, one set of plans and desires, or another set, or another set, and so on). Without plans and desires we should depart from the familiar framework of reality with all its rules that we accept without thinking. We sense ourselves subordinated to this framework, and "bound up" with it.

Described in the above way, the doubt which accompanies our existential course and the critical examination the doubt provokes are, both of them, instruments of *practical reflection*. They are both instruments set in motion by "life-crises." In crises of this kind, man — passively following his preoccupations — becomes disoriented. Critical examination of a crisis situation marks a turn in man's vital pursuit and its development. His pursuit must be rendered more expert at satisfying the concrete aims of the individual. However, it is not only during those moments when his existence seems "shackled" that man begins to reflect on the world, and his relations to others and himself. Clearly more general questions can come up than one provoked by a particular situation. A profound "sounding" of ourselves already occurs at the moment of adolescence, when we must find some "sense of direction" in our lives. We ask, "Who am I?" and "Where am I headed?" In our individual existences as children we submitted to and accepted certain standards, values, and lines of conduct. Now we suddenly discover that the human world of which we are a part leaps over the intimate world of the "family" which once regulated everything for us. We discover that in the large world, suddenly blowing open in our faces, previous "family" rules are not universally accepted, nor even respected. Those who used to decide for us and regulate the familial universe no longer have power over the world. An unexpected array of viewpoints appears. There is such a diversification of value and behavior that it is impossible for us to integrate ourselves into the world without taking a stand, without defining our attitudes and choices, without "falling back upon" ourselves. "But who are we?" This is the question which poses itself, since what we were is no longer valid without a concomitant definition. A task becomes urgent and indispensable: a systematic examination of all that we *find* in the world, of all that is *imposed*, and of the attitude we will *assume*. It is an examination which calls everything into question again. Its purpose is to allow ourselves to choose what we wish to be or even what we are without being aware of it. This conscientious grasp of ourselves as agents of selection, so that we are alone responsible for our existence, marks a crucial stage in our development. It is a stage in which man ceases to be simply an individual, distinguished from others only by tendency, capacity, and particular social development. At the new stage each of us rushes to form himself in a unique way as a "unique person" because personhood involves *deliberated* choices. From this time on, separated from the innocence of spontaneous integral participation in the "familial" world, separated in a way from our own being, we constantly proceed through examination of existential *données*. We examine the necessities of choice which impose themselves, and the references upon which choices hinge. In sum, as soon as the development of our "personhood" begins, our existence progresses through this development by way of *systematic examination*. Our advancement, if we must advance, demands such examination as its principal and indispensable instrument. This very examination transforms itself in its modes, means, and process, in relation to the results that the formation of personality obtains. In turn, any advance in systematic examination marks an advance in the formation of personality.

4. THE COMMON "COORDINATES" OF THESE TWO METHODS OF EXAMINATION; THE IRREVERSIBILITY OF FORMATIVE ADVANCEMENT

The properly human life thus seems to depend on two activities: (1) man's reflection on himself, and (2) an examination reaching probingly into individual universes of other beings belonging to the common milieu, "the lived world." World there then be something special in the *radical examination* we have proposed from the start? The specificity of our proposed examination is what must be defined. The two types of examination to which man submits himself seem to be differentiated by the respective conditions of their provenance. The one functions in *all* critical situations; the other operates *sporadically* in critical situations. The first is characterized by its relativity to the situation which brings it into play. Indeed, the "givens" that it analyzes as well as the problems to which its processes connect derive from the very "givens," either direct or indirect. These "givens" are manifested or even implied more by the situation than by the nature and the givens of the crisis itself. The analysis could reveal some basal features regarding our knowledge of ourselves (as we discover ourselves to be), or even concerning the permanent conditions of our existence in the world. These conditions are going to "post a guide" for an examination of the same kind when a situation of successive crisis sets in. Nevertheless, this examination will be determined from top to bottom by the concrete givens of the crises: failure of another project completely different from the first; disappointment in another love, or even a friendship on which we have founded other hopes and had invested different types of feelings and efforts.

The second examination manifests itself by a continuity which weaves its way in and out of conscious life while at the same time carrying that life. Both examinations form part of man's vital mechanism and serve his advancement, his development — while being formed and modulated by this progress and development insofar as these are profound resources.

Consequently, in their distinct nature, both examinations are forged by their connection to the *coordinates of human existence*. But what are these coordinates? It is fitting to call attention to them at the very beginning by a definition of their orientation and common scope. These two examinations present a particular model of process. They emerge in the context of the vital process which proceeds along many levels, and by means of many techniques. The two examinations make themselves known, even to each other, by their contrasting functions. Both of them unfold, however, in a series of completed and incipient tasks, from forms of life that are *simpler* to forms of life that are *more complex*, diversified, richer, and then — after a certain point — they regress. The two kinds of systematic examination are at the basis of advance on the properly human level, and their performances are inserted into this advance.

In sum, these examinations share proleptically in life's development, and in the successive and irreversible transformations of concrete existence. *Their purpose and their importance* consist sometimes in actually *creating* new stages of the development, and sometimes in simply *clearing the path*. The first sporadic type of examination — which acts to solve a vital problem, to cause concrete existence to advance to the more complex, to the more fully human life — thus serves what could be called "the natural *telos*" of man, a specimen of the kind.

The nature of incessant movement has been customarily seized "in advance" of human existence under the universal form of "time." We dare to situate these two examinations at the interior of the advance of human existence. Here we must point out their *temporal coordinates*. The systematic examination that we undertake (so as to restore a new order to a situation thrown into chaos by crisis) arises in a precise instant lived "at the present time." We adhere to this instant perfectly with our whole being, while actively concentrating there all our strengths. This instant enjoys the fullness of our attention, *the present stage of existence* which we call the "present instant."

The "present instant" of existence is at once a vital state and a privileged mode. It is privileged because it enjoys our complete adhesion: we *are*, we *exist in* the present instant.

It is fundamentally the *vital state* because it is in this fullness, in this adhesion alone that we fulfill our psycho-organic functions as living agents. The present instant is likewise that of our lived experience which, ranging over the whole of our lucidity, makes us conscious of ourselves as identical with our acts, our feelings, and our thoughts. Although the present instant of our existence is itself without continuity or duration, it is through the topical states of our entire being (always new and always renewing themselves) that this being shows itself as distinct and identical and conscious of itself *acting*. And even if consciousness of self — acting in this luminous circle of lucidity — can only function within the network of our confined existence, it nevertheless thrusts its probes deep into the reticulations of our vital system, and lets nothing escape. The individuated experience of the diverse processes of our being (a sudden toothache, for example) proves that we are not strangers to any of our functions. For this we must thank the holism of our being as a "present consciousness." Even this is only a nucleus whose carrying elements always yield before others, which in turn take the "center stage" of our attention. The links, as elements of individuated and distinct present experience, impose themselves on our attention and monopolize us. These links keep in touch with the rest of our functions, and secretly effect the rapports essential for unified experience. Such unity of experience (which edges its fleeting way through the transient present moments) can be compared to the self-identical image constituted by a succession of cinematographic images.

Nevertheless, it is thanks to those links which the *present moment* maintains with the whole functional network that our existence weaves forward. Each present moment is a manifestation of our whole being, yet monopolizes our attention in a manner so exclusive, so dominating, that we behave as if only it existed. Nonetheless, this very attention, with all the vivacity that Locke suggests, passes from one luminous moment to another, and shapes a perfect continuity. The reality that such continuity establishes is seamless, ongoing, and unified. There is thus a functional oneness and an operational continuity whereby man "in process" forms his universe around him.

All this shows that movement, and the time which measures it, are coextensive and are patterned and brought into play by *initial spontaneity*. The operations of this spontaneity are neither sporadic nor unique events. On the contrary, these operations only arise according to a constructive design, progressively deploying themselves from one present state of operation *in actu* to another which follows (the task of the first having been accomplished). Similarly, the lived experience (the level of its conscious *prise* of self, and of actions in their instantaneous present stages) is inserted into a network of progressive development in which each particular state bursts forth while transfiguring the field of lucidity. Having fulfilled its role in the process, each present stage gives way to the subsequent one which it has already heralded. However, although each "present" state loses its grip on the ensemble, the state's intensity dies down in proportion as other present states succeed each other. While fading in their vivacity and their associative value in the periodical development that they carry, the present states linger like the tail of a comet that each new present moment, while in the ascendancy, drags behind it irreversibly. Indeed, every present state — while leaving its site to the following one — has marked its role in personal development of man, and done so in a definitive way. Every state resides in the background of what we call "the past": no feeling, perception, thought, or intuition could be brought back to lived actuality again, having once lost its present force as fact. We can never restore the past moments to actuality and correct them when we believe that we have not done justice to the situation which vitalized them in the first place. *Occasions manquées* are lost forever. The advance of spontaneity translating itself incessantly into states of actuality is born at a crucial point and projects itself ahead in an irreversible manner. The current of existence produced never stops; it advances in instantaneous states, one following the other in an inalterable and irreplaceable way. As a matter of fact, although the horizon of possible deeds seems infinite, when a deed does happen, nothing can bring about its negation or turn back its advance beforehand in order to modify it or make a replacement. Concrete actuality, as opaque as it may be, thus takes an absolute value in relation to the retrograde thread of the past which it leaves behind, having laid that thread once and for all. And actuality also takes on absolute value in relation to the wide-open vista of its forward impetus, the future toward which it tends. This existential current has then a two-fold function — one *operative* and one *signifying*. Existence follows the current's spontaneous dynamism on ahead, and when critical reflection intervenes, inserts itself in the ongoing current and follows along.

5. THE COUNTERCURRENT OF REFLECTION ON THE PAST, AND THE PERSPECTIVES OF INTERPRETATION

We have previously identified the examination of the "state of crisis" with the situation of the "present state" which the crisis brings to an end, and all the while, we have considered the mutual involvement of the "state of crisis" and the "present state," for both require a clarification in relation to our "deep being," and in relation to the possibilities of the world which embraces us. Our reflection has been oriented and directed by present needs; its scope and its import have not only been determined, but have also been limited by these needs. Once our examination reaches a clarification of the givens which seems sufficient to permit us to escape from chaos, and once this clarification proposes another path for our vital growth, then the links which the givens maintain with the rest of our experience — more elaborate pursuit of these very givens, for example — are abandoned. The state of crisis caught our attention originally because our relations with other people and with the world were perhaps not such as we naively imagined them to be; these relations had been warped. However, already determined and accomplished in itself, this examination dictates an inspection of the attitude which remains veiled, hidden at the bottom of what we believe to "hold as the solution." In short, whatever may be the solution which asserts itself provisionally, some of us are of the sort which remain perplexed, and we ask ourselves numerous questions: "But what does that *signify*? Is there no *reason* to believe that the woman we love shares our affection? Then how could she ever leave us?" We might also ask, "How could this faithful friend and that associate, both so devoted, still be able to betray the common cause? Haven't they shared our own feelings?" We might even ask, "Didn't this feeling have the same meaning for them as it had for us? What then happened to cause their present deeds to belie, apparently, our solidarity? Perhaps they don't really militate against our common interest, but do they have another point of view not obvious to us?"

And thus, in order to discover the true meaning of these actions, and wishing to convince ourselves that there is just a misunderstanding and that the precious friendship can be reestablished, we cast ourselves — *à contrecourant* — into the prospects opening up before us. The lines of pursuit follow the entangled events to which the present action seems to relate. Although we have believed ourselves in possession of proof that tender sentiment toward us actually existed, did not "something" of an alteration join in? Should not this "something" — it seems to us from our present vantage point — have attracted our attention more, when it came to light "sometime" ago on the occasion of "such and such" an event?

We have been involved in a two-fold "sounding" of events, and of sentiments, attitudes, and thoughts which translate these events into their *"human meaning*" (that which relates to other people, and that which relates to us on the rebound). We see Proust describe this sounding of the past so as to reveal this "hidden" meaning in his efforts at interpreting the elusive Albertine. For have not our own feelings and attitudes contributed to the forming of "conviction" about the state of things? Since the "state of things" had its initial moment of "presence" in our own existence, we should climb back to that moment (or so we think); and we climb back to find out if, full account taken, an error was not already committed way back then — an error in our estimation of "meaningful" relations with others, the world, ourselves.

Therefore, in order to straighten out the confused events of "dense" actuality, we review the links that actuality maintains with the past elements which formed it. Instead of advancing along with actuality, we climb back "by countercurrent" toward its genesis. Are these steps less opaque in their past state than in the "present"? At first we believe, with Whitehead, that, passing from the present to the past, the elements of reality become like wax figures, stiff, and then perfectly precise and certain. By leaving the play of living forces which make present reality impossible to enclose, these elements settle into the past as into their well-determined and univocal truth. But sadly, as soon as we undertake our analysis of the inert forms which reality has left behind, we discover it is not possible simply to reclimb the chain of our present experience to find "true" meaning in the ensemble of its genesis. Each of the links that we question in turn does not itself offer any sure and precise point of departure. On the contrary, each presents a multiplicity of profiles and of shaded references which expand over our whole existence, and implicate not only past deeds but also our feeling in their regard. Indeed, no "raw" facts lie there waiting as the background of our present moment. Rather, *our feelings* toward the raw facts are what we are looking to evoke, to revive. And it is in our *present* sentiment that they vibrate as evoked. Feeling penetrates them through and through.

Thus while scrutinizing genetic links, we necessarily choose "such and such" a point of view, "such and such" a frame of reference and emotional stance. In sum, it is not a matter of seizing these links; all that we can do while asking ourselves the question of their significance, is to *interpret* them. In other words, to follow back along the chain of existence while picking up recognitions with each link seems to relate it to the incomprehensible phenomenon that we are trying to clarify. We are no longer possessed by a present fact but we "evoke" one at our leisure, as if we look at it as a pebble in the hand, able to turn it over on all sides. However, such activity is only an illusion, because to evoke a "fact" we can only make it "live again" (*revive*) like a shadow — by means of our present psychic action and by means of our present feeling. And the shadow is not empty, so one cannot "let it hang" or "pick it up"; the shadow must be filled with our felt psychic action, and structured according to its framework. The alleged "fact" rests in one affective nuance more than another, and is viewed from one point of view rather than another.

For example, we may ask ourselves the following: "How can our child seem to take a path contrary to the one he should have taken and appeared to commit himself to?" Always starting from the present given situation, that is, the "present" from which we interrogate ourselves, we reexamine the current of past life which has led to this situation without our being aware of it. We are sometimes incredulous and hope to find a repudiation of the child's present attitude; sometimes we are pessimistically resigned "to the worst," and the same deeds, colored differently by our attitude, present themselves in various forms. Nevertheless, as soon as a feeling of love arises in us, which calls to our attention some deeds of his childhood intimately fondled by our memory, we doubt everything, persuaded that this child is fundamentally as we have always felt him to be: good, generous, open, noble. An entire bundle of memories come to our attention as support.

Eventually a number of present deeds resists our attempts to save everything. But, then, must these negative deeds represent something other than the surface of the total ensemble — a badly turned corner, as it were? — so that "fundamentally" even some hopes can remain intact? And as a help to our reflection, this or that distant fact comes and seems to mask this bad turn. We resume our investigation of the ensemble biased by this distant, remembered "fact" while trying to understand how such could happen. How could this gesture, this action, this expression carry so much weight and significance? And what events, feelings, reaction could have contributed to the present evolution of his attitude? However, the question "Is he truly like this?" constantly returns, and memories with it. By means of their effective power, these memories take possession of our imagination and force us to reassess everything according to their "slant." In the light of this feeling about things, could we ever know anything precisely? On the contrary, as long as our interest remains unsatisfied, we will ceaselessly discover new elements in the past which will make the past live again while opening up new perspectives on it.

These perspectives multiply with each new feeling along the moment inspiring us, until we reach a range of interpretations of past events. Each interpretation proves capable of shedding new light on the genetic background of a particular action. Thus we believe ourselves to be approaching closer to its authentic sense. Nonetheless, we will never reach an *explication*, a *definitive* classification. This "sense" is really the mystery of the "other." But, does the "other" know it himself? Or, while trying to understand it, is he likewise lost in the labyrinth of interpretations?

Indeed, we follow this "sounding" of the past in countercurrent because it seems to us that reality is too complex, too poignant, situated as it is on the intersection of too many unknowns and uncertainties to be defined. It is in its already withered state, exhausted by this palpitation of uncertain life and exorcised of its monopolizing power, that we come to uncover the "real" meaning of the present. Above all, this revelation of meaning occurs in the light of the clarifications which emanate from the sum total of involvements. Indeed, this is how we discover the Real! Reality withdraws both from its movement in imminent action (in obscurity and relative inertia) and from the future projected before us with its *presumptive virtualities*. All reckoned, the friend leaves us, betrays our solidarity in such and such circumstances, etc.

For, while plunging into the obscurity of the retreating past, while trying to unravel what can well signify — by its results and its causes — what has already happened, we refer nevertheless to the future. But this is certainly not the real future — that is, the anticipation of what really is going to follow along the genetic lines of present existence. Rather, it is the future already achieved in the past, and the *presumptive future*, that is, what could have happened if. . . .

This presumptive future is that which in reality becomes our central interest, and consideration of it contains the affective instigator of our quest. Perhaps . . . if . . . our child had chosen as a friend an *X* whom we had suggested, then . . . . Perhaps if we had at one certain moment better understood the situation and had encouraged him instead of punishing . . . . The *presumptive future* — because it is not real, and we do not know if it has ever been a possibility (an opening for action that is not just our imagination) — remains present at each step, while showing us how everything could have been different. Creating regrets, remorse, and solace, the *presumptive future* is at once a landmark for the hypothesis of interpretation, and its object as well.

Whatever the "depth" with which we try to penetrate our deep being, clarifying the most primitive of our convictions, it becomes a matter of seizing on the meaning of factual reality (*réalité de fait*). It becomes a matter of answering the question, "What is the factual reality of ourselves and of our universe?" We never succeed in finding an answer, belonging as it would to the empirical condition of man. For examination shows that the "instantaneous present" or the "incontestable fact" is elusive in its nature first of all because it provokes us entirely without yielding space for the interval of reflection. Further examination shows that this fact itself – the "actuality" or "present" – is not in itself clearly distinct, because it takes part in a *spontaneous current, forming instantly on too many levels, and at the intersection of too many ramifications*. We can only look to it to find a rational nucleus which is approximate, abstract, and always subject to various interpretations according to the level or even the sidestream of its envisaged current. How powerless we remain before a current which is irreversible!

6. THE IDENTITY OF THE INTERROGATING SELF

While following the very evasive sequence of events in which our existence is carried on and becomes exhausted, we discover that everything vanishes without being retained. Everything escapes us; everything glides irretrievably between our fingers. Therefore, in this very fluid, elusive existence, are we not being transformed ceaselessly, at each lived instant? Does each person, while passing by, take away something of us (such as thought, feeling, desire, joy, sorrow) which can never be recovered? Does anything remain which can be called "ourselves"? Is there indeed anything which persists through these evolutions, other than an empirical conductive thread which we call the "self" (*le moi*), present in each feeling, thought, conviction, and desire? And is it the case that this "self" disappears one lived moment and returns the following moment without reason or purpose? Or, would this conductive thread only be a logical self, the unity of conscious actions – an empty self, indifferent to what it unites?

This cannot be so, however, because our constant solicitude about understanding what "we" are doing and what is happening to "us" indicates that this self, far from being an impartial agent in the drama, is on the contrary its principal hero. What is more, this self is at once the playwright, the hero, the actor, and the spectator. In our interrogations concerning the significance of our actions and passions, we are all these things at once. Instead of being an empty and indifferent "self," this real self enters into each of these actions through and through, filling each one and being filled itself by an emotive, affective, and "felt" current.

It is about this self, pulsating with each wave of our existence, that we wonder. Weary of losing the thread which promises the decisive answer to our quest, we ask ourselves in regard to this self: "Who are we really?" "What are we looking to accomplish while knowingly trying to direct our existence in some causal fashion?"

Without knowing the answer to these questions, will not our existence remain incoherent, arbitrary, and futile? Our views about existence multiply with endless interpretation. Is there then anything that remains which does not change, or which, changing completely, remains identical? Who are we then who exist, who are trying to understand? Wandering in the opaque veil of this current, are we merely the conductive thread of arbitrary acts, and do we just blindly endure them? Do we allow ourselves to be tossed about without choosing our actions, without trying to direct a course toward an end which would give it meaning? Always completely involved as we are in the actions which arise and fade, do we not fade with them? Or, on the contrary, do we persist through all events? Instead of abandoning ourselves to evanescence, do we not try to direct the complete course? In truth, trying to satisfy our need to know ourselves, we are accomplishing rather more than we apparently had hoped: we are establishing the *personal self*.

First, we try to find and identify the deep impulses which animate us in our reactions, desires, hopes — impulses which, after all, direct our examinations while trying to penetrate beneath those interpretations presently practiced or considered possible. We question ourselves about what makes us happy or sad, about our talents or our weaknesses. We try to establish a way by which we might investigate "objectively" our anticipated possibilities, our cherished ideals — which have appeared suddenly uncertain and arbitrary to us in the context of the life and world which supported them. Each option suddenly requires a foundation and final reason.

How is it possible that, in the flow of our occupations (so strictly and faultlessly bound one to the other by concrete logic), even though everything appears "natural" in the established order of things, one begins to doubt everything and to wonder — "and afterwards, what is there . . . ?" From this question arises *sua sponte* the irresistible desire to examine the individual and universal condition of our personal life, a life apparently so well-organized, so sure and untroubled . . . . This desire moves manifestly in the inverse direction, against the on-going structures of life, springing as they do from a natural spontaneity. The inverse desire sheds doubt on laws and forms. We used to share unreservedly in our actions; they carried us with them along the stream of life. Now a self becomes detached and departs from the current. This self no longer identifies with its own facts; instead, it endures them. The self subjects all our paths and customary involvements to critical examination, and they are all found to be narrow, futile and banal.

This impulse to question ourselves about life, the world and ourselves is manifestly one of the most profound, because it brings into play all our faculties, and our entire being is concentrated there. Is this examination imposed on each of us? Does it belong to universal human nature, or is it indeed the result of particular personal developments?

CHAPTER III

ON THE SEARCH FOR THE CRUCIBLES OF POETICS:

The Basis of Literary 'Scholarship' in the Philosophical Interpretation of the Creative Origination of Literary Art

1. THE INVESTIGATIONS OF LITERARY SCHOLARSHIP AND THEIR PHILOSOPHICAL INVOLVEMENTS

From the previous mapping of the locus of the literary work, it appears clearly why literary work, and all works of art for that matter, solicit various types of scrutiny, analysis, and explanation, briefly, of "scholarship," in order to be enjoyed and "interpreted," that is, to be recognized in their peculiar and unique "identity." It is, however, enough to look into the various conceptions of literary scholarship to see that it is explicitly or implicitly based upon philosophical assumptions, if it is not directly an application of philosophical theories to literary interpretation.

Envisaged from the vantage point of its origination, which opens all possible avenues leading not only to the unwrapping of the hidden conditions of "human truth," but also to all the possibilities with which to forge its expression, the infinite spread of the "variations on the same theme" that art is, calls for the delineation of its types. It is assumed that they are founded within the work itself.

The Literary Genre: Conventional or Spontaneous Distinction?

It appears puzzling that, in spite of the ever-changing styles in literature, the basic category of literary production called "literary genre" maintains itself throughout all the variations.

The theory of literary genre, which classifies in an essential way the types of literary works stemming from Aristotle's *Poetics*, maintains itself throughout the long history of attempts to come to grips with the appreciation of the individual literary work. In spite of a vast increase in approaches to the composition, style, theme, and purpose of the literary work – with this diversity epitomized in the liberties taken by contemporary writers, who have practically suspended all the traditional writing conventions – no literary theory has so far provided an alternative, valid distinction.

It seems inherent to the very appreciation of the individual literary text to recur to such a differentiation which, before we enter into its intricacies, gives us a lead by stating whether it is a poem, a novel, a drama, etc. Gérard Genette attempts to show the innumerable efforts throughout history to classify them, which have led to deeper insights into this basic division, but have offered no basis for a new principle of classification.[19]

And yet the question remains: Is the distinction of genre "spontaneous" or is it a convention? I propose that the innermost need for the classification reflects the *conditions for the interpretation* of the work, for making it relevant to the reader/recipient, on the one hand, and the *existential condition of the literary work*, or of *a work of art at large in the essential bonds with the creator*, that is, *the locus of art within the human condition*, on the other. Unlike the well-ordered, repetitive, vital meaningfulness of things and beings, which belongs to the basic needs of human existence within the life-world, *works of art appear unchartered within the life-world*. In order to establish their bearings within man's existential "mapmonde", it is indispensable to discover their intrinsic meaningfulness, and for this man needs a most complex network of efforts.

The question which arises in every theory of literature which aspires unavoidably to lay down such bearings, is the conception of the "genres" and of the principles of their differentiation. These questions become urgent in view of the nature of contemporary literature and its tendency to experiment with and to challenge the validity of all the previously accepted norms according to which the various types of literary works were established. The most striking blow to the established norms may be seen in the radical rejection in contemporary drama of the three basic unities: unity of space, time, and action.

And yet, in spite of the widespread trespassing of the respective territories of literary expression – leading to tragedies without plots, as in Ionesco or Beckett, and novels without a beginning or end ("*le nouveau roman*"), lyrical poems without a focus, etc., etc. – there is still a basic differentiation to be seen. In fact, we can maintain that the classic distinction of the literary genre into epic, lyric, and drama remains unmistakably visible. Only the question of the nature of consistency within each of these categories and its richness of elements on the one hand, and of the precision or vagueness of their contours on the other, remains debatable.

But in order to investigate the validity of the classic differentiation of literary works into the once established genres, we have to raise three questions. First, we must ask: "How is this differentiation grounded in the nature of the respective literary works themselves?" Second, we may ask whether they do not correspond to the three types of man's endeavor to endow his experience with significant forms. Finally, we may wonder whether the "epic," "lyric," or "dramatic" quintessence of corresponding literary works founding the classic differentiation of the dramatic, poetic, and epic genres merely expresses either (a) the composition and structure of the work or the message to be conveyed through them by the author or whether (b) it would essentially consist in *manifesting the primeval ways in which the human being unfolds his existence*.

The description of literary genres by Réné Wellek[20] represents the first mentioned tendency of seeking their differentiation with reference to norms intrinsic to the nature of literary work itself. In contrast, but in agreement with Aristotle and Goethe, Northrop Frye sees the differences between epic, lyric, and dramatic works as stemming from and carried by different "rhythms"; rhythms which express the rhythm of human existence itself.[21] According to Frye, there are different rhythmic articulations which carry certain modes of order of human existence and at the same time are the basis for the literary expression of man. He relates the "rhythm of recurrence" with the epos; the "rhythm of continuity" with prose; the "rhythm of association" with lyrics and the "rhythm of decorum" with drama — Blanchot would divide the rhythm rather according to the internal strife and reconciliation of opposites.

Searching back in time, we discover the conception of the genre resting in a profound connection between the natural condition of the poet and the very forms of his artistic expression. Romantic poets like Goethe see in the differentiation of literary forms and types, the differentiation of the universal Nature, that is, they see the poet himself, the genius, as the discoverer of these supraindividual and suprahistorical laws and rules of poetry, as expressing the laws of *universal Nature* within himself.[22] As Lessing puts it: "*Jedes Genie ist ein eingeborener Kunstrichter. Er hat die Probe aller Regel in sich.*"[23]

Hölderlin, like most German Romantic poets, understands the division of literary works as essentially "poetic," that is, as belonging to the nature of literature itself and indispensable for its understanding, as well as for its creation by the writer. Lyricism, epic, and tragedy (tragism) — as the principles of classic differentiation — represent, respectively, personal feeling, exaltation of a collective aspiration, and exfoliation of an intellectual intuition — the last being "the unity of everything which is alive."[24] Thus, with Hölderlin's principle of differentiation, we go beyond the distinction between the work of literature as a distinctive entity and of the writer as an individual human being; in his thought they both stem from "the unity of everything which is alive," and are spontaneously differentiated in accordance with the different ways in which the writer experiences his life.

Seeking the spontaneous differentiation of literary genres at their point of origin in the human condition at the intersection of human functions I am deliberately seeking their *philosophical* status. The spontaneous differentiation of the literary genre investigated in the function of man's creative meaning – bestowing, which is convertible with inventing *new point of significance for his self-interpretation in existence*, reveals how each genre unfolds specific schema to *orchestrate the coordinates of the creative context from which these original significant posts emerge.*[25]

2. THE EPIC COORDINATES OF MAN'S SELF-INTERPRETATION IN EXISTENCE

We have thus proposed the elucidation of means by which man seeks to "reinterpret" his life to "discover in it meanings significant for his existence, meanings which establish the basis for his creative endeavor and also serve as a common ground, or essential link through which the writer and his work participate together in the abyss of Nature. In several of my earlier writings, I argued that the creator, in and through his work, accomplishes his own route of existence. His progressing quest constitutes, through his work, his own unique *self-interpretation,*[26] while he struggles with the *elemental forces*, with those of Nature and of his life-world, to delineate his own unique route out of the anonymous, universal human condition; his own "destiny" out of the repetitive life-cycle of the human race.[27]

The differentiation of literary forms is itself the differentiation of the significant elements, designs, and curves that this self-interpretation takes. Yet the literary work knows a point of detachment from its bearer, while the process of *self-interpretation in existence* never stops. It is a continuing quest for the meaning of human life, the meaning of its aims, of the aspirations which the human being chooses to pursue, and of the struggles into which man is thrown by a common doom. While the lyric genre brings us into the spring of the heights and of the abyss of these aspirations and elevations of man, while the dramatic genre seeks to discover the meaningfulness of the highest points of significance of human existence to be established in tragic/moral conflicts, the epic genre interprets the quest of the human being as such – in the individual as well as collective perspective – for the *life-significance in its entire course, as situated within a collective destiny*.

From the preceding remarks, it is clear that I agree with Tzvetan Todorov's penetrating analysis of the epic genre. He has shown from studies of the narrative (*récit*) through which the epic genre — whether in prose or rhyme — essentially proceeds, that there is no "primitive narrative" (*récit primitif*) of which we could presume to be in parallel with the formalists' conception of the "primitive tale,"[28] and to expect to find in it an originary ground for the unfolding of the passage from the conditions of experience and those of interpretation, in the nature of the literary form itself. Furthermore, the divergent yet complementary contemporary approaches to the literary work make every *direct* extrapolation of the objects, ideas, reflections, etc., which the work expresses irrelevant to the retrieving of its profound significance or message with respect to human existence and to the human condition.

Nevertheless, we may distinguish in the epic genre at least three *significant lines* in man's coming to grips with the human condition in both its individual and collective dimensions. They may be considered as the "key-triad" toward *man's self-interpretation in existence*.

The epic genre consists essentially in presenting man's existence as a confrontation between his survival drive and the seemingly insurmountable obstacles which impede it; it is from man's continuing efforts, and heroic persistence, in overcoming these obstacles that his individual destiny as well as the destinies of nations and humanity are wrought.

First, we may say, in fact, that in the old primeval sagas, especially the Nordic sagas, there is a predominant thread of the universal human struggle with the *Element* to survive which reaches its peak in the saga of the monster, Grindel. (We find the same unavoidable destructive factor of the human condition in, among others, e.g., a contemporary novel, Calvino's *The Ants*.)

Second, we may distinguish another major thread of the universal human concern, particularly evident in the Greek epos, the *Iliad*, and the *Odyssey*,[29] which consists in devising constantly new strategies to dominate the natural and social situations and reach an accomplishment of some significant end (e.g., Aeneas' founding of Rome). Not only the heroism, force, courage, and determination, but the cunning, plotting, and deceitfulness of men personified by gods like Venus or heroes like Ulysses are enhanced.

Finally, the modern epic (Balzac, Thomas Mann, Galsworthy), seeks to unravel and interpret the entanglements of individual existence within the social fabric, the mechanisms of historical meaningfulness, its ground in the irrational propensities of the human being in conflict with the rational systems of social situations and their demands upon the individual — their merciless and unavoidable strictures and sanctions which break the lines of the highest and noblest individual pursuits.

Briefly, the epic narrative is an expression of man's historical *self-interpretation in existence par excellence*: it brings forth these "knots of significance" which highlight its fabric. The cadences of the narrative link the series of events with man's quest not only to understand, but to give meaning to the flowing, changing, often haphazard and indomitable circumstances of life, both personal and social. Its peculiar design, which orchestrates the basic factors of the *creative context*, consists in weaving within patterns of causes, reasons, and motivations, the individual destinies of men within trends of their nations, extending them between the coordinates of their highest vocations at the one extreme and their deepest flaws at the other, and pinpointing the interpretative schema by the "aspirations," "ideals," and "dreams" which humanity strives to accomplish.

Bringing together these coordinates, man extends a canvas upon which humanity embroiders its history in the largest circles that nature, the lifeworld, and man, may encompass in the historical progress of humanity. We come here to the closing issue of *man's historicity as self-interpretation in existence: its significant "epic knots."*

These three distinct most salient *knots of significance*, the wringing of which from the trivia of man's factual life-course, the writer proposes to the human being as the three major keys to translate the transient and anonymous vital course into a meaningful and lasting script of human *historicity*. They are grounded in the process in which man prepares this "translating" groundwork.

In fact, within the vital mechanism of man's functioning, we witness a built-in and passively operating impulse to *seek the causes* of past events, relate them to the present, and project their significance into the future: to derive, to look for, to wish, to expect, to fear, to avoid. We are entangled in this incessant concern — subjacent to all our actual vital interests, captivating our present attention — with "understanding" of our life-course by seeking the "causes," "reasons," and "motivations" intrinsic in the facts themselves or leading to them.

We never arrive at a conclusive answer, since the perspectives in which we should or could envisage each point extend infinitely with the life-course; yet we never cease to ponder; and it is *this pondering which establishes a continuity of the self-interpretative* line. It becomes an intrinsic line of our life-progress; within the *vital line of fact*, it carries on a line of *significance* for these — otherwise transient — facts, a continuing line, which we never stop to revive in our memory. We endow them with our very own meaning "creating reconstructively" our "history." In virtue of this continuity of *meaning-seeking* and *meaning-bestowing*, by retrieving from brute events of Nature specifically human interconnections of reason and intermotivations, which are never definitively fixed and never cease to be an object of wonder crucial for the quest after the meaning of our life, man establishes his *history; this forms the undercurrent of his self-interpretation in existence*: the specifically *human* phase of his self-interpretation.

We have succinctly analyzed the "radical self-examination" of man as this undercurrent of man's self-interpretation establishing his existential historicity. This undercurrent carries also the subjacent vital mechanism of *the creative quest*.

3. DRAMA AS ORCHESTRATING THE "LIVED ENGAGEMENT" BETWEEN THE CREATED WORLD OF FICTION AND THE INDIVIDUAL WITHIN THE REAL LIFE-WORLD

Goethe tells us in *The Theatrical Mission of Wilhelm Meister* that he entered the "secret gate of artistic experience" through the theater.[30] It was in fact, the puppet theater, which he first watched and then participated in actively during his early childhood: by the multiple chords it played upon his childhood sensibility, it progressively revealed to him the various dimensions of reality disclosed through art — dimensions which must be discovered if one is to participate in the fictive world of art. At the same time, the individual's sensibility is stimulated to conjure their presence within him; last the impact of fictitious personages prompts one to invent them oneself.

Goethe's description of the genesis of this full-fledged "artistic conversion" also reveals the specific nature of the drama whose essence flies out like the butterfly from a chrysalis only when brought upon the stage. Other types and forms of art (e.g., a novel, a poem, a painting, or sculpture, etc.), stand passively before the spectator, ready to be beheld or *deciphered*.[31] But the drama, the comedy, or any type of performing art is neither a passively present entity nor an inert text; on the contrary, the drama as a segment of life — even if fictitious in its various differentiations — stands out, addresses us directly, and solicits our total involvement. It calls upon the spectator not simply to enter its canceled door, but to ENGAGE in its EVENT.

Aristotle's analysis of the dramatic art has brought to light the specific status of the onstage art among the other arts. This status appears more clearly when we grasp its specificity in the way it orchestrates the *creative context* upon which it is suspended, both in its genesis within the creative process and in its subjective enjoyment.

First of all, the dramatic work does not consist in and is not limited to a text meant to be enjoyed by being *deciphered* in solitary contemplation, silent or sonorous, the way a novel or a poem is. To bring the literary work which is not conceived in a dramatic fashion (e.g., Galsworthy's *Forsythe Saga* or Flaubert's *Madame Bovary*) on stage demands a profound transformation, not only in the form, but also in the very texture of the work. In the dramatic transposition of a novel or story new perspectives have to be established within the text to bring into focus the themes, central to both, that have to remain. New ways of onstage presentation have to be devised, replacing the descriptions within the text, so that the drama may convey the same ideas. Developments incarnated in the plot and the protagonists' characters, dispositions, and aims have to be reworked in terms of the modalities of their presentation. Reflection is replaced by an intensified characterization of situations leading to significant events, shortcuts, shifts of emphasis in presentation, and a new selection of events themselves, etc. has to be devised in order not only to evoke the presence of a *fictitious* reality in our imagination, but to *confront us with it directly in our current reality*. That is, a novel, epic or poetic dialogue, in order to be transformed from a reflective presentation into a dramatic stage performance, has to be transfigured into something that is evocative, catching, and captivating.

In its very essence the drama has been devised for *engaging* the spectator, or its reader as a substitute spectator, into an integral and direct participation in what is manifested and "going on on stage," so as to enter immediately into our presently current life and captivate us to the point of "suspending" it, making us "live" for the time of its performance the "life" segment of the fiction. Indeed, the spectator is not only watching the "happening" on stage. He is to be caught up by it and drawn into its labyrinth; he has to enter and follow the thread of the unfolding sequence of men who act and/or submissively suffer, dream and/or struggle. He is called upon to participate in their plight and rejoice in their victories, as in his own. He is induced to absorb them into the palpitating flesh of his own experiential flux, so that they spread throughout his entire experiential system.

The drama "plays," indeed, at the borderline of two worlds: the fictitious one it presents, and the real one, which it breaks into. When merely read, it remains self-enclosed within the fictive realm, and it is by the reader that its essence is drawn into the reader's experiential system by being translated from an inert script into a revived, embodied, imaginative experience. When brought on stage, when it comes to its own, the drama's *significant script* breaks in, in "body" and "flesh" of reality, into the real world, "as if" it were a segment of it.

The intricacies of the dramatic orchestration of this fictitious world in its various designs into which the "realistic" performance (despite the variety of styles) brings the natural forces of real life in their evocative power with the life-world elements upon which the fictitious world draws. They appear in their transmuted meaningfulness, as the ground for the dramatic situation and the emotive, passional and rational functions of the human being which bind the dramatic knots become embodied in real human actions happening within our most immediate reach. They draw us into the very heart of the universal human struggle for survival, as well as for a higher "vocation." They aim at installing the dramatic "life-simile" *within* the real life flux.

Thus they play upon our vital vibrations with a life-quintessence: at the same time they dissect the real life's forces and bring them together into a *higher order of human significance.*

Drama is an urgent appeal to the spectator to identify himself with this "life-simile" in a very special way; the way in which a two-fold engagement may occur. Unlike in real life, in a dramatic participation, *all* the protagonists call on the spectator to be absorbed by him and to identify himself with each of them, thereby expanding his one solipsistic self. With all their interests differentiating them and feelings and passions held in obeyance, each calls for his attention. Instead of remaining the one protagonist of his own life-drama, who has only a vicarious access to the partners of his conflicts in the real-life game, in the fictitious drama, the spectator is asked to sympathize with *all* of them, as each is revealed to him, whether they are sympathetic to his own attitudes or adverse. Not only is the hidden self of each protagonist exposed through various prisms of its self-expressions, and the way in which it incarnates this self, but in its bodily soul-presence each protagonist acts upon the spectator's own inner self, drawing him into the intricate web of its mutual accords or discords and engaging him to "enter into," "to play with," to "be one" with its existence, "as if it were his own."

To accomplish this, the drama ties together in a specific way the creator/spectator network. It amalgamates the experiential network of real life on the one hand with the "life-simile" on the other. It establishes a common ground for the actual life-world on the one hand and the fictitious universe, on the other. It moves with the in-and-out of the dynamic tension between the "fictional truth" of the drama and the "personal truth" of the spectator.

The concrete game of life manifesting itself "in front" of the spectator in a "life-simile" draws the spectator in, sparing him the reflective effort necessary to enter into the fictitious universe of epics or poetry. By touching directly upon chords of his sensibility in all registers – sensory, emotional, spiritual – affecting profoundly his anxieties, fears, passions personified on stage, it ties together the knots of his own innermost conflicts and brings to his attention his own vital concerns, worries, sorrows, joys, and hopes in a *new form*. The "life-simile" in the dramatic plot – tragedy, comedy, satire, monologue, etc. – is absorbed vividly by the spectator's *own schema of real life-experience*. But it is absorbed and retroactively affects the spectator in a special way.

In fact, the "life-simile" of the succession of events that the dramatic plot performs on the stage not only drills itself into the very flesh of the sensitive spectator, engaging him into identifying himself concretely with the protagonists, feeling their pain and trembling for their safety "as" for his own, but in an inverse movement of the concrete engagement it also endows his everyday life-course, in its otherwise pedestrian calculations, struggles, failures, and accomplishments, with a "higher" and "deeper" *significance*. No other form of art, indeed, exposes the hidden springs of human existence more deeply and more clearly.

In its quintessent form the "life-simile" of the dramatic unfolding, the events in their "inner logic" are brought out in succession, arranged along the spectrum of cause/effect, their line of "inner logic" ranging from the unavoidable, at the one extreme, down to the absurd, at the other. Vitally involved in similar successions in real life, we are so absorbed in the vital significance of each singular happening in its radius of causal interconnection, that we necessarily always fall short of following the "inner logic" even of some segment of the sequence. Only in retrospect do we ask ourselves: "How could it have happened?" And once resolved, the human conflicts are open to so many rays of associative reflection that we never find reasons for one or another of their solutions with any degree of certitude.

It is the gist of dramatic art to conjure insight into the otherwise hidden causes and consequences – unavoidable or absurd, premeditated or arbitrary, in or beyond our control – connections of real life events, which make out and decide about the course of life and the destiny of man.

Furthermore, it lies in the essence of the dramatic manifestation to expose the intimate recesses of human desires, tendencies, dreams, and anguishes, those very recesses of which man is only dimly aware in real life, but which actually prompt him in unforeseeable situations – the real turning points in his existence — to take stands, to withdraw, or to undertake to act in a way "true to himself" or, on the contrary, "in spite of himself", or "besides himself."

The dramatic pattern of "life-simile" distills the quintessence of these various perspectives into *the condition of human existence* and distinguishes itself from the real succession of events by intimating to us the *elemental* bends and pulsations of the human being against which, in real life, his "better judgment" remains powerless and would remain so, even if he had been able to recognize them in the "heat of action."

The dramatic conflict orchestrates the creative context by bringing together the extremes of dark passions indomitable by reason and of man's translucid rational self-projection in self-control over his ambitions, aspirations, nostalgia and higher strivings, into a coherent pattern of deeply experienced significance, calling on the spectator to evoke it in the spread of his own experiential system and to make *its* lived significance *his own*. Thus it exposes and projects great interconnective lines of the human life-course conceived, either according to the inexorable, irreversible law of the fatum, or that of the redeeming providence, or of the universal doom, which no cunning can divert from final oblivion, or as flowing totally haphazardly as a transient play of forces, which comes from the abyssmal unknown and steers toward the enigmatic nowhere; last, making the life-sequence appear either as transgressing the borderlines of man's mute submission to the natural forces, or as the unavoidable outcome of his unaccountable, *subliminal* depth, the drama-orchestration of the *creative context* brings the otherwise crudely trivial events of life to *universally human significance*. While the spectator with his vital concern gears into the fictitious sequence of events in the drama and identifies himself with its various elements, his own plights, sorrows, and struggles acquire their universally human significance. They acquire a dignity of being "truly" human.

4. THE "LYRICAL CADENCES" OF PURE POETRY AND OF HUMAN EXISTENCE

A. *"Pure poetry"* [32]

"What makes poetry poetic?" All poets have raised this question and there are as many answers to it as there are probings into the nature of poetry and as many great poets who have found an original access to the hidden source of poetry and a mode of communicating it.[33] The question itself is invariably concerned wih the norms and skills of the art and with the nature of what is called the "poetic language" as its medium. In this it touches directly upon the great philosophical issues not only of its own origin but also of the origin of all art. Indeed, most poets attribute to poetry a quality encompassing all arts. As Théodore de Banville tells us:

> La poésie est à la fois musique, peinture, statuaire, éloquence; elle doit charmer l'oreille, enchanter l'esprit, représenter les sons, imiter les couleurs, rendre les objets visibles et exciter en nous les mouvements qu'il lui plait d'y produire: aussi est-ce le seul art complet, nécessaire, et qui contienne tous les autres.[34]

Obviously he means not only that poetry needs all these means for its communication but also that it pervades them all.

And yet, it is not only the case that poetry differentiates itself from other literary genres discussed previously, that is, with the proviso that we are talking about lyrical poetry as distinguished from the epic and drama; it also holds when it comes to the question mentioned above, that the nature of poetry becomes mysteriously stripped of most of the paraphrenalia of intelligence, imagery and representation, and we are tempted to identify it with one specifically "poetic" element as the key to this mystery.

The controversies about the profound nature of poetry abound in history. The celebrated debate about "pure poetry" between Paul Valéry and Henri Brémond which, as Brémond recognized, continued the discussions of an entire line of literary theories by Edgar Poe, Baudelaire, Mallarmé, and which may be through l'abbé Dubos traced to Italian Humanism, was a significant contribution to simultaneously stripping poetry of all its accessories in order to grasp it in its unique "purity" and to introducing us immediately into the "locus" of the poetic genre.

Brémond brings us into the heart of the matter by emphasizing that after we have enumerated all the possible features of poetic beauty there still remains a "something" – "*un encore*" – to which it owes its unique power of incantation. It is this *ineffable* "pure poetry" which constitutes "mysterious reality."[35] By virtue of its presence, its unifying and transforming action and communicability, a work of poetry is poetic; no imagery, no elevated thoughts or sublime feelings could alone accomplish it. On this we cannot but agree with Brémond; in confronting a great lyrical poem, ideas might be totally confused, images vague and dispersed, rhythms in dissonance and yet the poem might vibrate like a flame and penetrate us through and through. Examples are innumerable. We could begin with the poetic folklore of each culture and finish with the mysterious poems of Gérard de Nerval or the *Duineser Elegien* of Rainer Maria Rilke. Hence we could draw the conclusion that this "pure" and "ineffable" element is not in intelligence but in expression. Equally innumerable are conceptions stressing the poetic expression propounded in present day poetry and criticism. However, by stressing expression the question is not solved but merely postponed.[36] Which elements of the expression itself may account for the ineffable factor of poeticity?[37] On the fringe of expression there remains musicality, which has always been brought forth by most poets as an essential, if not the crucial, element of poetry.

Does not Verlaine tell us in his poetics: *"De la musique avant toute chose ..."* and hence derives from the clues for the rimes, words, rhythms how to conjure the magic of musicality.[38] Contemporary poetry also recurs chiefly to the "purity" of musicality. And it seems indeed that a lyrical poem "signs", and a ballade is in its essence both poetry and music. However, in agreement on this point with Brémond, I must say that musicality — as indispensable (in a way) as it is in Occidental culture for true lyrical poetry, and emphasizing that the essential poetic genre is the lyrical genre — could not account for this incantation of the poem which vibrates within us in sympathy with our own being. No more does it suffice than the fact that a "good" musical work makes a symphony or concerto vibrate within us in the same vein. Thus we conclude that musicality is not enough. To answer the initial question, "What makes poetry poetic?" by the mystic element, as Brémond does, is merely to reopen the issue at an even more puzzling level instead of closing it. Yet with the issue formulated in terms of "pure poetry" we come to the heart of the discussion about the poetic genre.

B. *The lyrical cadences infusing with sap the coordinates of the existential experience*[39]

The question, "What makes poetry poetic?" is then not without a profound reason recurring ever and ever again for every poet, critic, and literary scholar. I am proposing that it is the *lyric element*, unique in its nature, that by its presence not only accounts for the uniqueness of poetry among other genres in literature, but also for the *poetic factor* crucial for all great art. Simultaneously its presence plays an important role as "poetry of life," that is, as the decisive and indispensable fruit of human creativity it is the gist of *man's experience of his existence*.

No wonder that in playing such a central role within art and human existence, the lyric is situated at the crossroads of all possible interpretations. In one of the most penetrating attempts, Hölderlin distinguishes between the "tragic poem" and the "lyrical poem" stressing in the latter especially the feeling in its *sentient mode*. The reason for singling out the sentient mode is that he sees the specificity of the lyric poem in its overall unity which transmits itself in the most easy and direct way and believes that from all the modes of feeling the *sentient mode* characterizes itself by these features. Carried by its "pure tonality" the "sentient" mode of feeling avoids an expression that would translate elements of every day reality, or seek gracefulness. Furthermore, it has the tendency to combine the opposites, (e.g., elevation of the spirit and factuality of life, the simple emotion and the ideal of the heroic, etc.) and maintains its *own* "profound life".

Hölderlin concludes that in the lyrical poem − in contradistinction to an epic one − the emphasis falls upon *its* language. The feeling transmits us *its* language. It is the profound emotion inherent in the feeling that comes to its own in the expression it releases.

This classically Romantic view brings forth three essential features of the lyrical poem: the profound emotional "life" running through it to which any type of expression is accidental; the unique unity of the emotion which constitutes this "profound life"; and, lastly, its position between the "elevation" of the spirit and the real life, between the "heroic" and the naive, between its concrete lived substance and its ideal universal orientation. It cannot be denied that Hölderlin's intuition in discerning a specific element within the lyrical nature of the poem that remains totally independent of all meaningfulness − in which it can be but does not need to be clad − is correct. This intrinsic element which constitutes the profound life of the lyrical poem, reposing in itself and prompting its own "language," I call the "lyrical moment."[40]

The self-reposing nature of the lyrical moment is evident in all great poetry, Occidental as well as Oriental.[41] All their other features, including rhythm and musicality, which are either different or lacking, meet in the presence of lyricism. It is the lyrical moment that gives poetry its ineffable quality and endows it with an all-pervading mysterious attraction and power.

It is precisely from its aloofness to the meaningfulness of ideas and expression and to particular associative and evocative feelings and emotions of a *life-significance* that its universal forcefulness springs simultaneously with its mysteriousness. From the most refined poetry of John Donne, William Blake, Byron, Keats, Coleridge, Burns, Villon, Aragon, Eluard, etc. and moving backward into the naive folklore poetry passed from mouth to mouth and sung on local village occasions, we witness this independence of the lyrical moment from any direct life-meaningfulness. Aragon's repetitive line at the end of each stanza of his otherwise historical poem "Vézelay, Vézelay, Vézelay," Coleridge's closure of 'Xanadu' " ... who drunk the milk of paradise ... ," Petrarch's concluding line of the sonnet on the death of Laura " ... e la speranza è morte"; as well as its opposite, Leopardi's outcry in the famous poem 'Infinito' " ... e mi ricordo delle morte stagioni ma la presente è viva ... " and an infinity of other instances in world literature display this.

Is it not the lyrical moment that, breaking through the intellectual, psychic, emotional molds of the life-constitution and the objective sets of circumstances within which our life-entanglement with its dynamism presses upon our sentient functions prompting them either to submission to the "force of circumstance" or into self-protecting reactions, runs through the conflicting forces on its own? Is it not fortified by all of them alike that it gathers momentum and dissolves with equal suavity anxiety and fear, pathos and tragic feeling, distress and despair, cheer and short-lived joys of success, victory and satisfaction, obsession of love and of hatred? Is it not the lyrical moment which is released in true compassion, in being moved, touched, in solidarity, faithful as we are to the other in spite of all? It appears in *sequences of cadences* and our first impression of it is in the form of a musical incantation.

It seems that Thomas Carlyle, in talking about the "truly musical expression" and identifying it with "poetic" expression, while emphasizing the special meaning of its "authentic" musicality, had this all pervading nature of the lyrical moment in mind. He explains his ideas by stating that a "musical thought" is spoken by the mind which has penetrated "into the most intimate heart of things" and which discovers its most intimate mystery.[42] This mystery consists in the "melody" hidden deep down in the "soul" of the thing envisaged poetically, that is, in its inner harmony. He goes further in attributing to the "melody" of the thing the reason for its existence as well as its right to exist. Lastly, he extends the role of the "melody" to all "intimate things," that is, to the entire universe of *human discourse felt poetically*.

We have here a musicality which is understood as differentiated from, and more essential than the audible music itself; a "thought" which is differentiated and more essential than the intelligible instance of thought, a principle of internal harmony of "coherence" differentiated from, and more essential than, the order of coherence itself. The lyrical moment which germinates in all human functions but distills slowly and gathers its nourishment from the intergenerative mixture of their operations is slow in gathering momentum —

as slow in measuring its maturation as the drop of honey which, while being taken from the juices of flowers, dissolves, melts, ferments and falls ripe into the honey-comb. It is not to be rushed, manipulated, calculated or prompted before its own time. Thus it breaks unexpectedly through the various dimensions of the human being, through the filter of consciousness and its molding by the *logos*, as an "inspiration of," as an "intuition of," an idea or a feeling, of a "tune" or of a "suggestive color." It breaks into our entire being, galvanizing it. At the same time, our being prompted to pursue it, to discover its implications, and its indications activates its curiosity and culminates in an urgency to grasp it and to elucidate it. It suggests we should leave everything aside in order to seek to its riches. It urges us, in other words, to seek its appropriate "cipher" for its significant communication. "Gift of the gods" or "inspiration of genius," no matter to whom it is attributed, poets say: "only the first verse is *given*, all the rest is literature." Similarly we could see in this visionary outburst of the lyrical moment prompting an "intuition," the crux of philosophical reflection, the unfolding and theorizing of which would be just an intellectual process of scholarship.

And yet although as indicated above the lyrical moment does not appear in a single instance, yet its force belongs to the fact that it surges and spreads in a sequence of cadences which neither those of the rhythm nor of the melody as such command. It is due to this cadence-form spread by its force that it makes our being "ignite" and through its rays vibrating through the poem reverberates further within our whole frame. As we may observe it in such exemplary poems as 'The Wasteland' of T. S. Eliot and 'La Chanson du Mal-Aimé' of Guillaume Apollinaire the outbreak within the poem of the lyrical moment takes place at discreet,[43] seemingly disconnected times and yet it makes the entire poem alive with a subjacent life-process. The outbreaks of lyrics stay to it in relation comparable to the drippings of juice from the life-growing process of a maple tree.

This life-growth process through which the lyrical streamlet flows is that of the human existence in its quest after creative fulfillment and its striving toward the *ultimate telos*. It carries through its subjacent life and its emerging cadences, which spread their vibrations of special *universal significance* to life throughout our entire frame, the very current of this universal significance into the functional coordinates of experience which the creator is then ciphering into specific experiential forms.

To conclude I propose that the nature of poetry does not need to be sought in a speculative notion of being nor in the relation of man to the divine, mysterious in itself. The lyrical cadences, which conjured by the creative function of the human being into the poetic appearance in ritual, dance, plastic art, myth and literature, lift the human life to a higher and authentically human level *of his existential experience of himself* and find their most condensed and strong outlet in the poetic creativity; they surge from the *subliminal ground of the* INITIAL SPONTANEITY *from the resources of which the human being generates his life and existence.*

CHAPTER IV

THE RADIUS OF POETICS IN PROBING INTO THE EXISTENTIAL SIGNIFICANCE OF LITERATURE AND HUMAN EXISTENCE

Now it is time to situate poetics among the other approaches to the investigation of literature.

Poetics has not only been a focal point of writers — who marveled about the mysterious source of their urge to create and yearned to find a perfect form for their inner aspirations — but has also been searched for clues to literary appreciation. What specific role may it fulfill, if any, among the other approaches to the study of literature (e.g., sociology of literature, history of literature, comparative literature, philosophy of literature, aesthetics, etc.) — which seem to dissect all the complexities of the literary work from all possible angles in all possible perspectives and are able to interpret it in its entire meaningfulness? Nevertheless, it has also been, and still is, the place where philosophy enters into literary scholarship by the front door. It is then up to the philosophical investigation to seek what place poetics rightfully occupies with respect to other branches of literary studies. Since in the present investigation we arrived at the conception of poetics as the *vehicle through which philosophy enters into the crucibles of art and human existence*, it is first the role of poetics with respect to philosophy that characterizes poetics itself as a philosophical reflection at its deepest. Since it is also, as shown above, within the frame that art — literary art — in its root-condition reveals itself, it is then up to philosophy to elucidate the inner ties which the complex nature of the literary work in its making may solicit as an appropriate perspective for study and how they stand with respect to each other in order to bring the work together, "alive" and *existentially significant*. We must then start by concerning ourselves anew with the philosophical significance of poetics, from whence the recognition of the pivotal place and role of poetics with respect to the fragmentary apparatus in literary work will flow.

The distribution of roles and the respective emphases put upon the different sectors of literary studies is in a subjacent way motivated by the views upon the above questions. Entangled and in solidarity with either the writers themselves or the literary theorists are the philosophical answers to them. Because, quite obviously, the understanding of the principles and modes of the "making" of the literary work in its form reflects the basic directives of the creative endeavor and naturally determines the place of these issues within the entire spectrum of the problems concerning the understanding of literature; they reflect in turn the philosophical presuppositions governing their formulation and treatment.

Therefore, poetics, which by the very nature of the question with which it starts should be the core of literary theory, plays in various conceptions a different role with respect to the various other sectors of literary investigation.

In fact, the particular conception of poetics stems in general from the implicitly philosophical approach to literary work itself.[44]

It is not the proper place here to give a complete account of the various conceptions of literary studies which abound in our times, particularly these inspired by Russian Formalism, the New Critics, phenomenology, and structuralism. To pinpoint our subjacent argument about the crucial amalgamate-in-contract of literature and philosophy in depth, which takes place in poetics, it is enough here to single out the two most comprehensive systems of literary scholarship those of by René Wellek and Roman Ingarden. They both aim at an "organon of method"; both are in the contemporary vein pervading the new trends of literary theory taking the "text," as the springboard of inquiry; both operate with a clearly defined philosophical apparatus. Yet, in spite of Wellek's initially heavy reliance upon the basic philosophical concepts of Ingarden's, they have quite divergent approaches in building their systems of literary studies, and they offer us intrinsically a case in point.

1. POETICS WITHIN THE ENTIRE SPREAD OF LITERARY INVESTIGATIONS[45]

The root differentiation of the literary genres seem to spring forth "naturally" and "spontaneously" when we find ourselves in front of a literary text. Yet seeking in a pre-theoretical attitude the first and simplest access to it, so that we may "read" it and enjoy it in its full right without being led astray from its authentic richness by preconceived expectation, is as we know, connected to an enormously varied effort on the side of the reader who needs to be educated to be capable of emotive, intellectual and aesthetic analysis, and to perform it adequately. This differentiation is already the fruit of the cultural unfolding of man and of the literary genre proposed above, which comes from the creative forge of meaningfulness and expression. With the influence of Heidegger, stressing the literary "text" as the ground for aesthetic investigation the complex of issues concerning enjoyment and appreciation is currently formulated philosophically as the issue of "interpretation." The concept of "interpretation" in Heidegger, Gadamer, Ricoeur and other versions might well stretch beyond literature into all modes of intersubjective expression and communication in its already fixed or progressing genetic forms (flux), but beyond its basis, whether established by respective thinkers in the layers of the text, transcendental consciousness or linguistic metaphor, lurks the *enigma of the origin of expression* as such.

Thus, with this root-differentiation of the literary genre proposed above reaching to the creative forge, we may pursue the questions: "What is literature as opposed to other types of writing?" and further on: "What is Art in contradistinction to other types of man's cultural manifestations?"

For a literary theoretician like René Wellek, it is the question of the "essence" of literature which is at stake. We are told that in order to find it, we have to become aware precisely of the specificity of literature among other cultural manifestations of man, in particular, among other forms of oral and written communication.[46] Along this line we have to distinguish from it the general study of cultural works, we cannot e.g. identify "literature" with the "great books" as highlights, taken in isolation and enjoyed in their remarkable form of efficacious communicability and expression. Unlike scientific language, on the one hand, and the ordinary language on the other, a special use of language may distinguish literature adequately in its "essence" from non-literary written or oral expression. But in view of the principle of the distinctiveness of the literary work as based on the modes of expression, Wellek returns to the nature of the literary genre, understood in the traditional sense of lyric, epic and dramatic. He does this for two reasons. First, because, according to him, the mode of representation in the literary genre is peculiar to the world of invention and imagination, as contrasted with the real world. Second, and with reference to Ingarden's by now classic distinction between the proper statements of the information of facts, and the *quasi*-statements or judgments in the form of which the literary text is made, Wellek sees the appropriate approach to the "essence" of literature through its representation, fictitious and inventive distancing from the factually informative representation of an analogical state of affairs, by way of sociological or historical interpretations.

These representational differences become the criterion of differentiation between literary genres. Accordingly, Wellek proposes a division of the study of literature into five sectors, complying with its intrinsic need for a many-sided appreciation. The *literary theory* identified as "poetics" comprises the philosophical problems which concern the nature of the literary work, its structure, its mode of existence, etc. *Literary criticism* is understood as the evaluation of literary work. Furthermore, while *literary scholarship* should provide the factual knowledge of direct and fringe elements entering into the scope of the represented objects, a *comparative study* of works among them is expected to bring forth their respective features and to enhance the points of their originality. Finally, *literary history* will reveal the unfolding "dynamics" of literary ideas, styles and forms in their inter-influences within the flux of the historico-cultural development, in contrast to the "statics" of literature exemplified in *literary theory* and in the *criticism* dealing with principles, norms and their application.

This "organon of method" is supposed to place literary work within all major perspectives from the combination of which its "essence" could be grasped. In this conception the role of poetics is central; however, it is understood as comprising simply the properly philosophical question: ontological and metaphysical. We may detect in this the inspiration of Roman Ingarden's formulation of issues concerning the nature of the literary work of art and, beyond that, the neatly structural phenomenological approach from which it stems. However, could we ever in this way — focusing one-sidedly upon the structure of the already finished product of the creative activity of man — reach the essential and unique links between philosophical reflection and the creative effort from which this product emerges? By identifying the core of the literary investigation with philosophical reflection taken in separation from its creative source in the Human Condition shared by both the philosophical reflection and the artistic process, do not we introduce an artificial cleavage into the very heart of human endeavor by means of which man aspires to reach beyond the anonymous course of his vital existence? By the same stroke, are we not threatening to cut the strings which carry the germinating literary work into the entire network of existentially meaningful, cultural, as well as profoundly and uniquely subjective existential threads?

Could these severed arteries, meant to convey the signals, signs, appeals, meanings, messages, the pulp as well as the forms of the fictitious world into the real one, from the creator to the recipient — a transmission of the highest existential significance, of which the literary work is a *virtual creative transmitter* — be restored by the prospective investigations of the particular sectors of literary studies?

Let us now turn to see how the philosophical preconceptions of Ingarden's approach to literature determine his conception of poetics and of its function. As a matter of fact, Ingarden delineates — although he does not elaborate —

the most comprehensive "organon of method," drawing explicitly upon his entire ontologico-aesthetic work and organizing it into a system.[47] In the first place it is founded upon a thoroughly developed ontological basis. However, the analytic, structural foundation of the ontological skeleton of the literary work is coupled with a transcendental structure of the modes of cognition of the types of literary works. Because of its clearly defined, detailed and nuanced differentiation and its fusion of philosophy and literature, it deserves special attention.

2. THE POETICS OF LITERARY CREATIVITY AND ITS PROPER PLACE IN LITERARY STUDIES

The spontaneous differentiation of literary forms within the interplay of the vital, subliminal, conscious and intellectual forces of the human being who creates them by projecting novel patterns of his functioning and, in so doing, frees himself from the bondage of Nature, is simultaneously an expression of man's effort to construct and create his own meaningful expression reaches that beyond his merely vital frame, thus transposing the secret of the significance of the literary work from its content to its *making*.

This creative origin reveals a unique configuration of the functions by which man, *orchestrating* them into a meaningful way, invents for himself a Nature-Transcending level of his *self-interpretation in existence*. No wonder that all creators have sought to understand the secrets of their creative endeavor by means of modes, principles and conditions of *making* as congenital with the fruit of the work. No wonder that we see the investigation of *poesis* as "making" of verses, of paintings, of dance, of myths, of scientific theories, technical innovations, etc., searching for its roots in the creative activity of its maker. After our investigation of poetics in literature by the question previously raised and treated, "What makes poetry poetic?" we now ask the direct question, "What makes a poem a poem, a novel a novel, a tragedy a drama, etc.?" According to Aristotle's intuition, they appear to be differentiated according to their "making." However, from the emphasis upon their intrinsic "make," that is, from their intrinsic content and form our emphasis shifted to the *conditions of the "making" itself*, that is, upon man's creative endeavor. Roman Ingarden's conception of literary work bars the door to this crucial question.

The crux of the matter, as I have pointed out elsewhere, resides precisely in the correlation between the structures — ontological, in the literary work, on the one hand, and that of cognitive process in the text, on the other hand. In their essential coordination resides the existential status of the literary work. Indeed from a mere meaningful skeleton, the literary work comes "to life" only when embodied within cognitive processes, which complete it and are also the conscious acts of the recipient. Within these processes it turns from an objective abstract formula into a subjective "concretion." But in the Ingardenian approach in which the ideal structural ontology takes on an overwhelming precedence over the *fragmentary* and scant inquiry into conscious processes, the emphasis falls upon the ontological structure of the work of Art.

Thus within his "system" of literary studies – or "organon of method" – first comes the *philosophy of literature* which would comprise the ontology of the literary work dealing ultimately with the question of what makes it a "literary" work and what makes it a work of literature. Correlated with it stands the *theory of cognition of literary work*.[48]

In correlation with the permanent rational core of the work's structure stays the range of cognitive contacts with the work in reading schemata which – falling into a coherent process – is sustained by the persistent core of structure which allows for correction, redressing of false impressions, etc. as well as leading through its coordinated phases of temporal succession, in spite of subjective fluctuations and distortions, in their once for-all-organized sequence. This subjective process of cognition aims at the reconstitution within experience which is strictly singular and relative to the readers or critics (as well as to the intellectual and cultural climate of the period), to the variants of the structural contents of the work as such, and to the evocative effect of the work upon the feelings, emotions and sensations of the reader. It leads toward the concretion of the universally determined literary universe within his strictly and intimately personal life-world. The climactic accomplishment of this concretion would be the emergence within the experience of the reader of a value-synthesis of all the elements of experience concretized in an *aesthetic object*. *Aesthetics of literature* is assigned the task of studying the aesthetic values to deal with the "sociology of literature" and philosophy of literary creativity. Then comes the *theory of literature* proper.

In contradistinction to the "philosophy of literature" defined above – "philosophical" in the sense that it envisages literary work not in its factual singular existence but in general, that is, as a distinctive type of entity relative to human consciousness, in its experiential expansion, and as its complement – Ingarden distinguishes a *science of literature*, which is concerned with literary facts, that is, with the actually existing singular works in their influences upon each other along with literary trends, etc. The evaluation of the aesthetic values of the literary work is attributed by Ingarden to *literary criticism*.[49]

This philosophical conception of literary criticism which, in fact, should grasp the literary work in the most immediate and essential way, appears so one-sided that it unduly prejudices the concrete literary appreciation orienting it formally. In its formal character it contrasts, on the one hand with Heidegger's hermeneutical method which is meant to be directly applicable to the body of the literary text. On the other hand, it contrasts with the Geneva Critics, who see the key to literary criticism — that is, to the direct analysis and appreciation of the literary text — in the subjective "experiential pattern" as latent within the literary text and at the same time, making the author "present" with his human experience and his literary intent within the text of his work as well, by unifying in an objective way the elements of the construct of his imagination. This experiential pattern is retrievable within the subjective experience of the recipient/interpretator of the text. As we know, both of the last mentioned approaches, although incomplete, prove themselves fruitful in literary analysis.

To *poetics* Ingarden attributes what he believes to be the most important role within this vast spectrum of perspectives upon literary work, namely, the role of interlinking them within an all-embracing synthesis. But it is clear that although Ingarden restores to poetics its full-fledged Aristotelian scope that seemingly includes *all* the aspects of the poiesis, the "make" of the literary work, yet by grounding it and delineating its ramifications by means of the ontologico-cognitive structure of the already-*made* work, as it stands in front of us in its accomplished form and in its strictly philosophical analysis, he establishes a framework of philosophical presuppositions how to approach literary work. Instead of opening it out for a far-reaching investigation of what it can teach us, bringing *into* philosophy crucial insights from the commerce of man and world, nature and consciousness, a tight philosophical seal is imposed upon it. From its calling to be "first," prior to any other decisive inquiry, relegated instead by this philosophical assumption of Ingarden to be "last," poetics is robbed of its true role. Irreplaceable in essence — not in term — by reason of its unique access to the springs of art and Human Condition, its misplacement bars the entrance.

In Ingarden's system poetics which alone could and should open up access to the labyrinth of human intercourse with Nature itself in the progress of man's origination in the womb of vital forces of mere passive survival toward his active *par excellence* self-devised and self-determined life significance, preceded by and drawing upon ontoloty, is relegated to a mere synthetic role of a wide spectrum of fragmentary researches.[50]

On Ingarden's ontologico-cognitive basis we may certainly answer the question: "What makes literature different from other cultural manifestations of man?" But we remain incapable of dealing with the basic wonderment: "What makes art, art?"; "Why does the human being create works of art?"; "What role do they play in his world and his destiny?" Turning now directly to literary criticism within the framework of the *Poetica Nova* it will appear that the scope of the literary work of art does not reside, as Ingarden claims, uniquely in the realization of aesthetic values, nor, as Heidegger proposes, in conveying the revelation of Being, last, not in the Geneva Critics' conception of communicating with the "moi profond" or of the author. In a reverse course I propose that the nature of the "critical" or "appreciative" process analyzing the body of the work itself rejoins the heart of the creative poetic investigation by elucidating the *existential significance* of art.

3. *POETICA NOVA* AS METAPHYSICS OF THE HUMAN CONDITION

In my perspective as developed so far the question of poetics centers upon the issue of its origin, that is, upon that of "How does the *poetic*, in particular the *lyric*, element emerge within *the human functioning*, the vital functioning subservient to life, and how does it find its way through the entire span of *man's self-interpretation in existence* until the peak of its *free meaning-giving* which takes place in the creative work?" Moreover, poetics appears as the proper field to assess the ways in which the widely spread differentiation of the creative works — that is, the differentiation within an emergent dimension of reality, which marks one of the main autonomous dimensions in the entire spectrum of "realities" — is linked together with its other dimensions. Last, *creative* poetics allows us to raise plausibly the metaphysical questions *par excellence*: "How is it possible that from the discreet and instantaneous acts of the human functioning a constructively organized and purposefully oriented continuing chain of the creative process builds itself up?" "In virtue of what factor does it occur that different modalities of the continuous-discreet purposive segments (biologico-vegetative, biologico-physical, psycho-intellectual and last, psycho-spiritual) are intertwined, encroach upon each other's functioning, mold each other and ultimately come together in one discreet, infinitely differentiated and yet constructively coalescing creative endeavor?"

Analyzing the spontaneous genesis of the *lyrical sense*, we come close to conjecturing that it is the lyrical element which, from among the *elemental* human endowment, works itself into the fabric of the functions playing out the human existential script that breaks the narrow borderlines of the life-serving functions, expands the vision and the horizon of life-meaningfulness, sets into motion the germinal *Imaginatio Creatrix* and brings the juices and the sap for the poetic dimension to emerge within the web of human self-interpretative guidelines. We are then led to conjecture from our analysis of art in its origin regarding some basic features of art's crucial factor: the *creative agent*. The *creative agent* — unlike the "human agent" or actor in real life — does not develop his virtual acting capacities with the passive spontaneity of his natural development. He constructs himself from the material of the SUBLIMINAL SPONTANEITIES by means of his own initiatives. Nothing is present in the actual life-world that he may choose as it is and rely upon. His inventive judgment alone guides the discrimination and selection. In the maintained sequence of ever renewed efforts to measure his inventions with his innermost exigencies, dimly glimmering from an unapproachable, distant horizon, there is nothing he may fall back upon. To use Nietzsche's expression, he is like a dancer on a rope in which each step is at the cost of a calculated struggle for equipoise; his rope is the intended masterpiece in progress.

And yet, this urge to find, to discover, to create is not a self-reflexive and self-centering act, it neither takes over the self for the sake of its object nor is it meant for the satisfaction of the discoverer or creator. It is simultaneously a desire to *express*, to *manifest*, to *show*; in short, to communicate to the Other. It contains a three-fold intention: one, directed toward the sought "truth of things," the second, toward its new form of expression within the life-world as a challenge to its present forms. Each intention is coupled with the desire to bring in a more adequate one. Both of these are anchored in a third, reversible intention bringing the two previous orientations back to their source in the natural subject, which consists of the urge to accomplish this quest *oneself*, giving it a communicable expression. Hence the *creative self* originates as an *agent*.

This three-fold intention delineates the *creative self* as a projector of an entire network of new interconnections between and among the familiar framework of Nature/human subject/society-culture-and human *telos*. The artist creates for the creative object's sake but he does so "for" the spectator, listener, etc.; it is to them that his effort is addressed; he is creating in order to fulfill himself and in that highest longing and aspiration, the *sense* of his entire being is put at stake. As an agent his meaning bestowing differentiates at the very generative junction of each new phase of his vital and existential advance.

From the way I have unravelled the great arteries at the crosscurrents of which the creative endeavor generates the common ground for philosophy and literature and outlines their intertwining, it is nevertheless obvious that our approach remains strictly within the phenomenological perspective. This statement calls for explanation. We identify with the "phenomenological approach to literature", usually either the use of the "phenomenological method" in stressing the aesthetic experience, or the "essentialistic" investigation of the structures of literary works (Theodor Conrad, Roman Ingarden, R. Wellek, etc.), or the interpretation of the literary works with reference to the "experiential pattern" of the work and of the writer developed by the Geneva Critics (Poulet, Staiger, Hillis Miller, etc.),[51] or a metaphysical interpretation of literature in which the "hermeneutic circle" is meant to lead to the revelation of Being (Martin Heidegger).[52]

These approaches unduly compartmentalize the field of inquiry through their peculiar emphasis. They twist the overall orientation, which should embrace the field, subsuming impartially the pivotal points, and their functional interrelations in a context, *before* singling out the most significant factors for a lead (either in the direction of the object to be created, or of that of personal destiny, or of a metaphysical telos).

Each of these approaches subordinates the remaining factors, distorting the entire picture. Each of them basically falls short of the fundamental aspiration of phenomenology. As I emphasized two decades ago,[53] the vocation of phenomenology which it has to a great degree fulfilled, is not only, as Husserl proposed, to give an unbiased, unprejudiced voice to all the types of experience and phenomena, but as I have attempted in my previous work, to show to all dimensions of reality mapping them by the extent and role of the human factor within them.

In the first of the phenomenological approaches to art, Moritz Geiger has undertaken to bring the pivotal points of the issues concerning art as the human enterprise *par excellence* to the "human significance of art,"[54] culminating in the enjoyment function of the already accomplished product. It is the aesthetic enjoyment then that gives the key to the significance of art. Phenomenological aesthetics then went in the direction indicated by Geiger, first, by interpreting art chiefly from the point of view of its function of originating aesthetic values and second, by seeing in aesthetic enjoyment its essential role.

In my own perspective, benefiting critically from certain sectors of phenomenological aesthetics that have been fully developed, as well as from my own contributions to the basic issues confronting the approach to the human being as such, the phenomenological inquiry has been considerably enlarged. The work of art appears in the full-fledged progress of its origination within the most complex functional network by means of which the human being establishes the significance of his existence — this existence itself. The work of art appears itself as working with a vertiginous strategy on the very battle field on which occurs the confrontation between the primitive, natural factors of human existence — with their elementary guidelines for the life-process — and the creative factors — aiming at the emergence of a specific type of human beingness. Hence from the narrow significance of "aesthetic enjoyment" of the work of art, the passage has to be made toward recognition of a crucial existential (in the strong sense of the term) significance of art as the telos of the creative process through which this unique type of human beingness emerges and is established: its *existence-significance*.

CHAPTER V

THE EXISTENCE-SIGNIFICANCE OF ART AND THE "DECIPHERING" OF LITERARY WORK

1. THE EXISTENCE-SIGNIFICANCE OF ART

Progressing thus far, we have reached the point at which the double-voiced fragments of close analysis, probing at strategic points of their congenital interplay, fall into line on the one side, yielding the key with which philosophy may open the way for us through the door to the puzzling situation of art in its connection with the alien world of Nature; on the other side, this calculated mapping of the *locus* of art within the Human Condition falls into an intricate design which should and does, indicate the key to the work of art's essential understanding and proper appreciation.

On the one side, we have in fact elucidated philosophically the foundational existential function of the human being consisting in his *self-interpretation in existence* as running through *all* the phases of his life-course.

On the other side, we have brought forth from oblivion the network of *coordinates of the existential* function projected by the human being within Nature and his life-world, which sustains the work of art in its generation and receptive exfoliation. One of our positive results is the discovery and laying down of the great lines of *phenomenology of man-and-of the-Human Condition* as rooted in *man's self-interpretation in existence*, culminating in its highest accomplishment in the *creative ciphering of sense*. The second indicates the *coordinates of the creative context* with reference to which the literary criticism may bring to life the work of literary art, *deciphering it in its full-fledged existential significance*. Indeed could philosophy conceived otherwise discharge itself from a task with respect to art, which is implicitly incumbent on it?

So far I have been stressing the issue of the nature of literature. First, by unfolding the spread of its generative tentacles. Then, by indicating the various perspectives of the appreciation which this first implies and which should circumscribe the radius of approaches to the great issue: "What makes literature a fine art and 'literature'?" In doing so, it appears that we encounter philosophical reflection and theory at every step as congenital with the issues concerning the generative process of the literary work and its nature, as well as a means for the clarification and establishment of modes and norms of cognitive and appreciative approaches.

When we talk about philosophical endeavor as the literary endeavor's essential partner in its attempts to unravel and grasp the enigmas of the human personal, social and natural existence, we already envisage this partnership and formulate the question of the enigmas in a way that either stems from, or implicitly entails, a set of philosophical assumptions as a foundation for a particular vision of the human being and of art, a vision comprising its role within the life-world and beyond it stretching into the intimately personal. Recapitulating our results: art in its role in human existence lifting the human individual from the narrow apparent determinism of vital progress into his own self-derived course, has appeared in its common originary phase of *human creative endeavor* as situated at the *cross currents of the Human Condition*.

Originating in the crucibles of man's creative function which itself is deeply seated in the Human Condition, questions concerning both stretch into all segments of human reality; to answer them, philosophy has to account for the multifarious and enigmatic threads which sustain and conduct the originative phases of their genesis. Only philosophy which has formulated this state of affairs can do justice to the all-embracing and singularly articulated philosophical dynamics there exposed.

No monistic approach tying up all the strings of human discourse within one metaphysical principle, no absolutism of a transcendental network of intentional structuration, nor that of an ontological realism fixating the dynamic current of life-genesis into sclerosed and artifical constructs of human intellect can even approach this existentially differentiated constructive progress.

A process philosophy which either seeks to account for the continuity of the life-progress in vitalistic terms at the expense of man's inventive function, or which sees in the constructive manifestation of life outbursts of haphazard vital élan on the other side of a spirit radically alien to it, misses the *originary situation of Man in his Condition*. Each of them assumes one single source of ordering and sees it either in the rational structures and rules of the human mind, or within a universal substructure of reality, or within the system of Nature, or stemming from an "objectivity" radically external to man, each of them perpetrating the traditional reductionistic assumption that all types of concrete segments should amount to one type and source of meaningfulness.

By contrast, following the creative progress of man it becomes clear that it is the real individual being who through his *vital-existential generative progress in its differentiations* functions as the crucial factor of order. It is not through his cognitive function alone, nor through the entire spread of his vital and existential operations that different modes of ordering are established. The human being is thus a meaning-bestowing factor, but in radical contra-distinction to the Husserlian conception of the strictly cognitive/constitutive meaning-bestowing-exclusively one, rational order, established by man as a rational being, he does it through *various* modes, characteristic of his vital existential progress, as a vital and *creative agent*.

2. INTERPRETATION: THE CREATIVE COORDINATES FOR THE EXISTENCE-SIGNIFICANT EXFOLIATION AND "EMBODIMENT" OF THE LITERARY TEXT

Each work of art, inert in itself, presents a challenge to the inquiring mind. The question which occurs in the encounter with a thing or being is: "What is it?" or "Who is it?" Instead, in the encounter with a work of art, when it is a work of plastic art, the question arises: "What does it represent?"; when it is a literary work, whether in a text to be read or to be performed on stage, the question occurs: "What does it 'mean'?" And yet, in opposition to the view that sees the literary text basically as a system of meanings, I propose that it is a *specific pattern of life-significance*, universal and uniquely personal at the same time. Whether, at one end, it culminates in a metaphysically significant "message" or whether, at the other end of its significance, it is merely setting up the elements for a peculiar "segment of human life" to receive flesh and blood from the experience of the recipient, the text is a *pattern of existence-significant signs* "ciphered" by the writer within a *creative context* and meant to be "deciphered" by the reader within the same context, although with a different emphasis upon its factors.

The work, the text, appears as a *configuration of signs* and not of meanings belonging to the current universe of human discourse and waiting to be identified in a stereotyped fashion. Access to this questioning derives from the nature of the sign: it indicates by its presence that it does not represent only itself but "stands for something else." Thus, it leads further to the question: "What does it stand for?" In fact, it entails a fivefold interrogation according to its five types of style: *visual configuration, sound-rhythm* (*cadences, sequence*), *meaningful configuration, poetic vibration*, and *experiential-emotive configuration*. These are exhibited through the configuration of signs, partly constructing it, partly intrinsic to it. Each of them is a construct in itself with its own most complex structural device establishing its significant network, calling for its opening/discovery with a specific key. Each of them points to its origin in the human *meaning-giving effort*.

Indeed, each points directly to two poles upon which the investigation of the text appears to be suspended: the significance of the text, that is, the inquirer as a human being fully endowed to receive the lead and to follow it on the one hand, and the human life-world within which the vision and the hearing with their *life-significant* relevance — the meanings with their objective correlates and the experiential configuration – on the other hand, have their resounding chords. They all draw upon the basic laws of the intersubjective world, and all its associative links with the cultural values of the given period; they exercise their respective functions in a varying but well-established order of the *existential significance* which they establish in common.

Although in direct encounter with the text, the human recipient-inquirer alone appears to bear the bringing to life of the work's significance — solely by his own means and upon his exclusive responsibility — yet, he is himself in the process of following leads as to *how* to give flesh and blood to the empty signs. Infusing them with HIS PERSONAL EXPERIENCE, developed according to the rules and order of the life-world. Moreover, by distinguishing among the ramified patterns of signs, while endowing them with five-fold experiential relevance, we establish this very relevance not with the solipsistic and world-withdrawn consciousness of the reader but, on the contrary, with reference to the entire spread of his life-world: we refer alternatively to various experiential perspectives which the work in its encounter with the experiencing subject as he confronts the challenge of the complexity of signs, triggers off, evokes, solicits, thus signling it out to come into focus from the world of the person which otherwise would passively remain in the shade of the all-enveloping active consciousness. It is within the re-creative forge of the intimately singular subject that the work of art — in particular the literary work — comes to life by calling up a specific experiential "shape" as a reverberation-schema in visual form, rhythmic cadences, poetic vibration, sensory, imaginative, intellectual and spiritual impulses, segments of life, meaningful objectivities etc., by means of the relevance of signs to the subject's experiential system. Ultimately, the subject seen as not only carrying the life-world but reciprocally carried by it, can accomplish the experiential relevance-endowing task of bringing the inert and lifeless skeleton of the meaning-discovering or analytic dissection of the text into full-fledged life. He does it by molding the received signals according to the universal meaningful system of the life-world. We may consider this full-fledged process of "interpreting" a literary text in the meaningfulness of its entire complex sense by incarnating the intentional skeleton of a structure with our own experience, "deciphering."[55]

3. THE THREEFOLD POLE OF THE CREATIVE CONTEXT

The two poles appearing in the inquisitive, exploratory encounter between the inert literary work and the living human being as its prospective "recipient" — the literary work which we have identified with the human being or the bearer of the cultural interpretation of human existence, from whom stems the meaningful system of the given type of culture at the one extreme, and at the other the life-world which is the all-embracing intersubjective network of the meaningfulness of every human subject, expressing his life situation, life condition and possibilities in general and in particular — would not suffice in our encounter for the "incarnation" or for the "deciphering" of the literary text. This latter as a work of art can never be adequately appreciated without reaching in a determinate way into the ground of its origin, the CREATIVE FORGE.

With the changing trends in literary appreciation, various perspectives upon the creative source of art, after having been abandoned, return in a different guise. Certainly the real-life biography of the artist or writer with his psychological processes can give no essential lead as to the origin of the significant system that the literary text has ciphered. Neither the social nor cultural situation within which the work has been conceived, even if as in a great many works dealing with the life of the period (like Lady Murasaki's 'The Tale of Gengji,' Balzac's *Human Comedy*, or Simone de Beauvoir's 'The Mandarins,'), can offer an essential lead to its meaningfulness.

Going deeper into the *creative forge*, we must also dismiss the "personal message" of the writer as a possible essential factor in the approach to the unique script of the work of literary art. Nay, even the "life-wisdom" peering in certain outstanding novels such as Manzoni's *Betrothed* or Tolstoy's *War and Peace*, etc. cannot as it is apparent from our analyses of the origin of both art and man's existence, usurp the claim to open the work's' uniquely-significant nature. The writer as the human being who projected into it his own unique experiential vision and the reader of this vision — established already into a text by which it is sustained — come together for a confrontation not restricted to the already constructed systems of meaningfulness but to be retrieved at its origin within both of them, a confrontation that alone would allow the appreciation of the work in all its depth.

In fact, the "experiential vision" presents the *existence-significant signs* as threads along which at certain junctures and tracts of the exploration it may play a role within an overall unified system. Their system of experience is constructed upon a highly developed critical level, at which the insights obtained from the investigatory encounter with the text is brought down to that of the *creative forge* and its depth within the Human Condition. It is patterned toward yielding the maximum efficacy in the surface interpretation of the work. Thus the experiential vision does not alone reach the convertibility of the signs with the *existential self-interpretation* of man. Undoubtedly all of the above mentioned perspectives (and a great many more) are necessary to enter sideways into the inquisitively exploratory "encounter" with the text. None of them, however, alone, nor all put any way together, can introduce us into the creative origin of the text in a way that could disclose its unique *existence-significance*.

Indeed, each of the possibly available interpretative schemata is already loaded with essential presuppositions that necessarily preclude the descent to the ultimate creative condition of the text, which tying all strings together, would expose them in their respective roles in the creative "ciphering" of its unique significance. Preconceived schemata misdirect the exploratory focus toward one single line of the meaningfulness of the work, proposing it as a uniquely significant "revelation."[56]

Any one-sided lead or focus of the critical attention will affect the response of the spectator, reader, listener as he confronts the work and is challenged by the question: "What does it mean?" "What does it stand for?" and reduce it to some lines of meaningfulness while disregarding others.

To answer this challenge he has to draw upon his entire experiential system in its specific cultural form already present, as well as upon the life-meaningful system of the life-world he is congenitally partaking of. Yet both of them are merely vehicles of the initial "deciphering immersion" with which he may establish his bearings in the initial no-man's land of the text. The main thrust of the deciphering process is directed to entering progressively into the *creative context* of the work.

The *creative factor* at work lurks within the multiple breaks of the otherwise consistent common meaningfulness of the "languages": sounds, rhythms, images, feelings, emotions, ideas, etc. Giving them an "uncommon" and enigmatic significance, he appears as the act of a magician, who transforms the ordinary life-subservient functions into arteries through which the new sap of life flows into remolded forms of experience. Within his text, the creator-writer himself is in no way present. In order to "bring his work to life," we have to be led with our initially cultivated self, present all the time, into these strange, unexpected, surprising, newly generated forms of meaningfulness: into their CREATIVE ORCHESTRATION. Creative orchestration, once we penetrate into its network, makes all our imaginative, aesthetic, moral, poetic, cognitive sensibilities vibrate, by appealing to functions which lead the creative factor in the exploratory process under the promptings of profound stirrings and high aspirations. We enter into the creative orchestration precisely because of the necessity to search within ourselves for the proper *chord* upon which the transformed, transfigured, transfused forms of experience which make an appeal to us could respond. Having once entered into this labyrinth, led by its intricate paths, no one is waiting for us like a Minotaur in its recesses; we advance into the creative forge to discover how the human being is, with supreme effort, working out the Human Condition of his own uniquely life-significant system of existence. The pursuer of the understanding of the literary work, "deciphering" within the creative forge the devices left by the creative effort within the work, comes to respond to them within his experiential system. It is "unique", surging from the unfathomable abysses of Nature and yet not totally life-subservient; it draws upon all the resources which man's roots in Nature may conjure up, and yet not making them passively flow into the current of life it puts to work the entire endowment with which he is fitted out for his vital survival, and yet does not submit it to its anonymous rules and aims. The existential significance *ciphered* within the creative forge is the root of the *recreative* deciphering of the literary work of art. It is also the root of the *specifically human existence*. With the effort of a demiurgos, freeing himself from pregiven laws and aims, *man-the-creator* orchestrates his inherited forces into new channels and ciphers them into an ever novel life-significant system.

According to his own unique aspirations, nostalgias, dreams, inclinations, strivings, man fashions a novel dynamic universe which depends only on his will. While compromising with otherwise vital challenges, the will is *here* inflexible to any; man's courage to believe, whose strength is usually measured by the obstacles in real life, is *here* indomitable.

NOTES

[1] Ortega y Gasset, J., *En torno a Galileo*, Madrid, 1933.

[2] Concerning the division of tasks between philosophy and literature Richard Kuhn writes: "Philosophy takes as one of its tasks an account and analysis of the grounds of experience; literature takes as one of its tasks the presentation and realization of specific experiences. What philosophy explains, literary art realizes or makes. Philosophy asks: 'What makes experience possible?' and 'What makes this kind of experience possible?' Literature establishes the realities for which philosophy must seek explanations ... enmity and ignorance have kept them apart." p. vii, *Literature and Philosophy, Structures of Experience*, Routledge and Kegan Paul, London 1971.

Although I agree with Kuhn about the role of philosophy in general, it is to misunderstand entirely the profound nature of the literary endeavor to restrict it to "establishing the realities" of "experience" and to relegate it to illustrations of philosophical theses. Should we concede to Kuhn that literature "establishes the realities" of "experience" it would have to be considered how these realities are being established — in contradistinction to the way in which it is done in current life for practical purposes and the way in which it is done by science. That is, we will still have to raise the crucial question about the *origin* of both philosophy and literature which we are proposing here as the key toward elucidation of their interrelationship.

Our initial statement about the most intimate interplay between philosophy and literature throughout history does not leave much room for "enmity and ignorance".

[3] I have analyzed the relationship of the philosophical style to its subject matter and to its specific approach in my essay: 'On the Philosophical Style', *Phenomenology Information Bulletin*, Issue III, 1979, *The World Institute for Advanced Phenomenological Research and Learning*, Belmont, Mass. I propose to develop this theme in a forthcoming work.

[4] In this section are expanded the main ideas of a study presented as the inaugural lecture at the Annual Convention of *The International Society of Philosophy and Literature*, Cambridge, Mass., in 1976.

[5] For the conception of "human Reality" cf. by the present writer: *Phenomenology and Science in Contemporary European Thought*, New York, 1961, Farrar, Straus and Giroux, The Noonday Press.

[6] Cf. Anna-Teresa Tymieniecka, 'Man the Creator and His Threefold Telos', in *Analecta Husserliana*, Vol. IX, 1979.

[7] Cf. Anna-Teresa Tymieniecka, *Eros et Logos, Esquisse de la phénoménologie de l'intériorité créatrice*, Nauvelearts, Louvain, 1964. I have established in this treatise on the creative experience the notion of the 'creative context' as sustaining it and allowing us to analyze creativity at large.

[8] The conception of the "Elemental" ground of man's pre-consciousness has been first introduced by me in my study 'The Initial Spontaneity ...', *Analecta Husserliana*, Vol. V.

[9] For the discussion of the role of spontaneity and free creative decision of the artist cf. my study: 'Imaginatio Creatrix', in *Analecta Husserliana*, Vol. III, 1974.

[10] For the notion of "transnatural destiny" cf. the pre-cited 'Man Creator and His Threefold Telos', in *Analecta Husserliana*, Vol. IX, 1979.

[11] Anna-Teresa Tymieniecka, cf. The pre-cited *Eros et Logos*.

[12] The question of "mimesis" is the most complex of questions concerning the literary

expression with respect to the external reality, inner reality of the writer, social reality, etc. and lastly the very process of transforming the germinal potential for a literary work into a literary 'project' or 'Vision' as well as its actual making. Should we with Elder Olson interpret Aristotle's notion of "mimesis" as "imitation" — as has been popularly done — understanding by it that "natural objects possess an intrinsic principle of motion and rest, whereas artificial objects, a chair, a table, being results of Art, do not possess this principle: they do not change due to the propensity of form, but that of matter," we remain in the crude perspective dividing between Nature and art. We would then say that art "imitates" nature, taking from it its form or that the artistic process is reminiscent of the natural process. However, through several centuries of pondering this issue we have greatly differentiated and refined it. In the phenomenological perspective we may equally say simply as Olson does that 'mimesis' — even in Aristotle — is not only concerned with the transposition into the aesthetic form of human reality but also as expressiveness of experience. Cf. Elder Olson, 'An Outline of Poetic Theory' in *Critics and Criticism, Ancient and Modern*, ed. R. S. Crane, Chicago, University of Chicago Press, 1952, pp. 65 and 588.

For a historical treatment of the issue of 'minesis', cf. Erich Auerbach, *Mimesis*, transl. W. R. Trask, Princeton 1953, Princeton University Press.

[13] This section renders in a condensed form the lecture given by the present writer at the opening of the annual ISPL Conference. The controversy about 'Pessimism and Optimism in the human Condition' has also been brought forth in her study: 'The Initial Spontaneity: Pessimism and Optimism in the Human Condition', *Analecta Husserliana*, Vol. V, 1978.

[14] Willi Fleming, *Bausteine der systematischen Literaturwissenschaft* (Meisenheim am Glan, Hain, 1965).

[15] Cf. Northrop Frye, *Anatomy of Criticism*, Princeton University Press, Princeton N.J., 1957.

[16] The conception of the "cipher" as the sense-forming moment in *man's self-interpretation in existence* has been proposed and explained by the present writer in her study: 'The Creative Self and Man's Self-Interpretation', *Analecta Husserliana*, Vol. VI, 1977.

[17] This significant function which the work plays within the author's *self-interpretation in existence* is particularly striking in the case of Søren Kierkegaard's plight throughout his entire spread of aesthetic, moral and religious reflection (which constitutes the phases of his work) in which he attempted to dig into the deepest source of his intimately personal life-problems in order to discover or invent the most significant and "authentic" form of their understanding and interpretation. This probing and this interrogatory effort to discover the "true" significance of the life-issues confronting him at the same time established the authenticity of his personal existence wrung out of otherwise trivial life-situations, and punctuates the phases of the written creative work.

The autobiographical aspect of creative work has often been pointed out e.g., in J.-P. Sartre, *Les mots*, and most recently William Earle, *The Autobiographical Consciousness*, 1972, Times Books.

In our analysis an attempt is made to show in what intrinsic mechanism consists such an inner link between the author and his work. I propose that it consists in the *tertium quid* of *man's self-interpretation* as expressing the Human Condition as its ultimate source.

[18] The examination meant here is "radical" in the sense that it reaches the frontier

of finitude while dying in the natural order of existence. There remains to be distinguished the examination of our past existence — and its interpretation *vis à vis* discovery of the destiny mentioned above — as well as the lived repetition of past existence and its expiation. Both appertain, however, to a "transnatural" kind of reflection.

[19] Gerard Genette, *Introduction à l'Architexte*, Seuil, Paris, 1979.

[20] René Wellek and Austin Warren, *Theory of Literature*, Harcourt, Brace and World, Inc., New York, 3d. revised edition, 1956, pp. 15–46.

[21] Cf. Northrop Frye, the previously cited *Anatomy of Criticism*. Northrop Frye distinguishes four rhythmic articulations which, on the one hand, express certain modes of order, and, on the other hand, are as such articulations of order as the basis for the literary expression of man. The "rhythm of recurrence" forms, according to Frye, the Epos; the "rhythm of continuity" forms the Prose; the "rhythm of decorum" forms the Drama; the "rhythm of association" forms the Lyric. Blanchot, who also emphasizes the role of rhythm would distribute them, I believe, according to their "internal strife and reconciliation of opposites." (Cf. Maurice Blanchot, *Livre à venir*, Gallimard, Paris 1959.)

[22] Johann Wolfgang Goethe, 'Naturformen der Poesie,' *Noten und Abhandlungen zum West-Oestlichen Divan*.

In *Wilhelm Meister Wanderjahre*, Vol. II, p. 9, Goethe tells us: "Was uns aber zu strengen Forderungen zu entstehenden Gesetzen am besten berechtigt, ist dass gerade das Genie, das angeborene Talent sie am ersten begreift, ihnen den willigsten Gehorsam leistet," 1949, Zürich, Artemis Verlag.

[23] Gottfried Ephraim Lessing, *Hamburgische Dramaturgie*, p. 35. "Jedes Genie ist ein geborener Kunstrichter. Er hat die Probe aller Regeln in sich ' . . . ' deswegen ist es frei gegen die Konventionalität aber fült sich gebunden durch besonderen Regeln der Gattung, in welcher er arbeitet . . . " 1958, Kröner, Stuttgart.

[24] Friedrich Hölderlin, *Bemerkungen über die Unterschiede der Poetischen Gattungen, in Sämtliche Werke*, 1972, Wiesbaden, ed. Volume 2.

[25] This is the further development of the analyses given in the pre-cited *Eros and Logos* by the present writer of the "Creative Orchestration" — a technical term introduced by her there.

[26] Anna-Teresa Tymieniecka, 'The Creative Self and the Other in Man's Self-interpretation', *Analecta Husserliana*, Vol. VI, 1977.

[27] Cf. Anna-Teresa Tymieniecka, 'Man the Creator and His Threefold Telos', in *Analecta Husserliana*, Vol. IX, 1979.

[28] Tzvetan Todorov, *Poétique de la Prose*, Paris 1971, Ed. de Seuil.

[29] The Homeric tales, which might appear to the modern reader mainly as an epic literary work, represent in fact the life-experience and wisdom of the Ancient prehistoric Greek nation. If they were considered by the Greeks as their cultural inheritance and passed from generation to generation in the oral transference it was because they were in a poetically inspiring and pathetic fashion presenting all the possible aspects of human life-struggle, its heights and its unavoidable abysses, its prospects, dreams and illusory expectations.

[30] Cf. J. Wolfgang. Goethe, 'Wilhelm Meister's Theatralische Sendung', in *Wilhelm Meister Wanderjahre*, loc. cit.

[31] I have introduced the "dyad" : "ciphering" and its counterpart "deciphering," endowing them with a precise sense with respect to the creative activity on the one hand

and the 'creative interpretation' on the other hand, in my above mentioned study: 'The Creative Self and Man's Self-interpretation'.

[32] The subject of "pure poetry" was the theme of the present author's doctoral dissertation which she kept developing further. This is the proper place to acknowledge the author's debt and profound gratitude to her teacher and mentor, the French literary critic Professor Pierre-Henri Simon under whose direction she prepared her doctoral dissertation in French literature and literary theory on the subject: 'Le débat sur la poésie pure entre Brémond et Valéry', at the University of Fribourg in the years 1949–1952. Her deep interest in the problems treated in this work have been developing further through her studies of Slavic literatures with Professor Alfons Bronarski of the same university and in the postdoctoral seminar in comparative literature conducted by professor Henri Brugmans at the College of Europe, Bruges in 1952/1953; maturing with her philosophical development it has found its final form in the present work. To all the above-mentioned mentors a profound appreciation is due.

[33] The problem of "What makes poetry poetic?" is also at the center of Eastern poetics. Not knowing the Oriental languages and consequently being in no position to discuss the nature of Chinese or Japanese poetry with the acuity indispensable to treat this issue I may only refer to the study by the well known literary critic and poet from the Korean University, Seoul, who presents an entire spectrum of possible aspects involved in this Poetics', to be published in *Analecta Husserliana* devoted to the literary studies East and West. It is the unfolding of the various meanings inherent in the concept of 'Ko' that seem to correspond directly to the Western analysis of 'pure poetry'.

[34] Théodore de Banville, *Petit traité de poésie Française* (Paris, Charpentier et Fosquelle, 1894).

[35] Henri Brémond, *Le débat sur la poésie pure.*

[36] Concerning the nature of lyrical poetry the expressiveness of language has always been emphasized and is often considered the decisive factor in its "ineffable" nature. Maurice Blanchot sees in the very nature of "la parole poétique" a self-sustained reality. Speaking about the nature of poetry he insists: "la parole poétique, la parole intransitive, qui 'ne sert pas'; elle ne signifie pas; elle est." If Blanchot can hold this view about the poetic language it is not that the poetic word stays distinctively as an entity for itself; it is because, discussing, the proposals of Barthes to "liberate" the literary style from its cultural sedimentations by returning to a fallacious 'immediacy' of a solely instructive experience, he himself sees the literary experience as an indissoluble wholeness: "L'experience qu'est la littérature est une expérience totale, une question qui ne supporte pas de limite, n'accepte pas d'être stabilisé ou réduite, par exemple, à une question de langage (à moins que sous ce seul point de vue tout ne s'ébranle). Elle est la passion même de sa propre question et elle force celui qu'elle attire à entrer tout entier dans cette question". The pre-cited *Le livre à venir*, p. 254.

But, then, the question still emerges: "In virtue of what human reality does this unique completeness arise?" This is the question that I am concerned with here.

The nature of language itself as the medium of expressiveness of literature has been subjected to a most thorough scrutiny in the most recent times. Between the apriorism of language, when considered with reference to the Cartesian schema of the universal philosophical discourse and the "innate ideas" as its center (as seems to be the position of Chomsky) at the one extreme, and at the other extreme, the total arbitrariness of linguistic forms when considered with reference to the pre-conscious, subterranean – as I call it – strivings, pulsations and drives within the human functioning which do seem to have some sort of articulation but infinitely looser than the rules of our established linguistic systems (as Jacques Derrida sees it), there seems to be a vast spectrum of positions putting into doubt the established naive faith into the power of communication that language has enjoyed.

This revolution erupting at the crosscurrents of several lines of reflection is, however, not only basically philosophical but it corresponds also to the deep-seated phenomenological problems concerning the passage from the so-called forms of consciousness which emerge in linguistic forms. The debate between Paul Ricoeur and Jacques Derrida on the subject of the metaphor ultimately deals with the role which the metaphor as a semantic form may play in this passage. Paul Ricoeur seems to attempt in his new analysis of the metaphor to account for this passage (cf. Paul Ricoeur, *La Metaphore Vive*, Paris, 1975, Seuil). Yet a serious doubt emerges: the metaphor is already a semantic structure, that is, as a fruit of the constitutive faculties, it is already carrying in itself the constitutive system of meaningfulness. The crucial question is: "How does the metaphoric configuration of meaning, basically constitutively pre-determined, account for the original SENSE which the pre-constitutive pulsations, strivings of forces, nostalgia etc. may contain?" The complementary question of primary importance is: "How does it come about that the semantic schema of the constitutive apparatus could transmit the pre-conscious pulsations in another way than that which is strictly stereotypical and repetitive?"

I would maintain that no semantic construct could accomplish this task without some special means serving as the crucial intermediary. I have proposed as such means the "cipher."

In his attempt to discover the fundamental means of this passage from the — what I càll — *subliminal* realm of the pre-conscious virtualities of man, to the poetic form of language, Northrop Frye recurs to "two crude, primitive archaic forms." One of them is the metaphor and the other he calls 'simile.' He explains: ". . . the first (simile) is by likeness to something else"; the second (the metaphor), saying "that is that" you turn your back on logic, because logically two things can never be the same thing and still remain two things . . ." These two, simile and metaphor, are for Frye the "archaic forms of thought," because his (the poet's) job is not to describe nature, but show you a world completely absorbed and possessed by human mind . . ." The motive for metaphor according to Wallace Stevens, is "a desire to associate, and finally to identify, the human mind with what goes on outside it . . ." p. 32–33, *The Educated Imagination*, 1964, Bloomington, Indiana University Press.

What is the role of the metaphor — and it is a basic one — in poetic expression, is not to be argued here further, since what I attempt to pinpoint is precisely the very *condition* and *virtualities* of the human being for metaphoric expression.

[37] Roman Jacobson goes into the very nature of language as such to find the source of "poeticity". To begin with, like Blanchot and many others he insists on the autonomy of poetic language. "Qu'est-ce que la poésie? Mais comment la poéticité se manifeste-elle? En ceci, que le mot ressenti comme mot et non comme simple substitution de l'object nommé ni comme explosion d'émotion. En ceci, que les mots et leur syntaxe, leur signification, leur forme externe et interne ne sont pas les indices indifferents de la réalité, mais possèdent leur propre poids et leur propre valeur".

However, we learn immediately that this poetic autonomy of language resides, according to Jacobson, in the "poetic function" of language as such: "La fonction poétique (du language) organize et dirige l'oeuvre poétique."

Jacobson's idea that one of the three fundamental functions of language itself is poetic merits great attention in my considerations. On the one hand, it attributes to language as such the role of conductor of the poetic 'fluid' throughout all modes of human linguistic expression and as unlimited to literature alone. In this it seems that Jacobson's intuition meets my proposal to admit the common streak of literary *creativity* and *man's self-interpretation in existence* in general. Furthermore, it seems to see this common streak in the poetic element which enters into human *self-interpretation in existence* through language. However, the question arises: 'In virtue of what is the "poetic" element one of the three basic functions of language and, as such, pervasive of all realms of man's discourse?' Furthermore we must ask: 'In virtue of what the poetic nature of poetry may be based in the facts that the poetic language has 'its own weight and its own value' independently from reality they may express?' Lastly, the question arises: "If we expand the poetic function through language into the entire universe of human discourse, should we not assume that it pervades all realms of human manifestation, that is also those which are not enverbalized in the discursive language?" It is in this sense that we speak about the "language of art", "language of Nature" etc. in general. All these questions point toward the direction which is being taken in the present work. Cf. Roman Jacobson, *Huits Questions de poètique*, p. 46, Paris 1971, Editions de Seuil.

Albert Hofstatter seems to come closer then other scholars to the probing of this investigation seeking the source of poetry in the nature of linguistic expression by seeing in language" articulation of human being", pp. 83–86, *Truth and Art*, New York 1965, Columbia University Press.

[38] Indeed, Verlaine makes his poem sing, yet the great poetic accomplishments of his art (e.g., *Le pauvre Gaspard* or "*Le ciel est par-dessus le toit si bleu si calme . . .* "), which are thoroughly "musical", and especially the last one carried by a "sing-song" tune, do not owe their deeply moving quality which brings the reader into a thoroughly melancholy, ('metaphysical') mood to this musical harmony. It is the lyrical substance which breaks through at the points of musical dissonances and transmits itself through the rhythmic and melodic discontinuities. The musicality alone, as masterly as it might be conjured up, does not yield more than a lulling aesthetic effect.

[39] The metaphysical function of the "lyrical moment" was the subject of my lectures at the Chinese University of Hong Kong in October 1979 and at the seminar of Professor Jiro Watanabe at the University of Tokyo in November 1979.

[40] Is not the lyrical moment an outcome of several elements combined, which could be calculated in advance? Can we recur for such a calculation to the power of imaginative associations and transformations? Obviously not. We may agree with Livingston Lowes (cf. *The Road to Xanadu*, fourth ed. London 1940, Constable and Co.) that there is an intimate connection between the memory of particularly striking images, events, etc. retained by the poets imagination from real life and his poetry. When Beaudelaire tells us: "J'ai tant de souvenir comme si j'avais cent ans", he does not indicate an inventory of a haphazard residua of memory obscuring his field of consciousness but a reservoir of striking items that his *imaginative memory* has collected and retained for future poetic use. Yet when it comes to the relationship between the retained images, ideas, feelings and emotions and their transposition into the poem we have to disagree with Livingston Lowes on several points. In the first place his stress upon the "associative hooks" and the role of imagination in "transmuting" the images from the memory into those of the poem in the process of being created is misplaced. Indeed, they do not

account for the transition of the trivial matter-of-fact-mode of the retained images to the emotionally charged intense and dense dynamic through which the poetic images "corresponding" to the bits and pieces of recollection presentify themselves in the poem. In his most penetrating and illuminating comparisons between Coleridge's logbooks in which images collected from the journals and reminiscences of sailors at the South Sea seem to relate to the corresponding *Ancient Mariner*, Livingston Lowes shows us the radical contrast between the modality of the *phantastic* description given by the sailor and the *poetic* modality of corresponding images; yet neither the vividness and strikingly appealing colorfulness of the poetic mode, neither the intense emotional charge of surprise, strangeness and mystery simultaneously appealing and frightening, nor the transcendent foreboding prophetism emerging from their chain may account for the "electrifying" current which sets in with the stanza introducing the "wedding guest" and runs throughout his story till the death of the albatross. It is not by the pathetic sequence of the inner development of the narrator, upon which the emphasis of the author seems to fall that we are captivated but by its *lyrical cadences* with their visionary outcome. The poet is a master in calculating the strategies of channels by which the lyrical current may be transported but as Paul Valéry says "Le premier vers est donné", which means that the onset of the lyrical outbreak is beyond poetic calculation: it breaks out from the abysses of the human *subliminal* realm.

(It is interesting to note that from among a host of various treatments of the relation of the writer and his work, Thomas Mann in *Lotte in Weimar*, through a series of interludes between the supposed "personage" for the sake of whom the young Werther has committed suicide and the author Goethe himself, twenty years after the *Sufferings of the Young Werther* were written, attempts to show how the real-life episodes and the "image" that the author took from the real-life person that he has cherished or loved, were transfigured and transmuted through the apparatus of his "creative forge", as I call it, to the point that the real life heroin of the *Sufferings* and her family, of which in the story such an intimately cherished picture is given, are twenty years hence almost beyond recognition. It would be also most interesting to study the points of this creative transformation of the real life depth of the author by comparing the loves of Rainer Maria Rilke and the role they played within his work. On the extreme opposite of such an assumed interconnection between the real-life images, feelings and experiences as they appear in the recording imagination of the writer is the poetic study by André Gide entitled *Paludes*. There we follow day by day and almost hour by hour the trivia of the life of an imaginary author together with the texts of writing which he produces simultaneously. No connection of any sort is apparent. I would venture a supposition that Gide does not grant his fictive author the capacity to analyze his personal experiences at any depth. Hence the discrepancy.)

[41] The incomparable power of incantation residing in the lyrical moment makes popular ballads, folksongs, minstrelsy (e.g., the ballad of Lord Randall, the English and French country songs of the XVIth and XVIIth centuries) as strikingly enjoyable in our times with mores so greatly changed from the times in which they originated. Enough to see how present day music is strongly inspired by it (e.g., the success of the Alfred Deller Consort specializing in the country music of "old days" or of Dietrich Fischer Diskau, contemporary minstrel of the German *Lieder* of the Romantic period). It is obvious that contemporary music of the new wave of popular composers seeks the renovation of music by looking for inspiration in African and

Old European folklore. However, even if we considered specific musically of poetry to be the effect of most of its other aesthetically powerful features like rhythm, sonority of language and its special sound selection, colorfulness of briefly appearing imagery, simplicity and directness of feelings expressed, the effusive nature of feelings expressed on the one hand and their vaguely evocative effectiveness on the other hand etc. – all this is not enough to explain by an 'alchemy' of elements the uniquely 'poetic' lyric current which flows through a poem even if this alchemy is lacking, if rhythmic assonances break at most points, feelings are unclear and unevocative, and sonority mediocre.

This unique lyric power of poetry has been long used in Oriental literature to enhance literary prose. Dispersed through the novel, short story and poignant pieces of poetry are the devices of the old Chinese and Japanese narrative. It is enough to mention the Ch'Uan-Chi stories of the T'ang Dynasty (e.g., *The Jade Kuan-Yin Carver*, which begins already with a long poem in the Chinese literary tradition and Lady Murasakis *The Tale of Genji*, in XIth century Japan).

Although music – the choir – played an important role in Greek drama creating atmosphere for the enhancement of the crucial points of the drama it is in Shakespeare that the lyric moment has been brought to its fruition.

In several Shakespearean tragedies the lyrical poems sang at strategic moments of the plot, like the song of Desdemona in Othello or of Ophelia in Hamlet, gives to the tragic knot of the drama a special significance which pervades the entire plot and lifts the drama to a unique plane. The opening line of Desdemona's song: "O willow, willow, willow..." by its own incantation brings us out of the immediate plane upon which the tragic situation of Desdemona is being worked out to its dénouement. The song, without any direct relation to it, through the tonality of its own profound life expressing Desdemona's loving acceptance of the dark foreboding of death from the hand of the beloved – which she cannot clearly guess – lifts us from the intricacies of the events themselves into the lyric moment in which the assuming of destiny makes love and death one. We may then conceive of such lyrics in Shakespeare as of the calculated stroke of the author's genius by which he brings us from the masterful presentation of the dramatic plot – which would otherwise repose in itself – into the very source of human existence. No wonder that the lyric moment, having been properly received, pervades the entire drama; the spectator or reader feels a generative stream of his existence released within her or his entire being.

The great classic tragedies of Racine and Corneille as well as the Romantic dramas of Goethe and Schiller, lacking this lyrical factor of the especially devised song do not – like those mentioned – transcend the peak of the dramatic accompaniment. In the contemporary drama Paul Claudel makes, in contrast, a great case for the role of music.

[42] Thomas Carlyle, *The Best of Carlyle: Selected Essays*, ed. H. L. Creek (New York, Nelson, 1929).

[43] The magic of Gauillaume Apollinaire's extraordinary poetic sequence *La Chanson du Mal-Aimé*, does not carry any sequence of meaning. The separate stanzas, each of which makes an appeal to the author's as well as to the reader's meaningful scraps of memories, become striking when followed by the next one, which begins with its own separate self-enclosed emotive charge, equally mysterious and life-remote as the preceding one (except for some which have a sequel in the next one), and the monotonous melodic line that each of them perpetuates would not yield even a musical effect due to the repetitiveness at the long stretch of the poetic sequence. Yet breaking off the emotive charge with its evocative significance in each stanza and picking it up again in the

next one quite differently modulated allows the pure stream of lyricism to come to its own. Due to these astonishing ruptures of the poetic units in their emotive modulations we release within ourselves the lyrical streamlet continuing from one to the next. From one rupture to another we are carried onward with our profound self being ever more deeply immersed in the subterranean lyrical current flowing within us.

[44] In fact, it should be stressed that the very conception of 'poetics' and its role stems from a particular approach to literary work itself. If we approach literary work as the basis from which all the directives for the investigation of literary should be drawn, we place the ontological structure of literary work of art into the center and, as happens in the case of Ingarden, which I will discuss next, poetics is relegated from its central role to that of synthesizing various sectors of literary scholarship.

If, with Roman Jacobson we approach literature from the point of view of linguistic rules, principles and modes of intelligibility, placing in the center the constitutive role of language, we will see 'poetics' as grounded in the theory of linguistics (cf. 'Poetics in the Light of Linguistics').

In my own approach I attempt to free the conception of 'poetics' from distorting biases and place it at the very heart of literary creativity and scholarship.

[45] This succinct presentation of the spread of literary investigations and of the role of poetics within it stems from my extensive lecture given at the Annual Phenomenology and Literature Seminar of the ISPL in Cambridge in 1980 and is repeated in a shorter version at the Ewha Women's University of Korea, Seoul in November 1980.

[46] Cf. René Wellek, the previously cited *Theory of Literature,* Chapter 1 pp. 17–21.

[47] Roman Ingarden, 'O Poetyce', in *Studia z Estetyki,* Vol. I, pp. 255–264, 1945, Warszawa. PAN.

Poetics presented by Ingarden is the unelaborated synopsis of all approaches to literature (pp. 306–315). This synopsis gives us a carefully differentiated spectrum of all the possible modalities in which the literary work can be envisaged – a real "organon" of investigation. What a pity, that even had it been eleborated it would by organizing the synoptic map around a wrongly understood axis, fail altogether in lying down the "organon of method".

As Ingarden himself points out in the preface to his *Studies in Aesthetics,* (Vol. I Warszawa, PAN ed. 1957) his conception of 'poetics' has been conceived by him as a complement to his work *O poznawaniu Dzieła Literackiego,* in which, as he says, he proposed: "an outline of a new foundation for general aesthetics: its complement and simultaneously a certain correction brings the study 'On Poetics' (O Poetyce), which has appeared in full for the first time in this volume" (in summary it appeared first in *Sprawozdania Polskiej Akademji Umiejętności* in 1945).

[48] I have brought out and emphasized this correlation in Ingarden's conception of literary work between its structure and cognition in my monograph, 'Beyond Ingarden's Idealism/Realism Controversy with Husserl,' in *Analecta Husserliana,* Vol. IV, 1977. For Ingarden's ontology cf. Anna-Teresa Tymieniecka, *Essence et Existence,* 1956 Paris, Aubier. It remains the only book on Ingarden's philosophy.

[49] As was pointed out long ago (cf. my *Phenomenology and Science in Contemporary European Thought,* Part I, New York, Farrar Strauss and Cudahy, The Noonday Press, 1961), Ingarden has established a structural basis for the approach to the literary work of art through his ontological investigation in *Das Literarische Kusstwerk* – now translated into English by Georges Grabowicz and published by the Northwestern University

Press, 1973. Concurrently with this investigation, he has worked out in his book *O poznawaniu Dziela Literackiego* the structure of the cognition of the literary work. These two investigations are however interdependent in their very conception and their interdependence, which I have presented as the "correlation between the ontological structure of the literary work and the transcendental processes in which it is cognitively retrieved by the reader" (cf. the pre-cited, 'Beyond Ingarden's Idealism/Realism Controversy with Husserl') is to be brought to the ontological structure of the work. Furthermore, both of these investigations are biased by Ingarden's strictly eidetic approach in treating these twofold structures in spite of the fact that in a truly phenomenological vein, he is clothing the abstractness of the eidetic formalism with many concrete examples from the psychic life and literary experience.

[50] This succint presentation of Ingarden's poetics as a 'system of literary studies' comes from the more extensive treatment of the classification and unity of literary scholarship presented at the annual Phenomenology and Literature Seminar, held by the ISPL in Cambridge, Mass. 1970. It has also been a part of the lecture given by the author at the invitation of Dean Holingsworth, at the University of Michigan State, in October 1980.

[51] Cf. Georges Poulet, *La distance intérieure*, trans. as *The Interior Distances* by Elliot Colemann, 1964 Ann Arbor, University of Michigan Press; Emil Staiger, *Die Kunst der Interpretation*, 3rd. ed. Zürich, 1961, Atlantis; Hillis Miller, *Charles Dickens: The World of his Novels*, Cambridge, Mass, 1958, Harvard University Press.

[52] Martin Heidegger, *Der Ursprung des Kunstwerkes*, Universal-Bibliothek No. 8446/47, Stuttgart, 1954, Reclam. For surveys of phenomenologically inspired aesthetics and literary theories, cf.: Michael Murray, *Modern Critical Theory: A Phenomenological Introduction*, Martinus Nijhoff, The Hague, 1975; For a remarkably acqurate and detailed account of the origin of phenomenological aesthetics, cf. Gabriele Scaramuzza, *Origini del Estetica Fenomenologica*, Padua 1976, Antemore; Eugene Kaelin, *An Existentialist Aesthetics: The Theories of Sartre and Merleau-Ponty*, Madison 1962, University of Wisconsin Press and *Art and Existence*, Lewisburg 1970, Bucknell University Press; Robert Magliola, *Phenomenology and Literature, an Introduction*, West Lafayette, Indiana, 1977, Purdue University Press.

[53] Cf., by the present writer, *Phenomenology and Science in Contemporary European Thought*, Farrar, Straus and Giroux, New York, 1960.

[54] Moritz Geiger, *Die Menschliche Bedeutung der Kunst*, edited by Klaus Berger, 1980, Freiburg in Br., Fink Verlag.

[55] This process of discovering and assimilating the *existence-significance* of art by the reader is a most complex one. We may see it in the writers autobiographical writings.

Johann Wolfgang v. Goethe in his fascinating work *Die Theatralische Sendung von Wilhelm Meister*, written by the young poet after *Die Leiden des Jungen Werthers* and before the two other volumes of Wilhelm Meister's *Wanderjahren*, in which — as in the two others — is presenting autobiographically the inward plight of the poetic genius working its way through the primitiveness of the a-poetic, or pre-poetic human nature. Goethe is unraveling the initiation to the aesthetic universe of values, artistic sensibility and the entire transformation of his inner self following it, through his childhood initiation to the theater. Showing himself as a child who easily isolates himself from the family circle retiring into himself, he shows how the intrusion of the puppet theatre into his life had been a turning point in his inward development. At the same time he outlines for us the way in which dramatic art breaks into real life and, making us assimilate its specific significance with our current means, transforms these very means and our entire sensibility. The first appearance of the theatrical presentation is according to the young Goethe preceded by: "hopeful expectations, impulsions" etc. and the first reaction to the first theatrical presentation on stage is: "the enjoyment of surprise and astonishment". At the second presentation of the same spectacle we enter into the state of "attention and investigation". After the spectacle the images, expressions, recitatives, feelings, experience return in differently arranged sequences within the experience of the spectator. They re-appear in relation to the participants. (Cf. Chapter 4.) Goethe describes in detail the twofold magic of the drama on stage: the one, brought about by the actors of the play, the other of the 'magic' incantation into which the spectators are plunged. This magic makes the child desire to be initiated into this great mystery: to be the one as well as the other. At the third step, he desires naturally to create himself the dramatic text and to direct all the magic motions. The author proceeds in describing how this 'magic' experience of the unknown and mysterious gives simultaneously an appeasement to the initial expectation while it deepens and intensifies the initial disquiet of imagining to have found and discovered something and simultaneously intuiting in this very experience not to know anything about it. Thus the spectator begins his initiation to the nature of art and is drawn deeper and deeper into marvelling in ever repeated attempts to penetrate the mystery of the dramatic performance.

Goethe exfoliates in detail this initiation which takes a decisive inward turn in the attempt by the spectator to put himself in the role of the protagonist of the play. He considers it the "initiation in the great mystery". There is no mode to imitate in this undertaking. The 'great actors', imitated by lesser ones or given as examples to study the 'trade', must have invented all by themselves. It is precisely from this necessity to invent our way of assimilating the role of a protagonist which creates the great fervor which animates the actor (or the spectator in the same attempt). (Cf. Chapter 8.)

In this inventive effort we simultaneously discover the significance of moral, aesthetic, spiritual values which the drama has "ciphered" and transmutes our own sensibility accordingly. Goethe tells us how, also by reading the great masters (*Gerusalemme Liberata* by Torquato Tasso), through the images received and turned around in imagination in infinite variations, the deepening of which has awakened within him the significant values of the story, he has learned to love Clorinda and hate Argante. (Cf. Chapter IX.)

[56] This argument holds against the model of a "stratified structure" with its correlative "phases" of the reading of the work by Ingarden, which yields the one, unique significance of the literary work as an "aesthetic object", an instrument of the aesthetic enjoyment, as well as against the Heideggerian conception of the literary work.

PART I

PESSIMISM AND OPTIMISM IN THE HUMAN CONDITION: THE LIMIT SITUATIONS OF EXISTENCE

A. M. VÁZQUEZ-BIGI

THE PRESENT TIDE OF PESSIMISM IN PHILOSOPHY AND LETTERS: AGONISTIC LITERATURE AND CERVANTES' MESSAGE

I. THE PESSIMISM-OPTIMISM CONTROVERSY AND THE POSITION OF LITERATURE

In beginning I shall recur to the fifth volume of *Analecta Husserliana*, in which papers and debate of the conference held by the International Husserl and Phenomenological Research Society in Montreal four years ago are reproduced, because its general subject, "the crisis of culture," seems to be related to our own subject here today. In particular the prologue by Anna-Teresa Tymieniecka could serve with few changes in stress but the same content to put us in an expansive, though proper, mental frame for the discussions that we will hold. The very first subtitle by Dr. Tymieniecka, "La crise de l'homme: Y-a-t-il encore une tâche pour la philosophie?"[1] could, without much effort, lead to a parallel question, "Is there still a task for literature?" Granted that programming future literature could be even more difficult, even unrealistic, than foretelling the development of philosophy; however, both the lovers of letters and philosophy are by no means immune to the quixotic infirmity, or heroic endeavor, or pretension if you will ... *voilà*, before knowing it I am arriving at my major paradigm. But let us continue with our approach; let us see our matter in perspective first.

Professor Tymieniecka situates the present cultural crisis historically: "a focus of interest since Spengler's *Der Untergang des Abendlandes*, and reappraised in relation to transcendental phenomenology by Husserl in his *Crisis*, has finally reached a culmination." The crisis has now reached the very core of the human being; his innermost tendencies are "permeated by the attitude of *radical pessimism toward the human condition*".[2] This note is stressed in the prologue: "the major and most profound issue upon which the destiny of humanity depends is the *controversy between the pessimistic and the optimistic versions* of man and life."[3] Professor Tymieniecka recalls that the controversy between pessimism and optimism is as old as human culture; she recalls the names of Aristophanes, Erasmus, Nietzsche, and Leopardi, to whom others could be added; I myself will have occasion to refer to the deep tide of pessimism at the turn of the century, the far-reaching effects of which cannot be clearly separated from the most destructive tendencies and somber cultural evaluations manifested in our day.

The quest after phenomenological guidelines for the investigation of man and the human condition has already led Dr. Tymieniecka through recent philosophy to contemporary literature. According to this line of inquiry, man's life appears as being "nothing else but a course of animal survival, a play of circumstances." Today's literature bears witness to a total devaluation of life; in the prologuist's words, "indeed, if everything is to be brought back to human survival, all the ideals and values (fairness, patriotism, self-sacrifice, family inheritance, loyalty, disinterested satisfaction from a well-accomplished work, great deeds, heroism, truth, justice, etc.) which have led mankind throughout the ages, inspiring, giving meaning to actions, pursuits, and struggles, seem to have been emptied of any intrinsic and transcending value."[4]

The subjects debated on that occasion and later published in Volume 5 of the *Analecta*, especially in the manner they are introduced in the prologue referred to above, may indeed be the germ of the idea informing the theme for our conference. Implicit in the quest after the heroic in literature and the artist's commitment is the question of whether some classics of literature, perhaps all the great classics, could serve as antidotes for the pessimistic and nihilistic venoms that are distintegrating the basic values of our culture. Ultimately, the daring question remains, "Is there still a task for literature?" A full commitment to such a task would make of anyone no less than a heroic artist or philosopher, and probably a quixotic hero (or, of a mere critic and expounder, a quixotic soldier of a humanistic cause) — a seemingly bizarre consideration evoking again the Cervantesque myth: the exalted, adventuresome, idealistic hero.

Professor Tymieniecka's formulation aims at the possibility of a literature that affirms values, but she closes it with a question mark, thus opening the way to philosophical exploration. Or — in our evocative vein — to the "adventures" of the spirit.

PROPHETS, FALLEN WARRIORS, AND CRUSADERS — ADVENT OF THE LATEST PESSIMISTIC TIDE

Chesterton, a great reader and re-creator of Don Quixote,[5] called Cervantes' book the best remedy for pessimism. Chesterton himself had experienced the terrible wave of pessimism that had gradually engulfed the European mind in the second half of the nineteenth century and reached its culmination at the turn of the century, to which wave he reacted in a work that in turn became an inspiration for many to emerge from those dark waters. Chesterton, whom Bernard Shaw called the greatest literary genius of his day, has not continued to attract in the Anglo-Saxon countries the widespread attention he has enjoyed outside them, not only by the Christian intelligentsia or more generally affirmative thinkers, but by agonistic minds such as Borges'. Although huge and fat and a humorist like his immortal Sunday, and a praiser of lunatics instead of being one of them, Chesterton was himself a sort of Don Quixote, a crusader of thought who could very well deserve special treatment in this conference; let me indulge in extending this mention as an ancillary topic, leading to the examination of a parallel cultural situation that may add perspective to the historical view of our own crisis.

In his study of George Bernard Shaw, Chesterton gives his testimony:

Bernard Shaw and I (who are growing grey together) can remember an epoch which many of his followers do not know: an epoch of real pessimism. The years from 1885 to 1898 were like the hours of afternoon in a rich house with large rooms; the hours before tea-time. They believed in nothing except good manners, and the essence of good manners is to conceal a yawn. A yawn may be defined as a silent yell...I meet men who, when I knew them in 1898, were just a little too lazy to destroy the universe.[6]

To that period correspond the gloomiest creations of English literature; on our side of the Atlantic we had to wait until the last decade, perhaps with the single exception of *U. S. A.*, Dos Passos' trilogy of the 1930s, to find a comparable predominance of pessimism in fiction. And the total collapse of faith in life among the British writers of the epoch was not limited to the world of fiction. J. M. Kennedy in his still authoritative study of English literature in the years between 1880 and 1905, notes that "Crackanthorpe, Adams, Lawrence Hope, John Davidson, and St. John Hankin deliberately took their lives. Charles Conder died insane. Overindulgence in drink led to the premature deaths, in deplorable circumstances, of Lionel Johnson and Ernest Dowson." The critic of that period concludes that this list is not exhaustive. "The last thirty years of the nineteenth century were gloomy all over Europe, and nowhere more than in England," where writers were "afflicted with the blackest of despair."[7]

Schopenhauer's philosophy was the most important ideological background of that pessimistic wave. Chronology will not give us trouble if we remember that Schopenhauer was not known in intellectual and literary circles before mid-century, and essayistic literature about his work and pessimism in general needed an additional quarter of a century to appear, when other factors for disillusionment – political, economic, scientific – had been felt on the Continent and the Isles. Thus James Sully, for instance, published *Pessimism* in 1877, and still in 1891, in the prologue to the second edition of his book, found it relevant to observe for his readers that the fashion of Byronic despair was past; pessimism was now a new philosophical attitude and vogue.[8]

In a different measure and with minor chronological variation these circumstances were common to the principal European nations, but the philosophy and prohetic thought of pessimism was provided to all by Germany. To the vogue of Schopenhauer succeeded or was added toward the end of the century that of Eduard von Hartmann, whose reputation rested on *Die Philosophie des Unbewussten*, a successful work that had already gone through six editions five years after it appeared in 1869. Its fortune among the wider intellectual public lasted until after World War I, when the Freudian conception of the unconscious — for which von Hartmann's is one of the antecedents — became in turn a general and literary vogue.

Von Hartmann's philosophy was a vigorous rationalistic effort to reconcile physical science with idealistic philosophy, Darwinism with metaphysics, rationalism with irrationalism, Hegel with Schopenhauer. His laboriously built system has received divergent interpretations, although according to the most commonly held view he is the culmination of *fin de siècle* pessimism and has contributed to the nihilism of our century. He criticized Schopenhauer's belief in the effectiveness of asceticism carried to its ultimate extreme — or of any individual striving — to achieve the annihilation of the will;[9] this must be the aim not of the individual but of mankind. "The *first* condition" in attaining it is the progressive concentration in humanity of the largest part of the unconscious Spirit manifesting itself in the present world; only then can the negation of will annihilate the total volition of the world, causing the whole cosmos to disappear at the same time.[10] Von Hartman makes here a distinction between the former process and a mere mass suicide of mankind, which would not be successful in order to attain its aim — here he is applying Schopenhauer's reasoning about the failure of suicide in negating the will. "The *second* condition" of the "success" is that the consciousness of mankind become thoroughly "penetrated" by the folly of volition and the wretchedness and worthlessness of existence. Once the cosmic-total negation of will is attained, volition and suffering will cease at one and the same time; consciousness will cast volition back into the unconscious absolute.[11]

It is not easy to understand how anybody could have interpreted von Hartmann's message as other than utterly pessimistic.[12] The literary minds of his day, in adopting ingenuously that message, never saw it otherwise. A clear example from the best literature is provided by Thomas Hardy. A theme recurring in Hardy's novels is that of frustrated suicide; characters with a suicidal bent abound in them, like the protagonist of *Jude the Obscure* (1895), or Henchard in *The Mayor of Casterbridge* (1887). The deaths of these characters are virtual suicides in their intense desire and consciousness of annihilation. Jude's biblical-Byronic accents on his deathbed, while the music and shouting of a sporting event reach him from afar (the instinctive will to life manifesting itself in gaiety), recall the Calderonian verses repeatedly quoted by Schopenhauer:

> for the greatest crime of man,
> is to have been born.[13]

But however diffuse or imprecise, von Hartmann's echoes resound throughout the novel, as when, after Jude's little son hangs his still younger brother and sister and himself, the doctor examining the dead children declares that it was in the older boy's nature to do it; children like him were "springing up amongst us ... the outcome of new views of life ... it is the beginning of the coming universal wish not to live."[14] Von Hartmann was understood in this manner, that is, as the prophet of the collective self-annihilation of mankind.

And he was not the only prophet of the "will to die" or philosophical follower of Schopenhauer's pessimistic vitalism. Philipp Mainländer (Philipp Batz), who was called "a new Messiah,"[15] anticipated von Hartmann in a general idea of a cosmic will directed toward its own absolute annihilation as well as conceiving its particularization in the multiplicity of the world (present), as a stage between original unity and future, ultimate nonexistence.[16] Mainländer regarded himself as a disciple of Schopenhauer, but refuted his master's opposition to suicide; for Mainländer it was a legitimate and effective way to negate the will, and he killed himself according to principle. It is interesting to remember that the hero of the homonymous novel *Mike Fletcher* (1889) by George Moore similarly disents from Schopenhauer's opinion and ends up committing suicide. And − staying with English fiction − the protagonist of Lucas Malet's *The Ways of Sin* (1891), also a follower of the old master's philosophy, does not actually commit suicide, but when falling or being pushed over a cliff, "with a great shout ... − a shout of triumph, of consummated warfare, of emancipation, of hope − that strong soul hailed Death, − the consoler, the restorer, 'delicate Death' − sitting waiting for him just this side of the white line of the slow breaking waves on the purple-grey shingle fifty feet below." These are the last words in Lucas Malet's novel.

In going through these literary testimonies we can still detect faint — occasionally resonant — romantic reverberations, but from them is absent the romantic *Weltschmerz* with its heroes predisposed to weeping or being the shoulder for weepers to lean on; it was a new epoch, with a new science and a new political economy. Likewise a clearly detectable change occurs from the *fin de siécle* novelists of doom and their successors during the following decades down to their counterparts in recent years — our century has already borne an awful portion of history! — and we shall endeavor to analyze some elements of the change. Still, there is a common basic element or factor in all pessimistic and nihilistic expressions of the last hundred years, a common element or factor that can be perceived primarily in philosophy.

The philosophers of the romantic age were Kant's followers prior to Schopenhauer. Kant had experienced a gradual change from confident optimism under the influence of Leibniz and Wolff, toward philosophical pessimism. But Kant had never considered himself a pessimist, and never was one beyond the phenomenal realm; his lifetime task was to save ethics and religion without rejecting the mechanicism observable in the empirical world. It was Schopenhauer who punctured the spiritual values, or illusions, as they were preserved in the Kantian system, and he was also a readable thinker who made Kant intelligible and accessible, in his own interpretation, to the wider intellectual public — with this mention we could not omit referring to the novel that incorporated into fiction the philosophical relationship between Kant and Schopenhauer in the development of Western pessimism, Pío Baroja's *The Tree of Knowledge*.[17] Schopenhauer became the great inspirer of Wagner and Nietzsche as well as the philosophers who continued in his pessimistic vein, and we have seen his original message being transmitted in thought and literature, directly or indirectly, during the first part of our century. And as a source he is still detectable in Heidegger's pessimism and the tragic side of existentialism; likewise his conception of the will to live as absolute and still absurd — since it has no origin or meaning — is at the root of the recent doctrine and literary expressions of the absurd.

Accordingly, although our century has witnessed several successive waves of pessimism and nihilism, it is legitimate to speak of the present "tide" of pessimism as beginning in the second half of the last century, the period of Schopenhauer's strongest influence. And if we focus intellectual attitudes, we observe during the same period, in its entirety, the full growth of the blind speculative one-sidedness pointed out by Professor Tymieniecka in the essay quoted above. In her words, "it seems, strangely enough, to stem from the *abuse of the power of reason*, that is, from extending the faith in its power beyond its legitimate reach and prerogatives."[18] At a much inferior level, even the farthest reaching manifestations of irrationalism — we know only too well — were the result of a proud *excess* of incompetent reasoning.

Philosophers have made the pertinent distinctions between individual thinkers and schools of thought. In this succinct presentation I will keep the focus upon attitudes in which I observe a general change from the pessimists of eighty or ninety years ago to those of the very latest waves — a perhaps profound change if we believe that attitudes may be more significant than theories. Many of the older pessimists and nihilists were at least capable of one strong belief that could reach heroic dimensions: the belief in their very pessimism. That strong belief, that heroic commitment to what appears as truth, I feel also in a work of literary art such as Thomas Hardy's *Jude the Obscure*; I do not find it in Kurt Vonnegut's pages, for instance. Naturally, we cannot summarily dismiss all contemporary artists and philosophers in this consideration. There is agony in Adamov, Genet, Beckett; some of Chesterton's best fiction can be linked to the Book of Job and successively compared to the best of Kafka's and Borges' stories in the common theme of the ultimate challenge to God — and not in vain the latest phase of American fiction has been called "Borgesian."[19] However, our times seem not to produce that paradoxical, heroic, uncompromising former commitment to unbelief.

It may be that pessimism has settled within our culture. Is it that we are being overcome by the feeling of the uselessness of it all, pessimism included? The belief in an absurd contingency of life from an even more unintelligible all-eternity may have been transformed into a vague, unreasonable awareness of the *nonauthenticity* of life itself. The representative writer of today may be precisely Vonnegut rather than — keeping our attention on American fiction — Pynchon or Barth; analogous comparisons could be drawn with examples from recent French literature, Argentine ... wherever there is a sizeable avant-garde literary production.

This leads to an observation of a more objective nature. Hardy's novels were intended for intellectual readers; Vonnegut's fiction is read by teenagers of our not precisely intellectual secondary schools as well as by college youth and their parents and professors. Let us remark that the terrible tide of pessimism at the turn of the century could be all but ignored by the society at large, the representative society of the conservative small bourgeois of little towns and big cities and more so that of the rich and powerful members of the high bourgeoisie, enjoying the *belle époque* and the Edwardian years, proud of their present and blindly confident in the future. Pessimism was then considered a sickness of a small intellectual minority. What cannot now be ignored, especially in the advanced Western societies, is a series of unexpected, unbelievable genocides, one monstrous world was following another, and the unsolved — perhaps unsolvable — threat of annihilation by prodigious scientific artifacts. Von Hartmann's speculation about an evolved future consciousness and its role in the needed negation of will and desirable cosmic extinction, now seems a mere rational exercise; what might practically cause the greatest conceivable destruction of humanity, even the extinction of nations, is certainly not an enhanced, spiritual consciousness but the will to power or merely the very will to live of unphilosophical and unimaginative, pugnacious supermen too attached to their possessions and too fearful of death — certainly not the kind of superman envisioned by Nietzsche either. And the new widespread pessimism can also reflect a feeling of uncertainty and helplessness, of disbelief and fear that is soaking through successive layers of the social fabric — a diffuse, not very clear feeling, at times colored by a sort of stoic acceptance, at times coloring inconsistent attitudes, like reverting in a manner of senseless bravado to popular nihilistic fiction and spectacle for entertainment.

III. LITERATURE FOR TROUBLED TIMES — THE UNIVERSALITY OF DON QUIXOTE AND CERVANTES' VOCATION

Great literary creations confront the tides of nihilism whether the heroes in them are tossed and thrashed by the dark waters — as in the best literature of the absurd — or emerge over the tide to utter their challenge, affirming transcendental values — in the manner of a Chesterton, a Berdiaeff, a Solzhenitsyn. The agonistic quest for the absolute may even seem most sincere and uncompromising when the combatant finds that it leads nowhere. Kafka's obscure parables and Borges' dazzling allegories can be modern theomachies; even Samuel Beckett's plays, which remain as the quintessence of the absurd, were linked by Ionesco to the Book of Job.

Most of the literature of universal interest and worth of the last fifty years has been one of unresolved agony. We may here take up where we left off in the introduction, with the question whether there still exists the possibility of a great literature affirmative of values.

It should be added: a vital theater, a great realistic fiction. Perhaps it is not a problem of the soul's temper but of literary genre or artistic expression;

at the level of popular entertainment, or in the revived genre of fantastic adventures, we have seen in the United States musical plays, television shows, printed romances extolling ideals and moral values. The fifties and sixties saw the overwhelming vogue of television Westerns with their varied specimens of heroes such as the Lone Ranger, Paladin, Matt Dillon, a genre descending from the knight-errant tales of old through Cervantesque transmission;[20] the following years witnessed the growth in popularity and serious recognition of Tolkien's beautiful fantastic tales, similarly related to medieval romance; some of the new popular detective shows, which the United States exports successfully, follow the former Westerns in a more modern, "realistic" vein. There is an evident public demand and perhaps a need for fiction of great deeds exalting faith and valor, fairness and justice, loyalty and self-sacrifice; upon applying this consideration to our questioning about the literary phenomenon, the problem is better delimited and defined, and the scope of our questioning centers on the artistic expression, the creator and his work.

There is a critical consensus that we find or hear from time to time — I remember having heard it from Leo Spitzer — that the most universal great character myths of Occidental literature are Hamlet, Don Quixote, Faust, and Don Juan; they are the ever-present paradigms that seem to appear and reappear as the four cardinal directions of the psyche. Out of the four, Don Quixote is the embodiment of heroism, of action with commitment to an ideal. I will devote the second part of my address to an evocation of the Cervantesque hero and the poet and man who created him.

We may assume — we have just observed it — that for a literary production to have significance in our times it will need to deal with the common man, in other words, with the deeply, universally felt human being, and be able to reach the many, transcending social barriers of class and nationality. This is one more reason that impels me to examine the heroic element and the author's commitment in the work of Cervantes. Critics agree that among the great characters of literature Don Quixote is notable for his realism and universality. In his case it is not only the extension of that universality, but its direction as well; perhaps the notion of "volume" would be a fitting image to express this phenomenon. It may be due to the fact that the book in which Don Quixote was born has been the best-seller among the classics, or because of characteristics of the hero himself; in any event, his fortune has spread not only through extensions of time and a space that we may, in search for an expressive image, call horizontal, but vertically as well: Don Quixote interests the learned and the ingenuous, the aristocrat and the commoner, the old man and the child.[21] My large English dictionary does not have the term "Hamletian"; "Faustian" is a beautifully suggestive word, *for intellectuals only*; other important characters have given origin to common terms, such as "bovarism," "tartuffe," "pantagruélique" applicable to meals in Latin cultures; perhaps "Don Juan" is the only term of that nature with a popularity comparable to that of Don Quixote, but in its usage it loses consistency with its original myth or any of the derived Don Juans of literature; only "quixotic" has become a common adjective in many if not all languages related to Western culture, is of frequent use in nonintellectual contexts, and has kept its meaning in correspondence to the original character. This portentous vitality, this universality of Don Quixote, his unique common appeal, makes him an easily applicable literary model for mankind in modern society, whose problems are common to its members at all levels.

In the consideration of Don Quixote's universality we have included the literary notion of "realism," seldom precise. The impression of apprehending reality rather than a fictional tale that one receives from Cervantes' narrative is what critics frequently understand by "realism" when referring to his creation, and is one of its salient qualities. Erich Auerbach has called it "the 'peculiarly Cervantean,'" that is, a vigorous capacity for vivid visualization.[22] This concreteness characteristic of Cervantes' creations, in the case of Don Quixote and Sancho, has expressed itself in a profuse although consistent iconography; it is only Don Quixote among creations of literature, by himself or with his squire, that we find in so many places, whether on the cover of a treatise about fiction, or on a poster for sale at a smoke shop reproducing Picasso's etching, or in the guise of a statuette in the gift department of Sears. It was only days ago that I met him in Washington, in the Museum of Natural History; there *he* was with Sancho, on a mural, illustrating not literary types but somatotypes! This life, this concreteness has also led essayists and poets, especially but not only in the Spanish language, to address Don Quixote as if he were a real entity, a historical figure, in doing which quite a number have fallen into the aesthetical sin of ridicule.

Likewise Don Quixote has been repeatedly hailed as a living symbol of the Spanish spirit and the moral values dear to it, and as a model for heroic deeds, when it would have been both easier and more sensible to erect a national myth of Don Quixote's creator, who as a creature of flesh and blood performed extraordinary heroic deeds and is himself a supreme example of courage in adversity and commitment to an ideal.[23] The very Spanish Cid, for another example, is a ready-made symbol of this nature.

Like any great work, Cervantes' creation is rich and complex; it is like life, full of apparent contradiction, and like life, it can only be understood — as far as it is possible — by the simple or by the subtle. And the subtle as well as the simple have found it to be a fountain of optimism and an affirmation of values and *joie de vivre* in spite of the misfortunes of its mad protagonist.[24] Don Quixote spends the whole tale trying hard to believe that he is a real knight errant, a real hero, and he almost never achieves it. A bookworm is what he is, and a teacher he becomes; he is the unsurpassed *teacher* of heroism in world literature, and in this respect we need him. But he is also a meddler who does a lot of nonsense and more damage than good, is a natural object of ridicule, and of course he is beaten every other day. And still, by a prodigious artistic process, he emerges as the quintessence of the heroic spirit. But referring to him as an example of behavior instead of principle is plain absurdity, and we would not need to say it if it were not for the circumstance that essayists and professors, especially in Spanish-speaking areas I think, have made that mistake.[25] And the mistake is further compounded when the poet and the essayist and the patriotic speaker refer to Don Quixote in terms befitting a real, historical figure. We may be more inclined to tolerate such interpretations bordering on the delusional if we remember and consider how great a realist, or illusionist, wrote Don Quixote's story.

Frequently critics and essayists of literature have drawn a parallel between Cervantes and his hero, in which the author himself has appeared as a Don Quixote figure. We need not overplay the parallel. As a man of arms, early in his life, Cervantes was what Alonso Quijano el Bueno could never be; a real hero who, wounded several times, fought beyond the call of duty in the battle of Lepanto in which the Christian League stopped the advance of the Ottoman Empire into Europe from the South — in Cervantes' eyes, as he describes it in the prologue to the second part of *Don Quixote*, "the greatest occasion that the past or present has ever known or the future may ever hope to see." For his heroism Miguel de Cervantes y Saavedra merited letters of commendation from the duke of Sessa and the commander in chief of the combined fleet himself, John of Austria, the son of Charles V. Later, as a captive of the Turks ruling northern Africa, he performed outstanding acts of devotion to his comrades and courage facing the enemy.[26] And no less courage he needed and maintained throughout a life of adversity, toward the end of which he created his marvelous world of fiction.

Now we may ask ourselves whether Cervantes was fully conscious of his vocation — more so, to what extent he was aware of commitment, *of the need* for commitment in life. It is a valid question for more than one reason. Among new Cervantesque studies, the latest tendency is to consider "ambiguity," which is seldom compatible with "commitment." And among old critical voices, there was an extremely vocal one according to which Cervantes did not even understand Don Quixote; although it has long been superseded, its dissonant echoes still occasionally reach our ears in academia. Moreover, as we shall see, the awareness and intention of Cervantes appear in his work and cannot be extricated from it; not only did he conceive Don Quixote's archetype in its baffling perplexity, but also our understanding of it increases upon examining Cervantes' awareness and commitment in the work of art itself.

IV. DON QUIXOTE'S HEROIC VOCATION AND CERVANTES' ART

I reread the last lines and seem to detect contradiction in my own words. A point has to be made: in another relation, Cervantes does not appear in his work, or appears very exceptionally. Erich Auerbach put it forcefully: "He does not take sides (except against badly written books); he remains neutral. It is not enough to say that he does not judge and draws no conclusions: the case is not even called, the questions are not even asked."[27] The exquisite Spanish poet and great Hispanist Dámaso Alonso pointed out Cervantes' masterly *mesura*.[28] Leo Spitzer made it clear that Cervantes' novel was not "didactic" in any simple way; moreover, the great German critic affirmed a spiritual relationship between Cervantes and Goethe that may have more than one element of truth.[29]

Only one secondary character, identified as a social type, is clearly condemned by Cervantes in *Don Quixote*, and even then the critics derived diverging and conflicting interpretations from the corresponding episode. In all other cases, that of Don Quixote included, the author never judges, not even implicitly; his ever-present irony might imply judgment though, but it operates in all directions, and even the very presence of the Cervantesque irony — being of a subtle kind — is occasionally questioned.

Two episodes, both in the second part, lend themselves especially to seeing this Cervantesque irony at work, nonetheless permitting Cervantes' commitment in life and art to shine through: one is the encounter with the gentleman — "caballero," that is, nobleman — of the green-colored greatcoat and the ensuing adventure of the lions, and the other, later in the story, the dinner of Don Quixote and Sancho with the duke and duchess and the stern churchman. Both episodes correspond to the peak of Don Quixote's career as a knight; commitment and heroism appear as theme or theory in the latter and as praxis in the former; we shall treat them in this order.

The duke and duchess are having fun at the expense of Don Quixote and

Sancho, treating Don Quixote with regal pomp in the manner of the old farcical games of "king for a day," "bishop for a day"; among other honors, they are having a special dinner with their churchman in residence attending, and the duke invites Don Quixote to take the head of the table and insists so strongly, that the latter has to accept. After witnessing a comic incident caused by Sancho, the grave ecclesiastic listens to the conversation, in which the duke goes along with Don Quixote's and Sancho's talk of knights-errant and their ladies, encounters with giants and persecution by enchanters, and he becomes indignant. In great anger — *mucha cólera*, this point is important — he reprimands the duke (it is interesting to remember that in the temporal order, and in the last account, the latter is his master):

"Your Excellency, *señor mío*, will have to answer to our Lord for what this good simple man does. This Don Quixote, or Don Simpleton, or whatever his name is, surely cannot be such a dunce as your Excellency would make him out to be by thus lending him occasion for carrying on his stupid and nonsensical acts."

Without interruption the priest turns to Don Quixote:

"And as for you, poor innocent fool, whoever put it into your head that you are a knight-errant and that you defeat giants and capture malefactors? For the sake of us all, go your way and follow my advice: return to your home and raise your children if you have any, and care for your property, and stop wandering about the world, gaping in the air and making yourself the laughingstock of all, whether they know you or not. Where in the name of goodness did you ever come upon any knights-errant, living or dead? Where are there giants in Spain, or bandits in La Mancha, or enchanted Dulcineas ...?"[30]

Rendering the characters' words *in extenso* will make us perceive "the 'typical Cervantean'" observed by Auerbach, their intense, spontaneous, independent life. The clergyman could not have been more truthful and convincing. At this point the reader even admires the priest for his boldness in addressing the duke — a boldness that still has to reach its culmination. Don Quixote has been listening very attentively, and when he sees that the clergyman has finished, disregarding the presence of the duke and duchess he springs to his feet, and trembling with restrained anger he speaks in a hurried, agitated voice. The respect for "the place and presence" where he stands and for the churchman's profession tie his hands, preventing him from redressing what would be an affront if it did not come from a man of the cloth, whose weapon is the same as that of a woman, namely, the tongue. Don Quixote will use the same weapon in doing battle with his reverence,

" . . . from whom one might have expected good counsel rather than infamous verbal abuse. Saintly and well-meant reprehension requires other circumstances and other arguments. Your having reproved me in public and so rudely has exceeded all proper bounds, for Christian reprehension is to be based on mildness rather than on asperity, and it is not right, knowing nothing of the sin being censured, to call the sinner, without more ado, a simpleton and a fool . . . "[31]

Now Don Quixote could not have been more dignified and correct and precise in his rebuke of the clergyman. It is Don Quixote's turn to attack:

"Is all that has to be done to make one's way by hook or crook into other people's houses to rule over their masters, and after in some cases having being brought up in some poverty-stricken seminary, without having seen more of the world than what is contained within twenty or thirty leagues, to take it upon oneself to lay down the laws for chivalry and pass judgment on knights-errant?"

What Don Quixote says is still pertinent and convincing. Even his sudden reference to knights-errant, his *idée fixe* to which he recurs at the end, sounds reasonable — although we know that, unfortunately, knights-errant do not exist. He continues his course of reasoning to end with a marvelous line: "A knight I am and a knight I shall die, if it please the Almighty." Don Quixote concludes:

" . . . led by my star, I follow the narrow path of knight-errantry, in the exercise of which I despise wealth but not honor. I have righted grievances, undone wrongs, punished insolence, defeated giants, and trampled monsters under foot. I am enamored, only because knights-errant must be. . . . "[32]

Beautiful, noble words. Words! But still the dominant note is there, extending into a coda like the summation of Don Quixote's vocation: "I am . . . must be."

Sancho enthusiastically approves his master's discourse, the clergyman recognizes Sancho, and the subject arises of the government of an island Don Quixote promised Sancho. The duke does not waste the opportunity of starting a new chain of practical jokes: in the name of Don Quixote he promises Sancho to make him governor of one of his *insulas*. Don Quixote instructs his squire to kiss the duke's feet for the favor received; Sancho does as he is bidden. When the ecclesiastic sees this, he rises from the table, fretfully, and speaks:

"By the habit I wear, I am about to say that your Excellency is as much of a crackbrain as these sinners. How could they not be mad, when those who are sane canonize their follies! Let your Excellency stay with them, but as long as they are in this house I shall remain in mine, and I shall excuse myself from reproving what I cannot remedy."[33]

He leaves the room, despite the supplications of the duke and duchess — although the duke did not say much, as he was laughing hard at the clergyman's "impertinent anger."

This exchange, full of Cervantesque life, has been given surprising interpretations. The Marxist line of criticism, for instance, fell into a rut of misplaced sociology: the fun of the duke and duchess at the expense of the poorer Don Quixote and Sancho would be a clear example of corruption and abuse by aristocratic rulers; Cervantes, a victim if there is any of class struggle, would be (it should also be through the ecclesiastic's words) an inspired, early critic of that order.[34] This interpretation *ad hominem* is based on mere assumption while ignoring fact. The fact is that Cervantes himself has given very emphatically a social judgment of the stern critic of the duke's and duchess' attitudes; it appears exceptionally in the text on this occasion. Let us read the presentation of the churchman, when the duke and duchess, accompanied by him, meet Don Quixote in the dining hall:

> The duke and duchess came to the door to receive him, and with them was a grave ecclesiastic, one of those that govern princely households; one of those that, not being born to the nobility themselves, are incapable of instructing the ones who are noble how they should be noble [*cómo lo han de ser los que lo son*]; one of those who want the greatness of the great to be measured by the narrowness of their own disposition [*ánimo*]; one of those who, desirous to show the ones they direct how to spare their wealth, make them be miserly. . . . [35]

(Always there is the accent on "being" — *ser*, that rich Spanish verb so difficult to translate into other tongues.)

That vehement condemnation of the clergyman coinciding with a clear meddling of the author in the story — this is Cervantes' only condemnation in the whole novel — brought about an interpretative approach opposite to that of Marxists and equally out of focus: that Cervantes' outburst could only be an overflowing of bile against the church, presumedly originated in experiences like those he had collecting taxes and in harmony with the Erasmian *movement* in Spain when Cervantes was young. The fact that church censorship, particularly rigorous at the time *Don Quixote* was published, did not object to Cervantes' work in the least, should cast serious doubt upon such an interpretative approach; moreover, consideration of social mores and moral values of the late sixteenth century — the only ones properly applying to the matter at hand — would have been enough to understand this passage differently. Historical clues throw a revealing light on it: one of the problems created by relaxed habits of the clergy was the desertion of parishes for a parasitic association with the rich; there are church decrees condemning this mercenary attitude of clergymen wasting their call in diverse courts. The Council of Trent in its work of reforming the Catholic church insisted on this condemnation.[36] Therefore Cervantes, in this as in other passages of his novel, was a *divulgador*, a popularizer of the decrees of purification of the clergy, and could count on the approval by the ecclesiastic authorities. In this light some of Don Quixote's words to the clergyman acquire new significance as an extension of Cervantes' harsh judgment; thus the notable, to-the-point initial paragraph in Don Quixote's rebuttal contains lines paraphrasing Tridentine recommendations to the clergy.[37] Not in vain does Cervantes make Sancho call his master "theologian" – tologian (*tólogo*), says uneducated Sancho.

In refuting any view there is always the danger of incurring the opposite error. An Olympian freedom characterizes Cervantes, a freedom that informs his life – the same freedom that makes it so difficult now for critics to pin down his attitudes or his apparent message. It seems that Cervantes was genuinely interested in the Tridentine effort, but he did not give himself to Trent: he may have felt concerned with the purification of the church and informed about the corresponding Tridentine casuistic work, or even personally concerned with some clerical attitudes to the extreme of letting himself go and fulminating the ducal priest, but, on the other hand, any Tridentine anti-Protestant attitude, not to speak of militancy, is totally absent from his works. And the primary concrete concern of his life was to stand up to the Ottoman onslaught into Christian Europe, and he endured a cruel captivity in Moslem hands; nevertheless his consideration of Moslem characters or of the converted Moriscos left in Spain – suspected of relapsing at the time – is one of respect and deep human understanding. If any ideological attitude could be discerned in Cervantes, it would be one of anticipation of the spirit of Vatican Council II.[38]

In this light it is proper to point out a spiritual relationship between Erasmus and Cervantes. Like Erasmus, Cervantes was always able to see the two sides of all questions. Some critics, among those who have emphasized that affinity as well as among others who have underplayed it, seem to forget that Erasmus, in spite of official condemnations, was one of the great influences upon the Council of Trent. But Cervantes' spiritual affinity with Erasmus does not mean *erasmismo*. Even disregarding the chronological problem, that Olympian freedom I have just referred to plays also in all directions; we shall see how Cervantes' social ideals seem to differ from those of Erasmus.

What are Cervantes' tenets regarding social responsibility and moral values, particularly in Spain about the year 1600? Some are already apparent or hinted at in the terms of Cervantes' explicit condemnation of the ecclesiastic at the duke's court, or when Don Quixote questions the capability of a priest trained in a poor, country seminary to serve as counselor of noblemen, or in the merry attitude of the duke when the priest leaves the dinner table in "impertinent" anger. Let us add the fact that the duke's majordomo presents Sancho with two hundred gold crowns upon his and his master's departure,[39] a deed showing largesse — an aristocratic virtue — on the part of the duke. We should remember at this point that practical jokes and derision were common practices of the time, not necessarily regarded as excessive cruelty; there is a whole literary genre of *beffa, bernesco*, Spanish *burla*.[40] The extended fun made at Don Quixote's expense is a trifling fault, a peccadillo of the great; the gift of two hundred crowns is a show of largesse, a virtue of the great. Let the priests be good priests and the aristocrats be good aristocrats. The proper role of the latter is to help wisely in the government of the nation and defend it in war; it is not clear whether or not the aristocrat in this episode fulfills his mission. In another episode at the duke's castle Don Quixote implicitly denounces the uselessness of "the nobleman who never has set foot beyond the bounds of his district" and "the lazy courtier who is more interested in news that he may talk about and pass on to others than he is in performing deeds and exploits that others may talk about and set down in writing."[41] But it would not be safe, based on the text, to include the duke of the story among the criticized noblemen.

The definition of the role and function and commitment befitting a nobleman is found in the episode of the lions and the gentleman of the green-colored greatcoat in chapters 16 — 18 of part 2. Don Quixote and Sancho meet a man on the road who impresses Don Quixote as being of importance. His dress, as well as the trappings of his stead, are rich and elegant; his countenance is "half humorous, half serious." The impression he receives from Don Quixote's strange figure is certainly more striking than the one he gives. Don Quixote explains his identity and mission, which astonishes the other man even more.

Now it is the turn of the man in green to give an account of his station and occupation in life. His name is Don Diego de Miranda; his life-style, according to his own description — which is done in a dignified but humble manner — appears to be a model of judiciousness, sagacity, and common sense. He spends his life with his wife, children, and friends, and his occupations are hunting and fishing, although he keeps neither falcon nor hounds but a tame partridge (used as decoy) and a bold ferret or two. His behavior is proper in every respect and he fulfills his religious duties.

At this point Sancho Panza does something strange, although acceptable considering his simplicity: he slides down from his ass, seizes Don Diego de Miranda's stirrup and begins kissing his feet with a show of devotion bordering on tears. This naturally surprises the nobleman, who asks Sancho about the meaning of his action. Sancho answers that if he is not mistaken, his grace is the first saint riding a horse that he has seen in all the days of his life — the implication, humorous but not in the least disrespectful for the Christian mentality of Spaniards, is that all saints (allowing for St. George) ride asses. The reader not fully cognizant of the extent of Cervantesque irony, may also be somewhat surprised at Sancho's action. The gentleman assures Sancho that he is no saint but a great sinner. "It is you, brother," he says, "who are the saint; for you must be a good man, judging by the simplicity of heart that you show."

The gentleman goes on by telling his only worry in life: his eighteen-year-old son has been studying Latin, Greek, and the science of poetry at Salamanca for six years, and the gentleman is unable to convince him to undertake a worthier study, such as law or theology. Incidentally, this worry in a gentleman of high social standing, as studied in the sociology of art, is normal at that time. Don Quixote replies with great wisdom and modernity; his central contention is that he who possesses the gift of poetry and makes good use of it, will become famous and his name be honored among all the civilized nations of the world. The gentleman should let his son go where his star beckons him.

At this point a new adventure occurs and Don Quixote shows an unimaginable, mad courage. The group meets a cart in which are two caged lions that the governor of Orán is sending to court as a present for his majesty. Don Quixote, disregarding the reasons of the gentleman and the tears of Sancho, decides to fight the lions and obliges the lionkeeper to open the first cage. Nothing happens, because, in a comic twist of the action, the lion is in no mood to leave the cage and Don Quixote lets himself be convinced by the lionkeeper that the lion's failure to answer the knight's challenge gives the crown of victory to the knight.

The most significant part of this episode is in the contrast between the attitudes of the nobleman in green and Don Quixote, and the words of the latter. When Don Diego tries to reason with Don Quixote that knights-errant should undertake only those adventures that afford some hope of a successful outcome, for valor when it turns to temerity has in it more of madness than of bravery, Don Quixote answers, to the point — in spite of the madness of his undertaking: "My dear sir, you had best go mind your tame partridge and that bold ferret of yours and let everybody else attend to his own business. This is mine...."[42]

The one in green even considers resisting Don Quixote but concludes that he is no match for the knight in the matter of arms; then, too, it does not seem to him the better part of wisdom to fight it out with a madman.

The contrasting attitudes are analyzed in the conversation after the lucky happening with the lion, in which the Cervantesque theme can be detected of the performance of proper function for existing authentically as a man. The apparent understanding and concession on Don Quixote's part when he opens the dialogue fits a typical intellectual astuteness of paranoids consistent with his psychiatric constitution: "Who would doubt, Señor Don Diego de Miranda, that your grace must judge me to be a nonsensical and crazy person? And it would be small wonder if that were the case. . . . " In his discourse Don Quixote gradually accommodates his bizarre, incongruous behavior to proper, purposeful vocation — that of the knight-errant, naturally:

"... let him enter the most intricate labyrinths, let him attempt the impossible at every step; let him endure in desolate lands the burning midsummer rays of the sun and the harsh inclemencies of winter winds and ice; let no lions astound him, no monsters, no dragons frighten him, because to seek some of them, to attack others, and conquer them all is his principal and legitimate occupation. Consequently, since it has been my lot to be numbered among the knights-errant, I cannot fail to attempt anything that appears to me to fall within the scope of my duties; therefore, defying the lions that I have just defied, concerned me directly, although I knew it to be unreasonable temerity."

Gradually rationalization turns into sensible reasoning, here informed by Aristotelian ethics:

"For I well know what valor is, a virtue lying between two vices, namely, cowardice and temerity. Nevertheless, it is preferable for the brave man to reach the point of being rash rather than for him to sink into being cowardly: the same as it is easier for the prodigal than for the miser to become generous, it is easier for the rash one to become genuinely brave than for the coward to grow into true valor."[43]

All of a sudden, as it frequently happens in this book, the madman looks sane when compared to the sane. If we accept Don Quixote's mad interpretation of reality as a basis for judging his conduct, he ends up appearing admirable in his heroism and even in the wisdom of his doctrine of commitment and action.

In considering this episode, Américo Castro emphasized this exemplary character of Don Quixote at the expense of a categorical condemnation of Don Diego de Miranda's character and life-style; the distinguished Spanish critic equated the gentleman of the green-colored greatcoat and the ecclesiatic at the duke's house in their opposition to Don Quixote.[44] Erich Auerbach corrected Castro: even accepting that Castro may have been right in finding "a shade of irony in the manner in which Cervantes describes his [Don Diego de Miranda's] style of life, his manner of hunting, and his views on his son's literary inclinations," still "Don Diego is a paragon of his class, the Spanish variety of the humanist nobleman: *otium cum dignitate*," and "if there is possibly an undertone of irony in the portrait of Don Diego, Don Quixote is, more than possibly, unqualifiedly conceived not with an undertone of ridicule but as ridiculous through and through." Consequently, "It would be forcing things if one sought to see here a glorification of adventurous heroism as against calculating, petty, and mediocre caution."[45]

This discrepancy between reactions to the same episode by two eminent critics is not exceptional in relation to Cervantes' work; it may be why the most recent Cervantesque studies are not so much interested in arriving at interpretations as in describing the nature of the ungraspable in Cervantes, his "ambiguity." We may add the reaction of another great critic, Marcel Bataillon, to the same episode; the French Hispanist observed that Don Diego's life-style followed the Christian Erasmian ethics (an analogy that in my estimation is nearer to the mark than Auerbach's characterization of "*humanist* nobleman").[46] Bataillon's observation reinforced his view of Cervantes' relationship to Erasmus, but if there is any truth in Castro's interpretation of the same episode, Bataillon's valid observation would run against his own theory, that is, the Erasmian behavior of satirized Don Diego would indicate Cervantes' independence from Erasmus' dicta.

The critic aspiring to comment on Cervantes' creation must be prepared to detect more than one side to every question – any tendency to see things in black and white would be fatal! Castro may be basically right in emphasizing the principle represented by Don Quixote, but Don Diego is not at all a contemptible, lifeless figure to serve in contrast to the hero – nor is the ecclesiastic at the duke's in the least wrong in the substance of his admonition to Don Quixote – and Auerbach is particularly right in underlining that Don Quixote, in the text as well as his creator's intentions, is mad.

Mad he is – "a confused madman, full of lucid intervals" – as Don Diego's son, the young Salamanca scholar, diagnoses.[47] And as a result, the virtues that he embodies of faith and heroism and action with commitment to an ideal undergo a sort of crystallization, of distillation of principle beyond the imperfection and coarseness of reality. Is Cervantes, the author, as deeply committed as the young Cervantes, the soldier, was to his beliefs and ideals?

Incidentally, let us notice that for Don Quixote worthy ideal and commitment may equally consist of a genuine interest and occupation in poetic endeavors, as Don Quixote lengthily admonishes the gentleman of the green-colored greatcoat.[48]

In this episode Cervantes' involvement, although in a subtle manner, is still there. A special effort is evident in the text to define clearly the social station of the one in green: he is "hidalgo" in the old sense of the word, that is, essentially noble, and he is "caballero," actually belonging — although he may not be "titulado" — to the contemporary noble class with the right, he and his son, to be given the "don" treatment. Incidentally, the *merely* hidalgos were the typical snobs of Golden Age Spain, and more so the countless ones who claimed *hidalguía* without having it. Mere hidalgos did not have the right to the "don" treatment, and a royal decree was issued to remind the Spaniards of this fact. The conjunction of the words "ingenioso hidalgo don" in the title of the first part of Cervantes' book must have already prompted laughter in the contemporary reader.[49]

Just as in the episode with the churchman Cervantes hammers on the phrase "one of those," in this episode he hammers on the word "hidalgo" in relation to that man in green casually encountered on the road, as willing to indicate that he is an "hidalgo caballero," "hidalgo" in the pristine sense of the word, a son with *algo*, an aristocrat with stable property; later it is spelled out that he is of the farmer-knight class (doubtlessly noble) of old medieval tradition in Spain, and rich.[50] The action of Sancho kissing the hidalgo's feet because he must be a saint by the kind of life he leads according to himself, is Sancho's extreme simplicity but also Cervantes' fine, ever-present irony; Cervantes, of a race of ironists, is the supreme ironist of literature.

And the irony is even finer and therefore more intense in the cursory appreciation of the nobleman's reasonableness and prudence when he decides not to try to stop Don Quixote from freeing the lions and flees with the peasant Sancho. The point of sarcasm is reached suddenly — Castro appears to be correct in this — with Don Quixote's shouting, "let each one attend to his own business," "you had best go mind your tame partridge and that bold ferret of yours." As always, Don Quixote is prodigiously right in principle although he may be hopelessly wrong in action.

Cervantes' art of irony and parody takes care of making the latter clear, as when the lion "proved to be courteous rather than arrogant and was in no mood for childish bravado," and turns around, and presents his hind parts to Don Quixote. Auerbach is here the more convincing in his disagreement with Castro: he directs attention to a series of elements of parody in

Don Quixote's attitude after "conquering" the lion, and with critical sharpness observes that the whole episode is introduced by a presentation of the absurd pride Don Quixote takes in his previous victory over the Knight of the Mirrors — whom the reader knows to be the disguised scholar friend of the curate of Don Quixote's village trying to trick the deranged old gentleman into giving up knight-errantry; " . . . there is hardly another instance in the entire book," Auerbach presses his point, "where Don Quixote is ridiculed — also in ethical terms — as he is here."[51]

In refuting Castro's one-sided view, Auerbach is in danger of being equally one-sided in the opposite direction, thus contradicting his own dictum, quoted above: "Cervantes does not take sides. . . . " It happens to all of us in attempting to define Cervantes' art.

V. CERVANTES' MESSAGE

Lovers of Cervantes' creation are progressing laboriously in the task of understanding it. The more we learn about and are able to appreciate his creative work, the more formidable it appears to be. Perhaps we have progressed somewhat since the year 1948, but we cannot subtract from what Leo Spitzer then wrote about Cervantes' masterwork that "we are still far from understanding it in its general plan and in its details as well as we do, for instance, Dante's *Commedia* or Goethe's *Faust*."[52]

The recent studies focusing upon "ambiguity" — for which writers as diverse as Rabelais, Cervantes, Nietzsche, and Kafka are favored objects — may not add much to that understanding, especially when the critics are not fully aware of the range of diversity contained in the literary notion of "ambiguity." And before reading the treatises that have tried to define the term, we need to feel almost at home in the numerous ancillary disciplines of criticism, in order to be able to discard the ambiguity arising from our lack of knowledge — of which we have already encountered some examples in this discussion.

A distinction among others that may need to be established concerns the ambiguity observable in allegory with a perhaps hopeless, or baffling, or cryptic message, but still neatly wrapped-up in the allegorical fabric, as is the case with some of Kafka's tales or with Borges' stories such as "The Lottery in Babylon." Granted that Cervantes also wrote allegory, in *Don Quixote* his satirical material is irony from beginning to end; reading it we may be left with the feeling of a vital message implicit in the satire, but Cervantes' fabric is one of loose ends.

Some critics may be tempted to see in Cervantes one more precursor of existentialism, with the emphasis he puts on authentic character and life, actuality in being, genuine commitment and function. Particularly from the Christian existentialists, whose values frequently approach those of Cervantes' times, many pages could easily be applicable to Cervantes – for example, from Berdiaeff's early *The End of Our Time* (*Un nouveau moyen âge*) or works like *The Bourgeois Mind* in which he contrasts the knight, the monk, the philosopher, and the poet with man in the time of bourgeois ascendancy. Kierkegaard's ethics, concerned with being, in which moral choices confront us with actual being – *existenz* – and where the moral issue is one of opposing authentic to inauthentic living, could be illustrated – with the values of a Christian though still aristocratic society – in Cervantes' works. But let us not forget that Kierkegaard was a great reader of Plato and Aristotle, and so were Cervantes and his contemporaries. The conception of proper function rather than better function is a classical one, and we find it in *The Republic*, and in the *Ethica Nicomachea* and other writings of Aristotle.[53]

Auerbach felt the need to deemphasize exaggerated, tragic, or symbolic elements that had been read into the text of *Don Quixote*, and he may have been justified in affirming that the greatest merit of Cervantes' masterwork may not be in a transcending message but in its humor and humanity, in "the neutral gaiety the knight's madness spreads over everything that comes in contact with him."[54] That vivid, vital experience is itself an answer to the pessimistic philosophical conclusions. In the way Diogenes refuted Zeno's ingenious Eleatic arguments by walking, the very experience of reading Cervantes' book counteracts the philosophical denial of life: life cannot be understood rationally, but *connaturaliter*.

Nonetheless, a transcending message, a commitment of Cervantes to his fellow beings, to his society in particular, can also be felt with reasonable certainty, I believe. At a time when the very roots of Spanish being are withering, he is telling his compatriots: each of you mind your own business, and perform it well; above all, *perform it*. Let the priest mind the business of the priest, let the nobleman perform the hard duties of the nobleman, let the writer be a writer and be it in earnest, with authentic feeling – there *is* a great task for literature, there is something literature can do for mankind. And let us live: let us not worry about death; let us create; let us be ourselves. Let us *be*.

That a madman has to give the message to the guardians of wisdom and power may be the sign of a time of decay, when those who are supposedly sane lack the basic health of being authentic. Jaspers has a somewhat related reflexion when studying madness in Strindberg and van Gogh, Swedenborg and Hölderlin: how authentic an artist van Gogh proves to be when compared to the expressionist artists, who in sanity found it reasonable to appear insane.[55] This already applies to our times of crisis, of decay and pessimism — Cervantes' work is proving again its ability to give a new, fresh message of wisdom to every new age.

NOTES

[1] 'The Pessimism-Optimism Controversy Concerning the Human Condition,' *Analecta Husserliana*, vol. 5 (Dordrecht: Reidel, 1977), pp. 3–14.
[2] Ibid., p. 5.
[3] Ibid., p. 10.
[4] Ibid.
[5] Chesterton wrote *The Return of Don Quixote* (1927), a fictional tale about a Don Quixote figure in English politics. *The Napoleon of Notting Hill* (1904), one of his best-known novels, deals more closely with the theme of madness in Don Quixote through its quixotic character Adam Wayne. One of the principal and frequent themes in Chesterton is the meaningful madness – e.g., *The Poet and the Lunatics* (1929), *Lunacy and Letters*, ed. D. Collins (1958). Chesterton had also a powerful allegorical vein, which I believe influenced Kafka.
[6] *George Bernard Shaw* (London, 1910), pp. 255–56.
[7] *English Literature – 1880–1905* (Boston, 1913), p. 2.
[8] Madeleine L. Cazamian, in her classical study of this period in English letters (*Le Roman et les Idées en Angleterre: L'Influence de la science (1860–1890)* [Strasbourg, 1923]), has a passage relevant to this section of the present study: "Lorsque l'enthousiasme suscité par les découvertes et les théories de la science comença à décliner, l'oeuvre de destruction qu'elle avait accompli provoqua chez beaucoup la révolte et l'amertume. Cette réaction n'a pas eu, en Angleterre, d'expression théorique originale; la métaphysique de Schopenhauer a suggéré des études, que signale J. Sully dans son ouvrage historique et critique sur le *Pessimisme*; mais entre l'évolutionnisme de H. Spencer et le néo-hégélianisme de T. H. Green, il n'y a pas eu, outre-Manche, d'école ou de système pessimiste qui s'impose à l'attention. Pour connaître ce mouvement, c'est dans les oeuvres littéraires qu'il faut l'étudier . . . " (I, 451).
[9] "Diese Inconsequenz muss hier in der Kürze aufgezeigt werden. – Der Wille ist ihm das ἓν καὶ πᾶν, das All-Einige Wesen der Welt, und das Individuum nur subjectiver Schein, streng genommen nicht einmal objectiv wirkliche Erscheinung dieses Wesens. Aber wenn es auch Letzteres wäre, wie soll dem Individuum die Möglichkeit zustehen, seinen individuellen Willen als Ganzes nicht bloss theoretisch, sondern auch practisch zu verneinen, da sein individuelles Wollen doch nur ein Strahl jenes All-Einigen Willens ist?" (*Philosophie des Unbewussten* [Leipzig, 1904], II, 398).
[10] Ibid., p. 405.
[11] Ibid., p. 406.

THE PRESENT TIDE OF PESSIMISM 121

[12] In his effort to reconcile opposites von Hartmann was not always consistent, and this fact ought to be duly observed by anybody trying to point out a nonpessimistic aspect of his philosophy. Von Hartmann referred to himself as generally a pessimist — much of a pessimist at the bottom of his heart, as he says in the prologue to the seventh edition of his main work. A final exposition of his pessimistic tenets is found in *Grundriss der Metaphysik*, published posthumously in 1908 as a fourth volume of his *System der Philosophie im Grundriss* (1907–1909). However, in *Philosophie des Unbewussten* he expounded what he labeled as "Einheit des Optimismus und Pessimismus" (ibid, II, 403–4), in which an energetic and most strong impulse to effective action is not at odds with recognizing the nature of the final stage of illusion (ibid.), the truth of which completes the despair of existence here — first stage — and the despair of the hereafter — second stage — with the absolute resignation of positive happiness — "die Wahrheit vom dritten Stadium der Illusion war die absolute Resignation auf das positive Glück" (ibid., II, 402). Quoting out of context cannot render a complete, true picture of a complex conception; counting on the reader's caution, let me quote von Hartmann further: "Wir haben gesehen, dass in der bestehenden Welt Alles auf das Weiseste und Beste eingerichtet ist, und dass sie als die beste von allen möglichen angesehen werden darf, dass sie aber trotzdem durchweg elend, und schlechter als gar keine sei" (ibid., II, 396). At times, the Leibnizian facet of von Hartmann's theorizing sounds like a Candide *au sérieux*.
[13] "pues el delito mayor/del hombre, es haber nacido" (*La vida es sueño*, Jorn. I, Esc. II).
[14] *Jude the Obscure*, pt. 6, chap. 2.
[15] Max Seiling, *Mainländer, ein neuer Messias* (München, 1888).
[16] *Die Philosophie der Erlösung* (Berlin, 1876); 3d ed. (Frankfurt, 1894).
[17] *El árbol de la ciencia* (Madrid, 1911). To the alternative Schopenhauer-Nietzsche in the same novel I refer in "El pesimismo filosófico europeo y la Generación del Noventa y Ocho," *Revista de Occidente*, nos 113–14 (August–September 1972), 171–90.
[18] *Analecta Husserliana*, V, 11.
[19] Borges' influence on recent American fiction has been pointed out by several critics. There is an excellent study of this agonistic period by the English critic Tony Tanner, *City of Words — American Fiction 1950–1970* (London, 1971); concerning Borges' major influence see pp. 31, 33, 39–49, 419–20, and passim. Borges has acknowledged his debt to Chesterton.
[20] Heroes set out, on their own, to protect widows and maidens and redress all kinds of wrongdoings; their prodigious horses never tire and their magic weapons never need recharging. Cervantes' structure is seen in the ever-present pair of hero and comic helper, as well as in the locale — predominance of roads and inns (saloons), more reminiscent of *Don Quixote* than representative of pioneer Western environment. The Cervantesque transmission is a reasonable thesis; Cervantes' pervasive influence on Melville, Mark Twain, and other writers of the nineteenth century is well known. The genre relationship was intelligently illustrated in one of Paladin's episodes: the American hero has to solve the problems created by an old man who is trying to redress wrongdoings in California by riding in medieval armor (episode televised in December 1961).
[21] Leo Spitzer, in an article reproducing a lecture he gave at several American universities that we shall have occasion to quote again, remarked before his American audiences that in Europe *Don Quixote* is first of all a children's book ("On the Significance of

Don Quijote," *Modern Language Notes*, 77 [1962], 113). This "vertical" universality of *Don Quixote* is forcefully described by Paul Hazard in studying the *fortune* of Cervantes' book in the nineteenth and twentieth centuries (*Don Quichotte de Cervantes − Étude et analyse* [Paris, 1931], bk. 6, chap. 3); "il est devenu le livre moderne par excellence, comique et noble, populaire et hermétique, tout chargé de philosophie" (pp. 349−50).

[22] Essay 14, "Der verzauberte Dulcinea," in *Mimesis*, 2d ed. (Bern, 1959). The first edition of Auerbach's book (1946) lacked this study. I will use the English translation published by Princeton University Press in 1953, which incorporated "The Enchanted Dulcinea" at that early date; all references will be to that edition. Auerbach continues: " ... the vivid visualization of very different people in very varied situations, for the vivid realization and expression of what thoughts enter their minds, what emotions fill their hearts, and what words come to their lips. This capacity he possesses so directly and strongly ... that almost everything realistic written before him appears limited, conventional, or propagandistic in comparison" (pp. 354−55). The testimony of a great novelist is especially significant; Thomas Mann "sees" the Cervantesque hero in the "adventure of the lions" as he stands before the open cage of the first lion, full of heroic impatience; the German novelist perceives the scene as life in the words of Cervantes: " ... diese ausserordentliche Szene ist mir in den Worten des Cervantes wieder recht lebendig geworden ... " (*Meerfahrt mit Don Quijote* [Insel-Verlag, 1956], p. 45).

[23] In the Hispanic world, and to some extent outside of it, the Cervantesque myth has been the result of a "contamination" of the historical Cervantes by the Don Quixote literary myth − compare it with the French idealization *of the historical* Racine as "le bon Racine"; incidentally, it is interesting to note that while in the latter example the historical figure benefitted by the process, in the former it could hardly do so.

[24] Auerbach has closed his essay emphasizing this effect of Cervantes' humor: "So universal and multilayered, so noncritical and nonproblematic a gaiety in the portrayal of everyday reality has not been attempted again in European letters. I cannot imagine where and when it might have been attempted" (*Mimesis*, p. 358). Thomas Mann expresses this Cervantesque coexistence of humanity and humor in a beautiful sentence: "Der wackerste und dreisteste Eroberer im Reiche des Menschlichen was immer wohl der Humor" (*Meerfahrt mit Don Quijote*, p. 63).

[25] Leo Spitzer, in the brilliant article already quoted, has denounced such interpretations as well as the erection of Don Quixote as a *living* symbol of his people, "a national hero of Spain" (see above and n. 23): " ... was it in the interests of the moral regeneration of the Spanish nation to present an amusing fool in a novel as a true national hero?" ("On the significance of *Don Quijote*," p. 129).

[26] There is unquestionable proof of Cervantes' outstanding character and heroic bravery. See Richard L. Predmore, *Cervantes* (New York, 1973), chaps. 5−6.

[27] *Mimesis*, pp. 355−56.

[28] "La novela española y su contribución a la novela realista moderna," *Cuadernos del idioma*, no. 1 (1965), 36.

[29] "On the Significance of *Don Quijote*," p. 129 and passim.

[30] *Don Quijote de la Mancha*, ed. Martín de Riquer (Madrid, 1958), pp. 768−69.

[31] Ibid., p. 770. "The 'typically Cervantean,'" in this case "the vivid realization and expression of ... what words come to their lips" (see Auerbach, n. 22 above), is Cervantes' unique ability of having his characters say or do the right thing. This passage brings to mind Kant's dictum in *Beobachtungen über das Gefühl des Schönen und Erhabenen*:

"Dem Schönen ist nichts so sehr entgegengesetzt als der Ekel, so wie nichts tiefer unter das Erhabene sinkt als das Lächerliche. Daher kann einem Manne kein Schimpf empfindlicher sein, als dass er ein Narr, und einem Frauenzimmer, dass sie ekelhaft genannt werde" ([Leipzig, 1913], p. 42).

32 *Don Quijote*, p. 770.

33 Ibid., p. 771.

34 If the critic disapproves, or believes that Cervantes disapproved of the duke's and duchess' conduct, he ought to agree with the substance of the priest's frank, courageous reprimand to his aristocratic masters, and vice versa. Concurrently, political creeds should not affect the recognition of Don Quixote's madness. Marxist critics, among others primarily concerned – from different angles – with political interpretations of literary history, have found the way to be at odds about everything with everybody about Don Quixote. In the Marxist line, see the study by the professor at the University of Ljubljana, Ludovik Osterc, *El pensamiento social y político del Quijote* (Mexico city, 1963); concerning the episodes treated in the present article see pp. 69, 71, 73, 81, 123–25, 128–29, 134–35, 187–89. In the same book Marxist references and bibliography are included.

35 *Don Quijote*, p. 765.

36 See Paul Descourzis, *Cervantes, a nueva luz*. I. *El "Quijote" y el Concilio de Trento* (Frankfurt, 1966), pp. 126–27.

37 Ibid., pp. 132–33, 108–9.

38 Coincidences with the moral theology of our century are pointed out by the same author, ibid., p. 113.

39 *Don Quijote*, p. 949.

40 Paul Hazard in 1931 already devoted a section of his basic book to "la burla" (*Don Quichotte de Cervantes*, pp. 178–85).

41 *Don Quijote*. p. 809.

42 Ibid., p. 655.

43 Ibid., pp. 160–61.

44 'La estructura del "Quijote,"' *Realidad*, 2, no. 5 (1947), 145–70. Later included in *Hacia Cervantes*, 3d ed. (Madrid, 1967), pp. 302–58. In the final edition Castro maintained his view: " ... su abierta desestima [de Cervantes] por el Caballero del Verde Gabán, o por el Eclesiástico de la casa de los Duques ... " (p. 308; see also pp. 317–18). Castro enlarged his treatment of both episodes, especially the one about the gentleman of the green-colored greatcoat, in *Cervantes y los casticismos españoles* (Madrid, 1966), pp. 138–42, 147–60. In this essay Castro is more concerned with the history of ideas; he answered implicitly Auerbach and Marcel Bataillon (*Érasme et l'Espagne*) and seemed to tone down his own former stance with respect to the gentleman of the green-colored greatcoat: "Carece de sentido presentar a don Diego como un tipo ideal y apetecible para Cervantes (humanismo, renacentismo, etc.), o como un personaje desdeñable por no entender a don Quijote" (p. 153).

45 *Mimesis*, p. 356.

46 *Érasme et l'Espagne* (Paris, 1937). Bataillon corrected Castro's simplistic and perhaps slanted presentation of Cervantes' *erasmismo*: "Le Cervantès érasmisant de Castro, loin d'être en contradiction avec la Contre-réforme espagnole, s'accorde à merveille avec les grands hommes de ce mouvement. ... [Cervantes] Ce n'est pas un incrédule qui cache sa secrète pensée derrière d'onctueuses protestations d'orthodoxie. C'est un croyant

éclairé pour qui tout, dans la religion, n'est pas sur le même plan . . . " (p. 827). Francisco Márquez Villanueva, in *Personajes y temas del Quijote* (Madrid, 1975), has devoted a long chapter — entitled "El Caballero del Verde Gabán y su reino de paradoja" (pp. 147–227) — to the study of the episode we are considering; in his eclectic treatment, Márquez Villanueva learnedly discusses Erasmian ethics, Christian Epicureanism, and the changing Christian image of the ideal gentleman after the Protestant and Catholic Reformations.

[47] *Don Quijote*, p. 666.

[48] Américo Castro has pointed out the importance of Don Diego de Miranda's young son Don Lorenzo, poet and humanist, who represents the preeminence of letters over all human endeavors, and the changed position of Don Quixote regarding arms and letters in the part 2 of 1615: " . . . ella [poetry, poetics] se ha de servir de todas [las otras ciencias], y todas se han de autorizar con ella" (*Hacia Cervantes*, pp. 367–68).

[49] The meaning of *ingenioso* in the title was discussed by the learned Hispanist Otis H. Green, "El *ingenioso* hidalgo," *Hispanic Review*, 25 (1957), 175–93. The satire of the poor hidalgos as well as the fake ones is highly amusing, significant, and constant throughout the *Quixote* and the *Novelas ejemplares*.

[50] *Don Quijote*, p. 662.

[51] Thomas Mann coincided almost verbatim with Auerbach in observing the abasement of the hero in the adventure of the lions more than anywhere else in the story, but was able to see beautifully the Cervantesque complexity, the other side of exaltation of the hero in the same instance: "An keiner Stelle wird die radikale Bereitschaft des Dichters, seinen Helden zugleich zu erniedrigen und zu erhöhen, deutlicher als hier" (*Meerfahrt mit Don Quijote*, p. 45). Francisco Márquez Villanueva has added an erudite note about the symbolic meaning of the curds Sancho put into his master's helmet and of the action in which Don Quixote claps the helmet with the curds in it on his head precisely when he is about to confront the lion; this incident, lively and comic although not necessary — sort of intercalated in the total episode — not only ridicules Don Quixote (reinforcing Auerbach's and Mann's observation), but cheese on the head — as Márquez Villanueva observes — is a topos meaning madness (*Personajes y temas del Quijote*, pp. 185–89).

[52] 'Linguistic Perspectivism in the *Don Quijote*,' in *Linguistics and Literary History: Essays in Stylistics* (New York, 1962); quoted from *Essays in Stylistic Analysis*, ed. Howard S. Babb (New York, 1972), p. 150.

[53] *The Republic* (Loeb) (Cambridge, 1946), pp. 101–3). *Ethica Nicomachea* (Oxford, 1925), 1097^{b}25–1098^{a}19.

[54] *Mimesis*, p. 358. See Leo Spitzer's beautiful summation of 'On the Significance of *Don Quijote*' (pp. 128–29).

[55] *Strindberg und Van Gogh — Versuch einer pathographischen Analyse unter vergleichender Heranziehung von Swedenborg und Hölderlin* (Bern, 1922), pt. 2, chap. 6.

DAN VAILLANCOURT

IS THE BATTLE WITH ALIENATION THE *RAISON D'ÊTRE* OF TWENTIETH-CENTURY PROTAGONISTS?

I. INTRODUCTION

The purpose of my paper is not to answer the question but to show why this question, and related questions, can be asked. Literature watchers such as Blanche Gelfant[1] and Marcus Klein[2] have already argued the case for alienation as a major theme in modern American fiction. For example, Blanche Gelfant states unabashedly that "alienation is the inextricable theme of modern American fiction."[3] But is the battle with alienation also prevalent in non-American twentieth-century fiction? To show why this and related questions can be asked, I will turn to four outstanding protagonists who do battle with alienation: Antoine Rocquentin in *Nausea* by Jean-Paul Sartre (1905–1980), Ivan Denisovich Shukhov in *One Day in the Life of Ivan Denisovich* by Alexander Solzhenitsyn (1918–), Emil Sinclair in *Demian* by Hermann Hesse (1877–1962), and Tara in the short story *The Runaway* by Rabindranath Tagore (1861–1941).

Why choose these particular protagonists? Together they cut across much of twentieth-century literature. They represent different nationalities and cultures. Rocquentin is French, Shukhov Russian, Sinclair German, and Tara a Brahmin boy in India. They appear at different times, approximately twenty years apart: Tara in 1895, Sinclair in 1919, Rocquentin in 1938, and Shukhov in 1962. Their creators have all won the Nobel Prize for literature: Tagore in 1913, Hesse in 1946, Sartre in 1964 (declined), and Solzhenitsyn in 1970. (The relation of the Nobel literature prize to my paper will become clear at the end.) Rocquentin, Shukhov, Sinclair, and Tara constitute a representative foursome of twentieth-century protagonists.

II. ALIENATION: A DEFINITION

Most writers describe alienation as a separation between the human being (A) and some other entity (B).[4] If alienation is to be useful, the two terms (A and B) must always be carefully identified. A is most often the human being, and B can range from elements in my physical-social world such as food, work, other people, to elements in my psychological world such as inner forces of the self, to elements in my spiritual world such as God. I am purposely avoiding alienation from the self (self-alienation) and alienation from society (social alienation), the two most common uses of alienation, since in the end self and society must be further defined. For example, self-alienation can represent the alienation of the human being from his body, unconscious or consciousness, soul or inner self, and so on. Alienation should always be identified as a separation between two specific entities, the human being (A) and B.

But alienation represents a special form of separation. Although I am physically separated from my paper at this moment, I am not alienated from it. To be alienated from my paper, I must experience a sense of strangeness, or incompleteness, or a loss of unity in relation to the paper. In other words, alienation represents an experience of separation coupled with strangeness.

If alienation depicts a special form of separation between the human being (A) and B, the overcoming of alienation will represent a special form of relation between the two entities. The statement is easily said, but not all writers on alienation agree with it. One writer, Walter Kaufmann, does describe the overcoming of alienation in terms of relation: "We should use 'alienation' and 'estrangement' as antonyms of 'feeling at home' in or with B."[5] I overcome the alienation from my paper, for example, when I relate to it in a special way, that is, when I feel at home with it, when I experience a sense of oneness with it.

III. ALIENATION: ROCQUENTIN, SHUKHOV, SINCLAIR, TARA

With this description of alienation and the overcoming of alienation, I can now examine how Rocquentin, Shukhov, Sinclair, and Tara do battle with alienation. Rocquentin depicts the "completely" alienated human being. He is alienated from elements of his physical-social world, psychological world, and spiritual world. He represents the lower, ideal limit of alienation. The other three protagonists each battle successfully different aspects of alienation. Shukhov survives the bitter Siberian cold and the brutal camp guards to make himself at home with elements of his physical-social world. Sinclair turns inward to integrate his light and dark forces; he feels at home with elements of his psychological world. Tara sees God's truth in all the things around him. He is at home on a spiritual level with the world of nature. Taken together, these four protagonists comment on the struggle with alienation in the three major areas of human existence.

1. *Rocquentin*

Antoine Rocquentin battles unsuccessfully in making himself at home with the physical-social, psychological, and spiritual elements of his world. In the end, as Colin Wilson points out, Rocquentin is absurd, like an expensive car that "will do 90 miles an hour *in reverse*, and only ten miles an hour going forward."[6] Despite his absurd characterization, Rocquentin provides a striking example of the various areas in which human beings do battle with alienation.

(a) *Rocquentin Alienated From Food, Work, People*

Rocquentin is alienated from elements of his physical-social world, in particular, food, work, and people. All physical objects feel strange to Rocquentin. "Objects should not *touch* because they are not alive . . . . But they touch me, it is unbearable. I am afraid of being in contact with them as though they were living beasts."[7]

Needless to say, this attitude toward objects will not aid Rocquentin to have an at-home experience when he lunches with the self-taught man at the *Rendez-vous des Cheminots*.[8] His lunch is uneventful until dessert time, when he bites into a piece of bread with chalky Camenbert cheese. He chews the bread "with difficulty" and then cannot make up his mind to swallow it. What is worse, the taste of the cheese sticks to his mouth. Like a man who suddenly feels a fly in a mouthful of food, Rocquentin is alienated from his bread and cheese.

He is also alienated from his work. For several years, Rocquentin has been researching and writing the biography of the marquis de Rollebon. His last entry reads: "'Care has been taken to spread the most sinister rumors. M. de Rollebon must have let himself be taken in by this maneuver since he wrote to his nephew on the 13th of September that he had just made his will.'"[9] Rocquentin can write no more, since he realizes that he too has let himself be taken in – by Rollebon himself. "He needed me in order to exist and I needed him so as not to feel my existence. I furnished the raw material, the material I had to re-sell, which I didn't know what to do with: existence, *my* existence."[10] Rocquentin is not at home with his research work because he wants to lose himself – literally – in the subject of the work. Not wanting to relate to his work, he attempts to disappear in it.

Rocquentin describes his alienation from other people in most of the diary entries. In only the second entry, he states categorically: "I live alone, entirely alone. I never speak to anyone, never; I receive nothing, I give nothing."[11]

Although Rocquentin has exchanges with others, he never feels at home with them. He fails to create any sense of unity between himself and another human being. For example, Françoise, the patronne of the *Rendez-vous des Cheminots* and Rocquentin's after-dinner mistress, "satisfies" his sexual needs without conversation, except to serve notice at one point that she will do it with her stockings on.[12] Françoise is no more than a convenient sexual object for Rocquentin. Even the cerebral self-taught man is no more than an object. Reading in alphabetical order every book in the Bouville library, the self-taught man is a receptacle of facts and ideas, a computer who responds only if the question does not require information past the letter L. Rocquentin is not at home with these human objects.

(b) *Rocquentin Alienated From The Past And Inner Forces Of The Body*

In addition to the alienation from food, work, and other people, Rocquentin is alienated from elements of his psychological world. He is alienated from all forms of his past: "My memories are like coins in the devil's purse: when you open it you find only dead leaves."[13] Memories do not exist by themselves; they are created by the present self. The past, as Rocquentin admits unwillingly, "is nothing more than an enormous vacuum."[14] He cannot relate to the reminiscences of yesterday, since they have no existence in themselves.

As a result of his alienation from the past, Rocquentin is "forsaken in the present." And in the present, he is isolated from his body. As he notes, "The body lives by itself once it has begun."[15] Rocquentin cannot seek refuge in any of the inner forces associated with the body. He is completely devoid of secret inner dimensions. The alienation from elements of his psychological world is absolute.

(c) *Rocquentin Alienated From Spiritual Elements*

Separated from the body, Rocquentin is his thoughts. "I think, therefore I am; I am because I think, why do I think? I don't want to think anymore, I am because I don't want to be."[16] This stream of consciousness characterizes Rocquentin's stance in the world. He does not accept the responsibility for his existence; he does not create himself each present moment. On the contrary, he attempts to stop thinking, to stop existing. Undoubtedly, Rocquentin views the task of self-creation as extremely difficult. First, he does not admit the possibility of special exhilirating experiences, whether they be adventures or perfect moments.[17] He exists amid the monotonous and the mundane: "Nothing happens while you live. The scenery changes, people come in and go out, that's all. There are no beginnings. Days are tacked on to days without rhyme or reason, an interminable, monotonous addition."[18]

Second, Rocquentin sees no unity to his life. He wishes the disparate and separate moments of his life could come together like notes in a melody. As he explains while reflecting on his favorite song 'Some of these days,' "what summits would I not reach if *my own life* made the subject of the melody."[19] But the wish is futile: "You might as well try and catch time by the tail."[20]

Third, to create himself, Rocquentin must be able to manipulate the objects around him. But they are alive, like living beasts. He prefers "to skip between them, avoiding them as much as possible."[21] Under these three circumstances, self-creation is extremely difficult, if not impossible.

Unable to create any meaning in his life, Rocquentin experiences one monotonous event after another. He fails to transcend individual experiences on two counts. On the one hand, he does not go beyond individual experiences to the level of their meaning in his life. On the other hand, he does not relate to any spiritual entity. On both counts, Rocquentin is alienated from elements of his spiritual world.

Antoine Rocquentin is alienated from elements of his physical-social, psychological, and spiritual world.[22] Yet, there is still hope for him. While awaiting the Paris train, he resolves to write a novel. At long last, he will act so as to create himself. After going backward at ninety miles per hour with Rocquentin, the reader can now anticipate going forward. But the story ends – between gears, as it were. Although I cannot complete Rocquentin's story, I will go forward with other protagonists, namely Shukhov, Sinclair, and Tara.

2. *Shukhov At Home With Food, Work, Fellow Campmates*

Amid the Stalin labor camp atrocities, Ivan Denisovich Shukhov succeeds in making himself at home with the simplest elements in the world around him — food, work, and fellow campmates.

As Shukhov explains early in the story, "Apart from sleeping, the prisoners' time is their own only for ten minutes at breakfast, five minutes at the noon break, and another five minutes at supper."[23] Eating for Shukhov is a solemn ritual, anticipated well in advance of mealtime and remembered long afterward. As he sits at table, Shukhov takes his cap off his shaved head: "however cold it is, he will never eat with it on."[24] Then he stirs the gruel carefully; he will not hurry now, "even if the roof catches fire."[25] He eats very slowly, because "food eaten quickly isn't food. It does no good, doesn't fill you."[26] And he eats with all his thoughts on the food,[27] first putting it cautiously in his mouth and then rolling it around deliberately with his tongue.[28] This moment is more precious to him than all "the years gone by and years to come"; it is more precious than freedom itself.[29]

Shukhov also makes himself at home with his work. Although imprisoned in a labor camp, he insists on working. Real punishment is "not being let out to work."[30] For Shukhov, work is an opportunity to use his hands to make something his own. During this one day, for example, Shukhov is assigned with another campmate to build up the walls of the power station with mortar and bricks. He so applies himself to the job that not only does the wall become his own, but he stays "overtime" to slap on the remaining bricks and mortar. And at the risk of having the dogs set on him by the camp guards, he climbs the ladder one more time for a final look at the wall. "The wall was straight as a die. His hands were still good for something!"[31]

Finally, Shukhov feels at home with some of his campmates,[32] despite the constraining hierarchical structure of the gang (each prisoner, from the squad leader on down, has his own place in the gang). At the end of the day, when the trusty in charge of the barracks orders everyone outside in the snow for a night check, Shukhov helps Caesar save the package of foodstuff he received from home. Shukhov does not expect or want anything in return. He does it because he is "sorry for Caesar." And later, after a conversation about heaven and hell with the Baptist Alyoshka, Shukhov gives him one of his two cookies. The problem with Alyoshka, says Shukhov, is that "He is always trying to please people but he never gets anything out of it."[33]

Shukhov battles alienation and wins. He makes himself at home with elements of his physical-social world, namely, his food, his work, and some of his campmates.

3. *Sinclair At Home With Inner Forces*

In the psychological domain, Emil Sinclair makes himself at home with his light and dark inner forces. At the early age of ten, he acknowledges the existence of two different worlds: light and dark. The world of light belongs to the "realm of brilliance, clarity, and cleanliness, gentle conversations, washed hands, clean clothes, and good manners."[34] And the world of dark is dominated "by a loud mixture of horrendous, intriguing, frightful, mysterious things, including slaughterhouses and prisons, drunkards and screeching fishwives, calving cows, horses sinking to their death, tales of robberies, murders and suicides."[35] Although Sinclair lives at times in this dark mysterious world, he is more often "a stranger to it." Unquestionably, Sinclair belongs to the world of light and righteousness; he is after all his parents' child.

To be at home with his inner forces, Sinclair must overcome two problems. First, he must internalize the worlds of light and dark. At this point, they exist outside him as two different worlds. Once he transforms the two worlds into inner forces, his second problem is to integrate them. Only after Sinclair overcomes both problems will he be truly at home with his inner forces. (I will not discuss, in this essay, the relationship between Sinclair's friends and Jung's figures of the unconscious.)

Sinclair's encounters with Franz Kromer and a woman he names Beatrice help internalize his dark and light worlds respectively. Sinclair crosses paths with Kromer on a half-holiday while roaming the streets with two neighborhood kids. The meeting leads to the telling by Sinclair of a tall tale about stealing a sackful of apples from a nearby orchard. Kromer, always on the prowl for money, seizes the opportunity to blackmail Sinclair in exchange for silence. That night, Sinclair realizes that the bizarre events of the day betray an evil omen for the future. The point is not that he has lied but that he has shaken hands with the devil: "For the time being I was not so much afraid of what would happen tomorrow as of the horrible certainty that my way, from now on, would lead farther and farther downhill into darkness."[36] A few years later, reflecting on his encounter with Kromer, Sinclair concludes: "What Franz Kromer had once been was now part of myself."[37] The dark would is now an inner force within Sinclair.

While away at a private boarding school, Sinclair sees for the first time the woman he names Beatrice. Although he never approaches her or speaks to her, his life and his imagination become completely dominated by her image. Everything he does is in worship of Beatrice. As he says, "I was trying most strenuously to construct an intimate 'world of light' for myself out of the shambles of a period of devastation."[38] One of the practices he undertakes during his cult of Beatrice is painting (in addition to feeding his awakening sexuality with a steady diet of cold baths). He must paint the portrait of Beatrice. Initially, his attempts fail. But one day, as he describes, while "idly drawing lines with a dreaming paintbrush ... I produced, almost without knowing it, a face to which I responded more strongly than I had to any of the others."[39] It is significant that Sinclair produces this painting with a "dreaming paintbrush" since the portrait will now represent the fumblings of his inner self. Indeed, as he discovers a few days later, the portrait does not resemble Beatrice or himself but his inner self.[40] Sinclair's light world is now completely internalized.

To integrate these inner forces and to be at home with them, Sinclair will need the assistance and guidance of Demian, Pistorius, and Frau Eva. From Demian, Sinclair learns that he must destroy his old worlds of light and dark in order to feel at home with his new inner forces.[41] While staring at an evening fire with Pistorius, he realizes that feeling at home involves a surrender, not only to the inner forces of light and dark but more profoundly to the indivisible divinity that is active in every human being.[42] After his experiences with Demian and Pistorius, Sinclair is finally prepared to meet Frau Eva, who as "a universal mother" welcomes him home. He greets Frau Eva with these revealing words: " 'I believe I have been on my way my whole life — and now I have come home.' "[43] For Sinclair the battle is over: "the outer world was perfectly attuned to the world within; it was a joy to be alive."[44]

Sinclair battles alienation from his inner forces and wins. He is at home with elements of his psychological world. And in the sense that the inward journey leads him to the divine, he is also at home with an element of his spiritual world. But the divine for Sinclair resides deep within the recesses of the self. Another way of being at home with God is to "see" the divine in all things, and this is the way of Tara.

4. *Tara At Home With The Divine*

As a young Brahmin boy in his middle teens, Tara is running away from home because "his spirit longs for the freedom of the mysterious outside world, unhampered by ties of affection."[45] Unlike most human beings who see themselves rooted in one place on this earth, "Tara is just a ripple on the current of things rushing across the infinite blue. Nothing binds him to past or future; his part in life is simply to flow onwards."[46] In his travels from village to village, he is stirred to his very depths by such things as "distant kites flying high in the midday sky, the croaking of the frogs on a rainy evening, the howling of the jackals at the dead of night."[47] Nothing escapes his keen glance. To Tara all things reflect God's wonder and mystery.

On one of his journeys, Tara hitches a ride on the boat of Moti Babu, who is on his way home to Katalia with his wife Annapurna and their daughter Charu. Although Tara intends to stay on the boat for only a few days, he remains with the family for two years. This is the first time Tara remains caged for so long. Perhaps it is because he is beginning to learn English, or perhaps Charu, with her constant teasing, is casting a "spell over his heart." In any case, Tara stays, "And the world which his imagination now conjures up is different from the former and less colorful."[48] Tara is losing his sense of oneness with the world. He is becoming alienated from the divine in the world of nature.

At this time, Charu reaches marriageable age, and her parents are looking for a suitable bridegroom. As might be expected, they choose Tara. Moti Babu as is the custom settles on a wedding date, and sends out invitations to Tara's mother and relatives. But to Tara, he says not a word.

On the day of the wedding, Tara cannot be found in his room. His story ends with this last sentence: "Before this conspiracy of love and affection succeeded in completely surrounding him, the free-souled Brahmin boy fled in the rainy night .... and returned to the arms of his great world-mother."[49] By returning to nature where he is at home with the divine, Tara wins his battle with alienation.

IV. CONCLUSION

Is the battle with alienation the raison d'être of contemporary protagonists? Of course, four protagonists dealing with alienation do not prove that all protagonists battle alienation. In examining how Rocquentin, Shukhov, Sinclair, and Tara do battle with alienation, my first goal is to provide evidence showing why, at the very least, the question can be asked.

My second goal is to show why another more specific question can also be asked: Is the battle with alienation the raison d'être of Nobel Prize literature? It is no accident that the authors of my four protagonists have all won the Nobel Prize for literature. But is it coincidence that the Nobel Prize was awarded to four authors whose unforgettable protagonists do battle with alienation?[50] Alfred Nobel, in his brief testament of 27 November 1895 states that prizes shall be given "to those who, during the preceding year, shall have conferred the greatest benefit on mankind."[51] Concerning literature, he adds that a prize shall be given "to the person who shall have produced ... the most outstanding work of an idealistic tendency."[52] These two statements clearly point in the direction of literature that grapples with the quality of human life in the twentieth century.

Finally, my third goal is to show why a more personal, perhaps more important, question can be asked: If alienation is a major affliction of contemporary society, and if our protagonists elucidate the successful battle with this affliction, then ought we not as scholars to study alienation in literature in order to piece together guidelines for the better, qualitative life? If we fail to pursue this question, at least at the speed of ninety miles per hour, it may be said of us that although we were learned, we did not understand.

NOTES

[1] Blanche H. Gelfant, 'The Imagery of Estrangement: Alienation in American Fiction,' in *Alienation: Concept, Term, and Meanings*, ed. Frank Johnson (New York: Seminar Press, 1973).
[2] Marcus Klein, *After Alienation: American novels in mid-century* (New York: World Publishers, 1964).
[3] Gelfant, p. 295.
[4] Walter Kaufmann, for example, makes this point in several publications. See as an example his introductory essay in the now classic work on alienation: Richard Schacht, *Alienation* (Garden City: Doubleday, 1970), p. xxiv. Even the two fathers of alienation, G. W. F. Hegel and Karl Marx, agree that alienation represents at the very least a separation between the human being and some other entity — the (social) substance for Hegel and work, product, nature, among other things, for Marx. In the 1940s and 1950s, many Marxist and existentialist writers began broadening alienation beyond separation to include such experiences as the self as object (Jean-Paul Sartre), sin (Erich Fromm), one-dimensionalism (Herbert Marcuse), etc. As a result of this overextensive use of alienation, writers in the 1960s and 1970s began a critical evaluation of alienation. Schacht's *Alienation* and Johnson's *Alienation: Concept, Term, and Meanings* are two outstanding examples of this evaluation. Due to the critical efforts of these and other writers on alienation, the term today is being used to designate at the very least a separation between the human being and some other entity.
[5] Walter Kaufmann, *Without Guilt and Justice* (New York: Dell, 1973), p. 146.
[6] Colin Wilson, *The Outsider* (New York: Dell, 1956), p. 296. In addition to an insightful work describing the alienation of modern man, *The Outsider* is a seminal work on alienation in modern literature.
[7] Jean-Paul Sartre, *Nausea*, trans. Lloyd Alexander (New York: New Directions, 1964), p. 10.
[8] Ibid. The lunch is described on pp. 103–23.
[9] Ibid., p. 95.
[10] Ibid., p. 98.
[11] Ibid., p. 6.
[12] Ibid., p. 7.
[13] Ibid., p. 32.
[14] Ibid., p. 64.
[15] Ibid., p. 99.
[16] Ibid., p. 100.
[17] Ibid., p. 150.
[18] Ibid., p. 39.

[19] Ibid., p. 38.
[20] Ibid., p. 40.
[21] Ibid., p. 122.
[22] The reader should be aware that Jean-Paul Sartre, as opposed to Solzhenitsyn, Hesse, and Tagore, does address directly the phenomenon of alienation. In *Being and Nothingness*, Sartre describes alienation in conjunction with the Look of the Other. When the Other looks at a human being, he sees that human being as an object rather than a free subject. As Sartre comments, "My being for-others is a fall through absolute emptiness toward objectivity . . . [and] this fall is an *alienation*." See *Being and Nothingness*, trans. Hazel Barnes (New York: Washington Square Press, 1953), p. 367. Sartre's later discussion of alienation in *Critique de la raison dialectique* owes much to Karl Marx. Indeed, Sartre describes alienation as a separation between the human being and his products, others, and labor itself.
[23] Alexander Solzhenitsyn, *One Day in the Life of Ivan Denisovich*, trans. Max Hayward and Ronald Hingley (New York: Bantam, 1963), p. 17. In the interest of consistency, I have changed some verb tenses to the present tense.
[24] Ibid., p. 16.
[25] Ibid., p. 17.
[26] Ibid., p. 27.
[27] Ibid., p. 54.
[28] Ibid., p. 88.
[29] Ibid., p. 151.
[30] Ibid., p. 7.
[31] Ibid., p. 125; for the building of the wall, see pp. 104–25.
[32] Although I can discuss only briefly in this essay the experience of being at home with other human beings, I wish to bring to the reader's attention an outstanding philosophical description of this relation. See Karol Wojtyla, 'Participation or Alienation?' in *The Self and The Other*, ed. Anna-Teresa Tymieniecka (Boston: D. Reidel, 1977), pp. 61–73.
[33] Solzhenitsyn, p. 202.
[34] Hermann Hesse, *Demian*, trans. Michael Roloff and Michael Lebeck (New York: Bantam, 1965), p. 5.
[35] Ibid., p. 6.
[36] Ibid., p. 14.
[37] Ibid., p. 41.
[38] Ibid., p. 67.
[39] Ibid., p. 68.
[40] Ibid., p. 70.
[41] Ibid., p. 76. As a commentary to Sinclair's painting of a sparrow hawk struggling to free itself from a giant egg, Demian responds: " 'The bird fights its way out of the egg. The egg is the world. Who would be born must first destroy a world. The bird flies to God. That God's name is Abraxas.' " In the pages that follow, Sinclair learns that Abraxas is both god and devil (light and dark). Again it is significant that Sinclair produces this painting with a "dreaming paintbrush." The bird represents Sinclair's inner self transforming the worlds of light and dark into inner forces and then attempting to integrate these forces (flying toward Abraxas).
[42] Ibid., pp. 86–88.

[43] Ibid., p. 119.
[44] Ibid., p. 117.
[45] Rabindranath Tagore, 'The Runaway,' in *Tagore Reader*, ed. Amiya Chakravorty (Boston: Beacon Press, 1961), p. 66; when appropriate, I have changed verb tenses to the present tense.
[46] Ibid., p. 69.
[47] Ibid., p. 66.
[48] Ibid., p. 76.
[49] Ibid., p. 78; the verb tenses are changed to the simple past.
[50] I have not demonstrated that all the protagonists of each author do battle with alienation. Again, my goal is to show why the question can be asked by analyzing the battle with alienation of a representative protagonist of each Nobel Prize author.
[51] *Nobel: The Man and his Prizes*, ed. by the Nobel Foundation (New York: Elsevier, 1962), p. 647.
[52] Ibid.

STEPHEN NATHANSON

NIHILISM, REASON, AND DEATH: REFLECTIONS ON JOHN BARTH'S *FLOATING OPERA*

Nihilism is the view that nothing has value. The nihilist knows, of course, that people strive after things and treat them as valuable, but he believes that such pursuits cannot be justified by reason, that they are vain and fruitless. Just as money is, after all, merely paper, so all the things which people think are valuable are ultimately inconsequential.[1]

Nihilism is a grim doctrine, whose bleak implications possess a personal as well as a theoretical sting. Though it is not clear that there have ever been any philosophical nihilists, the view has been highly influential, functioning as a philosophical specter, a conclusion which one wants to avoid but to which one may feel forced by the acceptance of other premises.[2] As a specter, nihilism has motivated and intensified much of the debate among twentieth-century philosophers, figuring centrally in the literature of existentialism and in discussions by analytic philosophers of emotivism and its variants.

Nihilism has also held a central position in the works of the American novelist John Barth. In the mid-1950s, while still a young man, Barth set out to write a trilogy of "nihilistic amusing" novels.[3] By his own account, Barth believed himself to be the inventor of the doctrine, but any one acquainted with the general philosophical issues will find familiar positions and arguments in the novels which grew out of Barth's project — *The Floating Opera* (1956), *The End of the Road* (1958), and *The Sot-Weed Factor* (1960). The positions, though not original with Barth, are developed with great acuity, and Barth has a fine sense for perceiving and drawing out the implications of philosophical views.

The central character of *The Floating Opera*, Todd Andrews, defends nihilism, and no answer to his argument is presented in the book. This is likely to lead a reader to see the novel as an assertion of the truth of nihilism, and the interpretation is apparently confirmed by Barth's own description of the book and its successors. In the first part of this paper, I shall argue that this interpretation is mistaken. The dominant position of Barth's fiction is skepticism rather than nihilism. *The Floating Opera* (along with *The End of the Road*) is a critique of a rationalist and intellectualist point of view, not a refutation of values.

Although Barth's novel does not affirm the truth of the nihilistic view, the work discussed presents us with a character and a set of events which help to illuminate nihilism as a position. In the second part of this paper, I would like to use the events of the novel as data and see what might best be inferred from them about the truth or falsity of nihilism. I shall try to show that they provide no support for a nihilistic point of view.

I

1. *Todd's "Nihilism"*

Todd Andrews is both the central character and the narrator of *The Floating Opera*, and the novel contains his presentation of the events of his life, as well as his conclusions about the nature and value of life in general. Late in the book, Todd formulates the following argument; (1) nothing has intrinsic value; (2) the reasons for which people attribute value to things are always ultimately irrational; (3) there is, therefore, no ultimate "reason" for valuing anything.[4] Because there is no reason for valuing anything, a decision which Todd had previously arrived at – a decision to commit suicide – is seen by him to be justified. For if, as he puts it, "Living is action," and "There is no final reason for action," then it follows that "There is no final reason for living" (223).

It is only *after* an unsuccessful suicide attempt that Todd realizes that this argument provides no grounds for suicide at all. If there is no reason for *any* action, there is no reason for taking one's life.

The rejection of suicide implies no rejection of nihilism. Indeed, Todd's new view expresses a consistent and more thoroughgoing nihilism. What would he do now? Would he live or die? These are no longer matters which demand decision, for it does not matter how they are settled. Todd sheds the role of agent for that of predictor and informs us that he would probably continue living as he had, much "as a rabbit shot on the run keeps running in the same direction until death overtakes him" (p. 246). Thus, though he lives on for a good number of years beyond the events related in the novel, continues his law practice, and works on the *Inquiry* whose outcome is the narrative of the novel itself, Todd's life is no testimony to human vlaue. There is no shred of affirmation, only the continuation of a dying animal's previous path.

The consistency of Todd's nihilism is only one of its impressive features. There is, in addition, his apparent rationality and the extent to which he has worked hard at finding answers to life's questions. His major pursuit is the writing of an *Inquiry* into both the reasons for his father's suicide and the causes of the defective communication between himself and his father. What this project amounts to finally is a classic quest for self-knowledge, whose thoroughness and duration alone would have met the most stringent Socratic criteria for the life worth living.

Todd prides himself on his logicality, his objectivity, and his self-control. On the very first page he tells us that, the opinions of his neighbors to the contrary, his "behavior is actually quite consistent" and his "life is never less logical simply for its being unorthodox" (p. 1). Moreover, the intellectual reflections in which he engages are directly related to his conduct, for, as he says, "I tend, I'm afraid, to attribute to abstract ideas a life-or-death significance" (p. 15). That Todd acts on principle rather than from habit is illustrated by his cultivation of habit breaking. Referring to this practice, he advises: "A good habit to acquire, if you are interested in disciplining your strength, is the habit of habit-breaking..... It will slow you up sometimes, but you'll tend to grow strong and feel free" (p. 122). Strength, freedom, and self-control are all qualities related to Todd's rationality, and his cultivation of them seems in many respects to be quite successful.

These features and the degree of objectivity and detachment which they give to Todd are also evident in his sense of humor and the skill with which he illuminates the frailties of his acquaintances. The connections between humor and nihilism are made evident when Todd, having made his friend Harrison Mack look particularly ridiculous, generalizes his fun-poking and "cheerfully" tells Harrison that he "mustn't take things seriously," that "No matter how you approach it, everything we do is ridiculous" (p. 40). For most of us, our humor is restricted, but for the nihilist, everything – no matter how serious or sacred – may be a source of amusement.

Though we enjoy his humor and though it gives him an air of superiority and detachment which may strike us as signs of strength, Todd's suicide attempt puts all this in a different perspective. While the Todd Andrews of the first, expurgated edition of the book attempts a private suicide, the real Andrews of edition two attempts his suicide by trying to blow up the entire showboat "Floating Opera," bearing some seven hundred crew members, show persons, and spectators. As Todd waits for the explosion and watches the show, he calmly contemplates the expected results, picturing to himself the "cracked, smoking, scorched and charred" bodies of his friends Harrison Mack, his mistress Jane Mack, and little Jeannine Mack, who may indeed be his own daughter. Todd remains unruffled by these prospects, completely indifferent to them.

When mere chance prevents success, he entertains no regrets and retains his nihilistic principles. Indeed, as I note above, the change which does occur involves a deepening of his indifference, as he recognizes that suicide is no more rationally justifiable than is living. That choice, too, is a matter of no concern.

Todd's suicide attempt brings into extraordinarily sharp focus the image of the rational thinker who is willing to follow his conclusions wherever they lead him. Equally highlighted, however, is the ghastliness of the particular conclusion to which Todd has been led and the action which, on his views, is a matter of indifference. There is a wonderful tension created between the repulsiveness of Todd's act and the compellingness of his argument. If the argument is correct, his action may repel one but is in no objective sense wrong. If his action is in some sense wrong in itself, then the argument for nihilism must be flawed. The novel, however, contains no assurance that the argument is unsound. Nor does it suggest how it might be rebutted.

2. *Todd's Life and Character*

In spite of the lack of a refutation of Todd's nihilism in the novel, it seems to me a mistake to understand the book as affirming the truth of nihilism. While it would be easy to recall most vividly the bleak but powerful view affirmed by Todd, to do so would be to recall but half the story. For Todd Andrews is not a mind in a vacuum, spinning out arguments about life's lack of value. Rather, he is a man who has suffered greatly during his life and who might well be described as an emotional cripple. Moreover, the nature of his emotional sufferings is closely linked to the path taken by his reflections. His full story, then, might be said to undermine his conclusions by providing an alternative (nonrationalistic) explanation of how he arrived at his views and by suggesting that his inability to find a basis for value in life resulted from his own defective nature. Should Todd turn out to be a nonideal observer, we might feel justified in rejecting his observations.

As an argument against nihilism, these thoughts are blatantly *ad hominem*. As a reminder that the whole story has not been told, they are quite important, for whether or not Todd is refuted by his history, we need a fuller picture of what occurs in order to assess the overall conclusions of the book.

What kind of person is it, then, who can arrive at such an extreme state of indifference? The answer is that Todd is a person who has long lived in terrible isolation and detachment from others. He is a person whose emotional involvements have been so disastrous that he can only live by desensitizing himself, by becoming calloused. His detachment is made evident by his own description of himself as "incapable of great love for people" and as never having "understood personally what love is and feels like" (p. 34).

By themselves, these facts are not very remarkable, but together with other things that we learn about Todd, they give us clues about the origin of his views and the nature of his needs. In fact, we have quite a lot of data, for Todd has spent years collecting notes in peach baskets and trying to explain both the lack of father/son communication and the change of mind which led him from suicide to living. His pursuit of self-knowledge has not been in vain. And yet, interestingly, his perceptions completely falter in his assessment of his early life and his relation to his father.

Todd's mother died in his seventh year — that is all we know about her — and as a result, he was brought up by "a succession of maids and housekeepers." His father "always expressed concern over" his welfare but "seldom gave ... a great deal of personal attention" and "was incurious" about his activities (p. 114). In his youth, Todd indulged in elaborate fantasies about constructing a great boat in which he would sail off to the "endless oceans." In all this we can see the emotional barrenness of his environment, the lack of concern for him, the resultant desire to escape, and yet, Todd tells us, that he never "regarded my boyhood as anything but pleasant" (p. 57).

Life does present him with opportunities for relating to other persons, but these turn into traumatic failures and leave him feeling that he is an animal, incapable of either dignity or worth, a suitable object only for scorn and ridicule. His first sexual experience is marred by his catching sight of himself in a mirror and succumbing to hysterical laughter at what he sees. The mirror provides a demonstration of his own animality, and even thirty-seven years later, he cannot "expunge that mirror from my mind," cannot hold back the "tears of nervous laughter" (p. 121).

His battle experience in World War I provides him with another demonstration of his animal nature and another missed opportunity for emerging from emotional confinement. Having been reduced to the state of "a mocked, drooling animal" (62) by the terror of battle, Todd is joined in a foxhole refuge by a German soldier in a similar condition. The two embrace in recognition of their common condition and despite linguistic barriers, achieve a relationship unique in Todd's experience. "Never in my life," he says, "had I enjoyed such clear communication with a fellow human being" (65). But the intense feeling of connectedness gives way to fear that the German has only pretended, that he will kill Todd at the first opportunity. Todd decides to sneak out of the foxhole while his companion sleeps, but the German opens his eyes. Todd instinctively lunges at him with his bayonet and kills him.

There are other horrors in Todd's life, and Todd returns to them again and again in his reflections. There is the army doctor who informs him that he has a heart condition which could strike him dead at any minute. There is, too, the memory of his dead father hanging from a joist in the basement, his clothes "perfectly creased and free of wrinkles," his hair "neatly and correctly combed," his eyes popped and face blackened (178). Todd's father did think to leave behind five thousand dollars for his son (which Todd gives away to the wealthiest man in town), but he offered no explanation for his suicide. His real legacy to his son was a vivd and ghastly memory, as well as a sense of betrayal and isolation.

All of these experiences leave Todd a wounded and disillusioned person for whom the world is a continual source of anguish and disappointment. Underneath the cynicism is an idealistic purist who cannot accept the indignity of a sexual posture, the arbitrary dependence of his life on his heart beats, the suicide of a father for reasons of mere financial loss. It is the facts of life that are, in the end, what Todd cannot accept, and since they do not live up to his standards and ideals, it is the ideals which he comes to reject. We see this in his law practice, as well as his boat building, for in both activities he has become a skilled practitioner who does not really care about the end result. This contrasts with his prewar indulgence in dreams which could not be made real. His wartime experience, he notes, "cured" him of his tendency to daydream, left him with low expectations of himself and his "fellow animals" and no desire for the "esteem or approbation" of others (67). Todd is left without the strength to sustain either ideals or personal relationships because doing so would risk further disappointment.

Indeed, the succession of masks and life strategies which Todd adopts are all ways of desensitizing himself, of shielding himself from the pains of life. This single purpose underlies the intense debauchery of his college days at Johns Hopkins, the "misanthropic hermitism" of his "saintly" period, the principled indifference of his cynical stage, as well as the deeper nihilism whose beginnings are related in the book. His whole life, as he tells us, "has been directed toward the solution of a problem, or mastery of a fact." The fact which he refers to is the fact of his own mortality, and the problem is that of coping with the prospect of impending death. His strategy throughout is to attempt to become indifferent to what he dreads. Each of his masks is an attempt at achieving this indifference. His decision to commit suicide appears to be the perfect solution because it treats life so cavalierly, because it appears to establish his personal supremacy over the problem, and, finally, because it will forever free him from the problem. And his final nihilistic rejection of suicide is an even more perfect solution. One achieves mastery and solves the problem of mortality by recognizing that the achievement of mastery is unimportant, that solving the problem is no better than not solving it. The problem dissolves in indifference.

The net effect of the very full picture which Barth gives us of Todd is that Todd's credentials as a rational, objective thinker are undermined. His nihilism is as much the outcome of his personal despair as it is of his reflections. Indeed, the reflections may be epiphenomenal, rather than causal, factors. Todd himself describes them as "rationalisings ... the *post facto* justification, on logical grounds, of what had been an entirely personal, unlogical resolve" (p. 219). The final page of the book describes him as wondering whether his emphasis on *absolute* values might not have been misplaced and whether "values less than absolute mightn't be regarded as in no way inferior and even lived by" (p. 247). Even Todd, then, regards his reasoning as inefficacious and finds his conclusions less than compelling.

The result of all this is a supreme inconclusiveness. We are left with three uncertainties: a doubt about the status of values, a doubt about the powers of reason to discover values, and a doubt about the value of reasoning. Perhaps the skeptical Barth is suggesting that the greatest mistake we can make is "to attribute to abstract ideas a life-or-death significance" (p. 15). The novel ought to leave one unsettled about reason and values rather than certain of a nihilistic conclusion. It is supposed to leave us ripe for "another inquiry and another story" (p. 247).

II

What we have in *The Floating Opera* is a working out of a nihilistic point of view and a vivid portrayal of a character who is driven to adopt it. What we do not have is an answer to the question of nihilism's truth or falsity. Nonetheless, the novel may be helpful in assessing nihilism as a doctrine.

My strategy for the remainder of the paper is based on the following thought. At an intuitive level, the plausibility of nihilism seems to be linked to the plausibility of the view that Todd's suicide decision was rational. It is no accident that discussions of nihilism and suicide go together, even though Barth is correct and perceptive in indicating that the truth of nihilism in no way implies suicide. But the connection might be stated by saying that if nihilism is true, then the indifference to life which seems to be connected with Todd's suicide attempt is not inappropriate or irrational. It would appear, then, that we can approach the question of nihilism's truth by asking whether Todd's suicide decision was rational. I shall argue for the view that it *was* rational — in the sense that it was *not irrational*. It is allowed by reason but not required. Although I shall have to qualify this conclusion, I hope to show that even if Todd's decision can be considered as rational, it does not follow that nihilism is true. Indeed, the most promising arguments on behalf of Todd's rationality presuppose that nihilism is false.

1. *The "Cool Moment" Theory of Rationality*

One of the strongest reasons for calling Todd's decision "rational" is that it appears to be the product of careful deliberation. Todd is no raving maniac. He is in control of himself, and whether it comes prior to or after his resolve, he is able to marshall arguments which support his view and which are not obviously flawed. Even if we could show that his argument was flawed, we might still want to say that his decision was rational because the flaws might be subtle and missing them would not reflect negatively on his rationality. At any rate, one argument for Todd's rationality is based on the manner in which he reaches his decision.

This approach to the assessment of the rationality of decisions has recently been attacked by Bernard Gert.[5] Arguing against what he calls "cool moment" theories of rationality, Gert claims that there are certain things which a person might decide to do in a "cool moment" but which would still be irrational. Furthermore, an act may be irrational even if, in a calm and deliberate manner, a person decides that the act in question is the one which he desires to perform more than he desires anything else. According to Gert, there are certain things (death, pain, disablement, deprivation of freedom, opportunity, and pleasure) which it is irrational to desire without reason. One may, Gert acknowledges, rationally choose to die if, for example, this is the only way to spare oneself from great suffering or to save a loved one, just as one might rationally choose to undergo great pain in order to avoid greater pain, death, or deprivation of freedom. So, one can have reasons which make such choices rational. However, lacking these reasons, such choices are necessarily irrational, and there are a limited range of reasons which can perform this justificatory role. Indeed, the only reasons involve either the avoidance of some other member of Gert's list of evils or the attainment of some good. One can only be rational in choosing death or disablement or pain or deprivation of freedom, opportunity or pleasure if one thereby significantly decreases one's chances of suffering these same evils or if one thereby significantly increases the chances of attaining some good.

This view, if acceptable, would lead us to conclude that it is irrational for Todd to choose death on the basis of his nihilistic argument. While a person might rationally choose death so as to avoid some evil, it is not rational to choose death simply because an abstract argument yields the conclusion that nothing is of value. This is not the sort of thing which, on Gert's view, counts as a *reason* for choosing death. No rational person would make this choice.

Gert's attack on the "cool moment" theory is strongly supported by the following example. He imagines a person who decides in a cool moment that his most important desire is the desire to kill himself in the most painful manner possible. Though he has other desires, including a desire to seek psychiatric assistance, he thinks that the desire to kill himself is of greatest importance. According to the cool moment theory, it would be rational for this person to kill himself and irrational to see a psychiatrist. Gert's claim is that this implication of the cool moment theory is obviously false and that whatever plausibility the theory has depends on our overlooking the possibility of desires which are in and of themselves irrational.[6]

The lesson to be learned is that a theory of rationality which takes into account only the manner in which decisions are arrived at must be false. Some desires are irrational to act on even if the decision to act on them is made in the favored kind of circumstance. Applying Gert's point to Todd, we get the result that Todd is irrational in choosing death, even though his manner of choice has the appearance of rationality about it. Moreover, if Gert is correct, there is a rational presumption in favor of living and against death. If this is true, the nihilistic conclusion that both are of equal value (because nothing has value) is false.

This argument for the irrationality of Todd's decision can be countered by arguing that Todd in fact meets Gert's criteria for rational action. This is because the real reasons for his suicide are that life has been exceedingly painful for him, that he has reached a point of despair at which the prospect of further anguish becomes intolerable to him, and that death provides the escape from the psychological torment which had defied mastery by other means. These, one might argue, are the real reasons for his suicide decision, and these are the sorts of considerations which Gert acknowledges as constituting reasons for bringing about one's death.

My agreement with this explanation of Todd's decision should be evident from my discussion of the desolation of Todd's life. Moreover, the novel is explicit in recording Todd's despair prior to the decision and his relief afterward. Prior to his decision, he is gripped by an overwhelming sense of futility, pierced by a private hopelessness, attacked by "a battery of little agonies" (p. 222). The next morning, he wakes at six, begins his washing routine, and suddenly, when the cold water hits his face, "all things in heaven and earth" come clear to him. He knows he will kill himself that day and reacts with exhilaration (p. 10, p. 233). Todd's elation is close to incomprehensible when described early in the book. Redescribed later as the means of relieving despair, it is perfectly understandable and apparently rational.

Here, then, is a means of showing Todd's attempted suicide to be rational, even within the standards stated by Gert. What is crucial, however, is that this evaluation in no way supports Todd's nihilism. Indeed, to say that suicide is a rational means for escaping suffering is inconsistent with nihilism. This judgment presupposes that not suffering is better than suffering, and we judge Todd's decision to be rational because it will bring about the end of his suffering. A consistent nihilist could not use this argument as a means of defending the rationality of Todd's action. If nihilism is true, the cessation of suffering has no value, and there is no reason to prefer suffering's end to its continuation. It should be clear, then, that if one wants to defend Todd's decision in a way which is consistent with nihilism, then one cannot defend him by saying that he is relieving himself of suffering and hence doing something desirable and rational.

2. *Nihilism and Indifference*

In order to see whether a defense consistent with nihilism can be constructed, let me return to my original sketch of Todd. According to this view of him, he decides to commit suicide because of his belief that nothing is of value. It is clear that Gert would not count this as a reason. Could it, nonetheless, be a reason? And what would this show about nihilism and values?

Recall Gert's example of the person who, in a cool moment, wants most of all to die in the most painful way possible, though he also has some desire to see a psychiatrist. This is an effective counterexample against the cool moment theory. But does it really establish that the desire for death (apart from approved Gertian reasons) is always irrational? Perhaps not. It strikes me that additional factors in the example make the person's desire irrational quite apart from the status of death as an evil. One might argue that this person is irrational to choose death because (1) the painful manner of death will inevitably lead him to regret his decision and (2) his desire to see the psychiatrist indicates anxiety about his desire for death.

That is, the painful manner of death chosen is irrational because, even if at the time of the decision he feels indifferent to the prospect of great pain, it will be impossible for him to be indifferent to it at the time of infliction. At the time of infliction, he will want the pain to stop, and he has enough knowledge at the time of the decision to realize that this will be so. Therefore, this aspect of his choice is irrational because he ought to know that he will live to regret it.

The wish for psychiatric consultation is relevant in a different way. If he kills himself, he will be dead and thus in no position to regret his failure to seek a psychiatrist. However, this desire indicates that, at the time of the decision, the person realizes the extremity of his choice, senses its possible pathological nature, and desires not to be subject to the desire to die. Part of him, so to speak, wants to live. He is not really indifferent to life.

Suppose now that we abstract these two features from Gert's example Death is chosen, but painless means are to be used, and the person making the decision has no conflicting tendencies, thinks himself to be acting perfectly rationally. Todd, though he does not perfectly exemplify it, suggests what the perfect nihilist would be: a person who believes that everything is lacking in value and whose attitudes are, correspondingly, indifferent to everything. If such a person chose death because he believed it was neither better nor worse than life and because he had become completely indifferent to living, would that be irrational?

In such a case, I would argue that the choice of death was rational, even though death was not being chosen as a means of escaping some significant evil. Richard Brandt, in the context of a discussion of euthanasia, argues that we must distinguish between injurious and noninjurious deaths. If, for example, one comes "upon a cat that has been mangled but not quite killed by several dogs and is writhing in pain" and one puts it out of its misery, then one has "killed the cat but surely not *injured* it."[7] There are two reasons here why death is noninjurious. First, death brings relief from suffering. Second, and perhaps more importantly, the animal is almost dead and has no prospects for survival. Its continued suffering is gratuitous. Were it capable of survival and recovery, death might well be an injury.

Brandt continues with another case in which, he claims, death would be noninjurious. He asks us to consider "the case of a human being who has become unconscious and will not, it is known, regain consciousness. He is in a hospital and is being kept alive only through expensive supportive measures."[8]

Though he recognizes some exceptions, Brandt wants to say that there is no moral obligation to maintain the life of such a person because he is "beyond injury." He is beyond injury because all of the capacities for living a meaningful life have already been taken from this person. For his heart and lungs to cease functioning at this point will deprive him of nothing further. He has no more to lose and hence is not injured by death.

Though the analogy may be extreme, it seems to me that Brandt's second case provides a model for the perfect nihilist I have described. If a person arrives at the point at which all the possible satisfactions of life have become indifferent to him, then, for that person, death would not be an injury. It would deprive him of nothing. In saying this I am, of course, assuming that this state of indifference can be known to be permanent, just as Brandt assumes that the hospitalized person will never regain consciousness. This is important because a person who painlessly killed himself in a deep but *temporary* state of depression has suffered an injury in losing the opportunities for meaningful living. Death can be an injury, even when it does not involve suffering.[9] However, death need not be an evil in the way Gert's view asserts.

If my argument is correct, the nihilistic strategy has once again backfired. I attempted to defend Todd's decision by suggesting that it could be rational if he were, in fact, completely and permanently indifferent to life and its prospects. This conclusion, however, is a far cry from nihilism. Nihilism, the view that nothing has value, is a completely general doctrine, an "ism" with a purportedly universal truth. Todd's decision (as reconstructed here) is rational, however, only because a very special condition obtains – namely, total indifference. If a person has no goals, no significant desires, no prospects for deriving satisfaction from life, then indeed, death is no injury, and life has no value. This description, however, applies to hardly anyone.[10] For most people, death is an injury, and life, as the precondition of satisfying their goals and desires, is something of value. Once again, then, the attempt to defend the rationality of Todd's act yields a negative conclusion about the nihilism which, on this interpretation, provided the basis for his act.

3. *Todd's Argument*

What about Todd's stated argument for nihilism? This has not yet been dealt with. Might it provide the rational basis for both Todd's suicide decision *and* for a general doctrine whose truth is independent of the conditions of an individual's life?

Todd begins by asserting that "Nothing has intrinsic value." He offers no support for this premise, and there is a degree of plausibility to the claim of hedonists that pleasure and happiness have intrinsic value, while pain and unhappiness have intrinsic disvalue. Presumably, Todd's view is that these are experiences and states to which people *attribute* value, and he would deny that we *find* them to be valuable. Further, as his second proposition asserts, "The reasons for which people attribute value to things are always ultimately irrational" (p. 218). Todd's thought seems to be that attributing value to things could only be rational if value were an intrinsic property. Since value is not an intrinsic property, our attributions of value are without rational basis. What this overlooks, however, is that values may be both relational properties and yet objective. This is revealed in the earlier discussion of injuries and benefits. What is injurious to one person (death, for example) may be beneficial to another. It is not irrational for a person suffering a painful terminal illness to "attribute" to death a different value from that "attributed" to it by a healthy person whose prospects for a satisfying life are good. It is an objective fact that death solves a problem for the sufferer and that it deprives the nonsufferer of opportunities for satisfaction.

Todd is, in a sense, correct when he claims that there is "no ultimate reason for valuing anything." However, the truth of this claim has no bearing on the rationality of choices for human beings. What is true here is that *if* we take an "ultimate" or "cosmic" point of view, a view which abstracts from all natural human needs and desires, then there is no reason for valuing anything. However, given that we make our choices within a context provided by our own natures and the nature of the reality which confronts us, there are rational grounds for choices and evaluations.

Todd's mistake, then, is to take too seriously the ultimate point of view, for that perspective excludes the very factors which make possible the rational assessment of choices. Likewise, the perfect nihilist is an inappropriate choice of "ideal observer" when we are seeking a basis for values. For, as I have described him, he is a being who is totally indifferent to all things. He is lacking in the concerns which human beings characteristically have. Hence, while it is rational for *him* to draw conclusions about the lack of value in his life, it would be irrational for others of us to draw these same conclusions. The perfect nihilist is a freak rather than a paradigm.

III

In this paper, I have examined the status of nihilism both within and without

The Floating Opera. I have argued that Barth's book itself is skeptical rather than nihilistic, and that it provides materials which tend to show that nihilism is false.

There are, I have argued, two plausible defenses of the rationality of Todd's suicide attempt. According to the first, the suicide is rational as a means of escape from an intolerably unsatisfying existence. According to the second, it is rational because a state of complete indifference to life has been reached. Neither of these defenses, however, provides any support for nihilism, because the first assumes the truth of a value judgment (that not suffering is better than suffering) and the second makes suicide rational only under circumstances which are extremely unusual (total indifference).

What these arguments do is to make explicit some of the conditions under which suicide could be rational, but they do nothing to cast doubt upon the value of human life or human endeavors in general. Moreover, although I have developed them with the intent of defending the rationality of Todd Andrews' suicide attempt, reflection on them shows that Todd himself *may not* meet the conditions of rational suicide as I have described them. That Todd sinks to the depths of despair is undeniable, but what hints we have about his later life suggest that his despair was a temporary state, that a tolerable life may have remained possible for him. Moreover, Todd does not come close to approximating the perfect nihilist whom I described.[11] He may believe that all is without value, but he himself has both interests and concerns, engages in projects and extended activities, and derives satisfaction from them. Because of this, his suicide attempt may well be regarded as a mistake, the irrational product of a series of unhappy events which culminated in his bleak vision of a life containing no possible satisfactions.

Whatever one concludes about Todd's decision, the case for nihilism remains a weak one. In saying this, of course, I do not pretend to have said the last word about this disturbing doctrine. The last word would consist of a full-fledged theory of value which made clear the conditions under which particular value judgments are justified by reason. Lacking such a theory, I will content myself with two concluding comments which may further diminish the force of nihilism. First, as Barth correctly recognized, nihilism implies nothing whatsoever about how one ought to behave or act. In particular, it certainly does not imply that suicide is better than life. Second, though nihilism may strike us as a doctrine which formulates a deep, disturbing, and important idea, the nihilist himself can make no claims for the importance of his message. Nihilism only matters if it is false.*

NOTES

* I would like to express my deep appreciation to Bernard Gert and John Kekes for stimulating correspondence and continuing encouragement of my thinking on these matters. I am indebted, too, to Thomas Benson, Donald Cohen, Walter Fogg, Edward Hacker, Marvin Kohl, and Ursula Tenny for helpful comments on earlier versions of this paper.

[1] Though I take nihilism to be the doctrine which denies the existence of value, the term "nihilism" is used in a variety of ways. My understanding appears consistent with the explanation given by R. G. Olson in 'Nihilism,' in *The Encyclopedia of Philosophy*, ed. Paul Edwards (New York: Macmillan, 1967), V, 514ff. For a broader range of nihilisms, see C. I. Glicksberg, *The Literature of Nihilism* (London: Associated University Presses, 1975), pp. 9–33; and R. C. Solomon, 'Nietzsche, Nihilism, and Morality,' in *Nietzsche*, ed. R. C. Solomon (Garden City, N.Y.: Anchor Books, 1973), pp. 202–9.

[2] The philosopher most often cited as a nihilist is Nietzsche, but even in his case, the interpretation is controversial. For an attack on this view, see R. Schacht, 'Nietzsche and Nihilism,' in *Nietzsche*, ed. Solomon, pp. 58–82.

[3] Barth's comments on his work are from 'John Barth: An Interview,' *Wisconsin Studies in Contemporary Literature*, no. 6 (Winter–Spring 1965), 3–14.

[4] *The Floating Opera*, rev. ed. (New York: Bantam Books, 1972), p. 218. Subsequent page references to the novel will be made in the body of the paper.

[5] Bernard Gert, *The Moral Rules* (New York: Harper & Row, 1973), pp. 26–39.

[6] Ibid., p. 30.

[7] 'A Moral Principle About Killing,' in *Beneficent Euthanasia* ed. Kohl (Buffalo, N.Y.: Prometheus Books, 1975) p. 109.

[8] Ibid.

[9] For an illuminating discussion of suicide, see Brandt's essay, 'The Morality and Rationality of Suicide,' in *Moral Problem*, ed. J. Rachels (New York: Harper & Row, 1975), pp. 363–87. For an account of how death can be an evil though it involves no suffering, see Thomas Nagel's 'Death' in the same volume.

[10] A possible case of this sort is discussed in B. Williams, 'The Makropulos Case: Reflections on the Tedium of Immortality,' in *Moral Problems*, ed. Rachels, pp. 410–28.

[11] Edward Hacker has suggested to me that a perfectly indifferent being would have no motives and, hence, would be incapable of *any* action. Barth develops a similar idea in connection with the character of Jacob Horner in *The End of the Road*.

HUGH J. SILVERMAN

BECKETT, PHILOSOPHY, AND THE SELF

Samuel Beckett's trilogy (*Molloy, Malone Dies,* and *The Unnamable*) can be regarded as an examination of the self and its various modes of expression. In each novel, a central role is given to the quest for knowledge of the self — the "I" or ego that speaks or writes. In this paper, I (Beckett allows for a juxtaposition of this "I" along with those in his novels) propose to show how, in spite of the fragmented and multiple appearances of the self (with the concomitant negation of personal identity), language does nevertheless establish the presence of the self. Formulating the human as an absence or loss presents a pessimistic outlook on the possibility of self-knowledge. However, since images of the self do occur in the intuition of signs, names, and attempted self-indexes, some optimism arises — particularly if the self-hood of the principal character does in fact achieve parity with that of both author and reader.

From one point of view, the trilogy is a Cartesian project, an attempt to assert the existence of the self. Since Descartes' *cogito ergo sum* is a *dubito ergo sum*, skepticism prevails to a certain extent. While Descartes reaches a resolution through a method that translates doubting into thinking as an affirmation of self-existence, Beckett's ego continues doubting, turning skepticism into pessimism. In fact, the ego which Beckett portrays doubts itself right out of existence, i.e., into death — the theme of *Malone Dies*. By the time that it is the Unnamable, the self is pure mind — *not* a full living being in the Aristotelian sense of a fully actualized potentiality. The self of the Unnamable is like a Hegelian *Geist* which can no longer be individuated.

From another point of view, the trilogy is a Socratic project in that the "unexamined life is not worth living." But for Beckett, even if living is placed in doubt, there is still the optimistic attempt to examine the self. Unfortunately, although the self is continually under scrutiny, in Beckett the self never actually achieves what Locke, Hume, and others have called personal identity. The self becomes a multiplicity of names, until it no longer has a name — but always it remains ungrasped, and therefore elusive.[1] Nevertheless, the self talks about itself. Hence the "language of the self":[2]

What tedium, and I call that playing. I wonder if I am not talking yet again about myself.

Shall I be incapable, to the end, of lying on any other subject? I feel the old dark gathering, the solitude preparing, by which I know myself, and the call of that ignorance which might be noble and is mere poltroonery. Already I have forgot what I have said. That is not how to play.[3]

At least two possibilities arise from this text: (1) the possibility of self-reference and (2) the possibility of self-deception.

I. THE POSSIBILITY OF SELF-REFERENCE IN SPEECH

"I wonder if I am talking yet again about myself." It is only a possibility, suggesting uncertainty. Unlike Descartes, who attempted self-doubt but who thereby achieved certainty through clear and distinct ideas, Malone is not certain. His uncertainty can be outlined according to epistemological, literary, linguistic, and temporal considerations.

Epistemological. His uncertainty arises because there may not be identity of the "I" wondering and the "myself" referred to. Whenever the self (consciousness) attempts to reflect upon itself, it may not posit itself fully, if at all. Such a view would be similar to Sartre's assertion that the ego can never be transcendental; it must be, upon reflection, a transcendent ego. According to Sartre, there is no transcendental pole or source from which all conscious acts emanate. For Sartre, the ego which is taken as an object of consciousness is not the same as the consciousness of that object.

Furthermore, the uncertainty might be the result of what Shoemaker calls the lack of an "ability to self-ascribe those predicates whose self-ascription is immune to error through misidentification."[4] Shoemaker argues, however, that "anyone who can self-ascribe any predicate whatever thereby shows that he is potentially capable of self-ascribing some P*-predicates"[5] (i.e., "psychological predicates which can be known to be instantiated in such a way that knowing it to be instantiated in that way is equivalent to knowing it to be instantiated in oneself").[6] Malone is questioning here whether he is in fact actualizing this potential of self-ascription.

Literary. Malone's uncertainty arises because when one talks, one is not always certain of the referent. I can talk about someone else, giving this other a name and a series of experiences, but, in fact, be talking about myself. One might say that this is what Malone is doing. This literary technique (an apparently nonself-referential narrative) creates the ambiguity. Malone tells stories about characters such as Saposcat, Lambert, Macmann, and Lemuel

— but these may be versions of the same person. They may all be perspectives on himself — his self.

Linguistic. Malone's uncertainty arises because language may never actually reveal one's self. One can talk, speak, even write, but it may be that in each case language cannot quite grasp or reach the very self that speaks or writes. The self, in this instance, escapes reification. Although language goes on, the self retains its special status. This would be the Cartesian and Husserlian view, where the *ego cogitans* or simply the transcendental ego stands behind all languages, all speaking, all positing. For Husserl, intentionality requires that an ego perform the intending.[7] Shoemaker also wants to assert that there is a sense in which "each person's system of reference has the person himself as the anchoring point."[8] But Shoemaker would want to claim that this "anchoring point" appears in language.

Malone, however, is having difficulty because of the nature of language. When Malone appeals to "talking," he means what Saussure called *la parole*, i.e., "speaking" or the individual use-element of language (*le langage*).[9] But Malone is not speaking, he is writing in his exercise book. So speaking here must be a special kind of *langage*, employing, of course, the *langue* English in this case. Malone's speaking or talking is written speech — written speech narrated in the novel. If falls under the general category of what Saussure called *signification* — the act or process of relating the signifier with the signified (the word with the concept) in a value system. Signification is a relating of a relation. Can we say that the signification here is the "I" (*qua* signifier) relating to the "myself" (*qua* signified)? If this relating were fully successful, Malone would not have his doubts.

Since the situation described is actually somewhat different, Malone continues to doubt. The problem is that he cannot be sure whether the signified for the "I" (*qua* signifier) is the same signified as that for the "myself" (*qua* signifier). Jesperson and later Jakobson have discussed this issue in terms of "shifters."[10] The "I" which Malone writes is a shifter because it may not be the same linguistic "I" as that which he wrote down earlier — that is, as he *reads* it now. The shifter then has to do with the "I" as written by Malone and the "I" as read by Malone. He now asks whether they are the same.

Temporal. The linguistic perspective has already pointed to Malone's uncertainty from a temporal point of view. Malone might be stressing the "yet again." If so, the question is whether the self mentioned is the same or a different self from the previous one. The very first words of the novel read: "I shall soon be quite dead in spite of all."[11] The question now is: can Malone say that the "I" of "I shall soon be quite dead in spite of all" is the same as the "I" of "I wonder if I am not talking yet again about myself?" Is there continuity of self? Does the self persist through time? Does the self remain the same as time passes? Kierkegaard had claimed that subjectivity is *becoming* (in the tradition of Heraclitus). If the self is always becoming (i.e., a protean self), it cannot remain the same. Hence the self of the beginning of the novel is not quite the same as the self wondering whether it is talking yet again about itself. As time passes nothing remains the same. Even if one is always waiting as in *Waiting for Godot*, enormous changes occur in the individuals who are waiting. In *Malone Dies*, perhaps the only aspect of the self that persists is the *uncertainty itself*. But this would bring Beckett back to an unresolved Cartesianism, a pure pessimism in the search for self-identity.

II. THE POSSIBILITY OF SELF-DECEPTION

"Shall I be incapable, to the end, of lying on any other subject?"

Capability. At least four interpretations of Malone's capabilities with respect to lying become apparent. (1) Malone could be incapable of lying about anything including both himself and others. (2) Malone could be incapable of lying about himself, but capable of lying on other subjects. (3) Malone could be capable of lying about himself but not concerning other matters. (4) Malone could be capable of lying about any subject whether it be the self or other subjects. The question posed is essentially an ambiguous one, but given the context, the suggestion is that Malone is incapable of lying about the self (in addition to being unable to lie about other subjects).

Whenever the self is discussed, the claim is that only truth can be spoken. Hence whatever Malone says about himself must be the truth — even his uncertainties and doubts. All the stories that he tells cannot be other than manifestations of the self. This point is particularly clear when one considers the very characters of his stories: Saposcat — Sapo (*homo sapiens*), Macmann (son of man), Lemuel (related to Lemuel Gulliver[12] in Swift who, like Beckett, was an Irishman from Dublin). They are all manifestations, allegories, or representatives of universal man — of *the* self as everyman. This would explain why whatever Malone says about himself would be universally true.

Living and truth-telling. Malone could conceivably lie on other subjects —

although he would probably also be incapable of doing so. Since Malone's memory is not very reliable, he might not always be telling the truth about other matters. We know that he is probably an octagenarian, and although he can remember when he was born, he cannot remember how much time has elapsed since then. But memory and its powers are not generally equated with willful truth-telling or lying. If facts are incorrect because of faulty memory, this does not necessarily imply lying.

To the end. The final suggestion in the sentence is that Malone will tell the truth about himself, about the self — "to the end." But the crux of the issue in the novel is that Malone is not certain whether he has reached the end. He often indicates that he "can go on," but he is not quite sure whether he is still going on. This becomes particularly evident in the *Unnamable*, for example at the end, when we read (the sentence is three and one-half pages long, from which I select the following):

... I know that well, I can feel it, they're going to abandon me, it will be the silence, for a moment, a good few moments, or it will be mine, the lasting one, that didn't last, that still lasts, it will be I, you must go on, I can't go on, you must go on, I'll go on, you must say words, as long as there are any, until they find me, until they say me, strange pain, strange sin, you must go on, perhaps it's done already, perhaps they have carried me to the threshold of my story, before the door that opens on my story, that would surprise me, if it opens, it will be I, it will be the silence, where I am, I don't know, I'll never know, in the silence you don't know, you must go on, I can't go on, I'll go on.[13]

In *Malone Dies*, Malone does, it seems, eventually die. Lemuel, who has killed Macmann and others, now becomes Malone with a pencil or his stick, "who will not hit anyone any more":

This tangle of grey bodies is they. Silent, dim, perhaps clinging to one another, their heads buried in their cloaks, they lie together in a heap, in the night. They are far out in the bay. Lemuel has shipped his oars, the oars trail in the water. The night is trewn with absurd

absurd lights, the stars, the beacons, the buoys, the lights of earth and in the hills the faint fires of the blazing gorse. Macmann, my last, my possessions, I remember, he is there too, perhaps he sleeps. Lemuel

Lemuel is in charge, he raises his hatchet on which the blood will never dry, but not to hit anyone, he will not hit anyone, he will not hit anyone any more, he will not touch anyone any more, either with it or with it or with it or with or

or with it or with his hammer or with his stick or with his first or in thought in dream I mean never he will never or with his pencil or with his stick or

or light light I mean

never there he will never

never anything

there

any more [14]

In the middle of the novel, Malone says:

> Then it will be all over with the Murphys, Merciers, Molloys, Morans, and Malones, unless it goes on beyond the grave. But sufficient unto the day, let us first defunge, then we'll see. How many have I killed, hitting them on the head or setting them on fire? Off-hand, I can only think of four, all unknowns, I never knew anyone.[15]

One might assume that the Unnamable is "beyond the grave." Furthermore, Malone has not killed off all these people. Malone is not the author of the novels in which these other characters appear. Rather Beckett is the author of all the other novels. One may want to ask whether the author in the guise of Malone is the same as Beckett — but this is a further problem which is only hinted at.[16] The crux of the uncertainty is that the "I" that says "How many have I killed?" cannot be Malone, since the other characters are in other Beckett novels. Or perhaps it is Malone and all the "I's" are the same fragmented and multiple self — *the* self.

Beckett's theater is often referred to as theater of the *absurd*. The same applies to his novels. It has some basis in Camus' view of life as absurd, e.g., the *Myth of Sisyphus*. What is absurd? The absurd is generally regarded as that which is incongruously ridiculous. But a *surd* is also "a voiceless sound in speech." Hence *ab surd* is "away from" the "voiceless sound," i.e., a voiced sound in speech. Thus Beckett's characters are always talking, but in an ununified fashion. There is constant vocalization. The absurd is also the lack of a rational center: αλογος.[17] Beckett's characters in *Waiting for Godot*, for example, are always going out to the circumference. The multiplicity of selves mentioned in *Malone Dies* are all expressions of a self which does not seem to have a center, a focus, a rational, unitary base.

Malone's tedium arises from his inability to find this rational base, this center. If he had it, he would have a standard by which to judge when he is alive and when he is dead, when he is himself and when he is outside of himself.

Malone may even be capable of saying: "I am dead." Roland Barthes claims that such a sentence is neither in the active voice (e.g., I am going), nor in the passive voice (e.g., I am beaten or I am buried), but in the middle voice.[18] Yet Malone never actually says "I am dead," he simply goes on dying.

"I feel the old dark gathering, the solitude preparing, by which I know myself."[19] Malone is alone (*mal*, i.e., sick, and *alone*). He knows himself by that solitude, which is the essence of his name. But this means that he does not distinguish himself from others. *Solipsism* seems to prevail. In Hegel, self-consciousness arises only by the desire of the other (the master—slave relation). Malone relates the experiences of others, but he is unable to *distinguish* them from himself. He interweaves their experiences with his own. The "they" is an amorphous they – Heidegger's *das Man*.

"I call that playing," Malone says. But what is at play? Language is at play. Malone's storytelling is a series of what Wittgenstein would call language-games. The playing is the play of life. Shakespeare's Macbeth claims that life is a stage with poor players on it. Derrida is concerned with the play of language – *l'enjeu*. As Rey calls it, *l'enjeu des signes*.[20] The self qua "I" is the sign that is continually in play – in operation – operating itself – trying to find itself: the lost self. The enterprise is as optimistic as that of Voltaire's Candide, but the results are as pessimistic as those of Sartre's *No Exit*. Beckett's conclusion however is not that "Hell is other people," but rather "Hell is knowing the self."

NOTES

[1] See Frederick J. Hoffman, 'The Elusive Ego: Beckett's M's,' in *Samuel Beckett: The Language of Self* (New York: Dutton, 1962).
[2] This conception of the "language of the self" is present in Hoffman's aforementioned study, but it is also the theme of Jacques Lacan's research. See, in particular, Anthony Wilden, *The Language of the Self* (Baltimore: Johns Hopkins University Press, 1968).
[3] Samuel Beckett, *Malone Dies*, in *Three Novels* (New York: Grove Press, 1956), p. 189.
[4] Sidney S. Shoemaker, 'Self-reference and Self-awareness,' in *Philosophy Today*, ed. J. H. Gill (New York: Macmillan, 1970), pp. 83–84.
[5] Ibid., p. 89.
[6] Ibid., p. 86.
[7] For an elaboration of the Husserlian position on this question see my essay 'The Self in Husserl's *Crisis*,' *Journal of the British Society for Phenomenology*, 7, no. 1 (January 1976), pp. 24–32.

[8] Shoemaker, p. 90.
[9] Ferdinand de Saussure, *Course in General Linguistics*, trans. Wade Baskin (New York: McGraw-Hill, 1959).
[10] Roman Jakobson, *Shifters, Verbal Categories, and the Russian Verb* (Cambridge: Harvard University Press, 1957) and Otto Jesperson, *Language, Its Nature, Development and Origin* (London: Allen and Unwin, 1922).
[11] Beckett, *Malone Dies*, p. 179.
[12] Gulliver is a sort of universal man discovering the limits of possible human society in order to reflect upon his own. Beckett's characters can also be considered as instantiations of an archetype that searches out the ranges of the human psyche in order to return to some center, which may be Beckett himself.
[13] Beckett, *The Unnamable*, in *Three Novels*, p. 414.
[14] Beckett, *Malone Dies*, pp. 287–88.
[15] Ibid., p. 236.
[16] In following out this possibility, consider Beckett as a *deus ex machina* appearing with every reference to the self. Like the "deity" Socrates, who, according to Nietzsche in *The Birth of Tragedy*, spoke through Euripides and rationalized Greek tragedy, Beckett is present as an absence, a phenomenon that announces itself through the selfness that is other in his characters.
[17] The Greek αλογος is here interpreted as meaning "without reason" – the absence of a *definitive* word or center. What is particularly absent is the specific place from which the authorial voice speaks.
[18] Roland Barthes, 'To Write: Intransitive Verb?' in *The Structuralist Controversy*, ed. Richard Macksey and Eugenio Donato (Baltimore: Johns Hopkins University Press, 1970), p. 143.
[19] Beckett, *Malone Dies*, p. 189.
[20] Jean-Michel Rey, *L'Enjeu des signes* (Paris: Seuil, 1971).

GILA RAMRAS-RAUCH

PROTAGONIST, READER, AND THE AUTHOR'S COMMITMENT

I propose to discuss the complex interrelation between the protagonist, the reader, and the author – in the light of three models. The first is the classical model of that interrelation, embodied in ancient drama and epic poetry. The second is that embodied in the traditional representational novel. The third is that embodied in the contemporary "hermetic" (or obscure) novel or play. I shall only sketch my views concerning the first two. The main discussion will concern the third.

Classical authorship is imbued with a sense of continuity. We cannot imagine Sophocles without Aeschylus, nor Euripides without Sophocles. The earlier authors serve as paradigms for the later – and the imitation was organic and meaningful, because whatever differences in world outlook there may have been between these authors, all such differences were embraced within the framework of traditional and inherited values. Despite such interrelatedness, however, we may point to a certain distance as characterizing classic literature: First, there is the distance between the writer and the world he depicts (see Homer's invocation to the gods); second, there is the distance between the audience and the world depicted; third, there is the distance between the author and the audience. While listening to the rhapsodist declaiming the tales of Homer, or watching the unfolding of a tragic drama, the distance between audience and protagonist was tangible. You could identify with Oedipus or Antigone, and feel pity and fear. But above it all you perceived a certain *telos*: what moved you was the workings of fate, embodied in the human predicament. This predicament was seen under the aspect of eternity – again, the distance – just as Zeus, in the Olympus pediment, looks upon the fate of the house of Atreus. Whatever closeness there was, was between the individuals in the audience as members of one *polis* – a closeness that does not exist between the readers of a novel, no matter what society they "belong" to.

Let us now contrast that three-sided relation with another triangle. I refer this time to the novel of the nineteenth century and earlier, in the light of the triangular relation between author, reader, and protagonist. There are cases – perhaps the best known being that of Flaubert and his contemporaries – where the author and readers do not share the same feelings toward the protagonist (Emma Bovary). Another example is that of Balzac's Vautrin. Yet I daresay that these are the exceptions, even in Flaubert and Balzac: if the approximately ninety novels and stories of the *Comédie Humaine* form a united whole, this is because they are set in one world (the Paris of the 1820s and 1830s), and because the social values of that world are shared by Balzac and his contemporaries. The interrelation between author, reader, and protagonist can here be characterized as dynamic and viable. The three partners in that relation share the same moral code, even while criticizing it. Balzac may have Vautrin looking down upon Paris and declaring war, but Balzac himself does not. (If anything, it is Paris that does the attacking.) For both Flaubert and Balzac the world accepted by their readers was the world that was valuable and real. Flaubert says of these ordinary men and women, "Ils sont dans le vrai" – and this includes their morality. That is, there are such things as transgression and recompense, and sinners against the code are thrown into moral crises. Despite clashes of interest, the code was never really threatened – and this holds true from the vantage point of the author, reader, and protagonist. The moral code bound them together; and although it was questioned throughout the nineteenth century, it was not significantly shaken until the twentieth.

I referred to the relation of imitation in classical writers. This does not hold true in the same way for the early novelists. Thus Cervantes is unique – not imitating and himself inimitable – as is Sterne, and even Jame Austen. That is, their "imitation" does not concern literary form, but is only an imitation of the world around them. This is their common denominator, despite the uniqueness of each. We can even generalize and say that in regard to most middle-class novelists, the closeness between them is not in regard to form, since the novel is in many ways an open prose medium. In the nineteenth century there is a further closeness – between the author and the world he depicts, between the reader and the world depicted. The novelist presents his fictional reality against the backdrop of a nonfictional Paris, St. Petersburg, or London. Authors such as Balzac, Dickens, and Thackeray take us by the hand and lead us through their inner and outer worlds.

Thus we are exposed to characters in certain situations, and we follow their progress toward their individual resolutions. But the point we must stress is that the characters are activated within their complex social fabric, and their conflicts and clashes register those social complexities. We are not dealing with private heroes, each of whom in his individual fate represents the dialectical nature of society at large.

One cannot understand the fate of Emma Bovary or of Anna Karenina without taking into account the social world they lived in. Indeed, these books do not exist without their respective social frameworks. In effect, therefore, the protagonist (of the representational novel) serves as the *place de combat* of the norms of his/her society. It is the social framework that contributes the tone and tenor to these novels, and this accounts for the numerous "closenesses" I alluded to. The omnipresent and ubiquitous novelist knows the ins and outs of his protagonist as he knows the streets of the town the protagonist lives in. It is this factor that gives the reader a sense of security: just as we know where we are, so we know what happens and why.

The contemporary novel, on the other hand, reflects acts of desertion and separation. The author has deserted both the protagonist and the reader. The reader is in search of a guiding hand to lead him in his senseless meanderings through landscapes he does not know and cannot name. In Kafka and Beckett, the plot need no longer make either sense or difference – even when it does. The reader goes through endless pages involving existing/nonexisting castles, wastelands, and characters. The plot is not complex in form, yet it carries an ominous and inexplicable weight, an undissolved heaviness. We face another sort of complexity, therefore, in the simplest of plots – be it that of *The Castle* or *Waiting for Godot*. We are inside an inverted fairy tale, and the Garden of Eden spells death for us. The feeling of doom, of the end of all aspiration, reduces the plot to its most rudimentary form.

But this is interpretation. Where does true complexity lie? Is it in the colossal worlds of *Ulysses* and *Finnegan's Wake*, works of legendary scope and depth? A Joyce work is cumulative in essence, whereas a Kafka novel (such as *The Trial* or *The Castle*) gives one the sense of losing the little one knew at the starting point. Where, then, does complexity lie? Is it in the role of the irrational as a *bona fide* component of the narrative *durée*? Is it in the freedom of an unmapped road? What is the power of the contemporary novelists and fabulists to state the complex in the simple? What the power of linear artists such as Kafka, Beckett, Ionesco? (Baroque fabulists such as Nabokov and Borges are not our concern for the moment.) Can we suggest that the complexity we have experienced in the plot of the full-blooded representational novel is not to be found in the obscure twists and turns of the contemporary novel?

The author's desertion of the protagonist means the desertion of the reader, to his own cell of meaning. Sometimes we go unwillingly, especially when a certain typical order is reversed: Usually the reader's understanding is the basis for interpretation; understanding the text leads to a deeper consideration of its implications. In many contemporary novels, however, we must begin with interpretation, and this serves as a basis for an understanding of the text.

This desertion of the protagonist and the reader turns both into acrobats hanging from a loose fabric devoid of frame and limit. The literary critic hangs on those loose ladders as in a Stoppard play. To the reader this spells ambiguity, ambivalence, a noncumulative quality which bars familiarity. Moreover, the element of unreliability enters. The unreliable narrator, for example, is disconcerting to the reader, who in any event is seeking a poetic truth in fiction. Something is presented, a cloth is embroidered without being put into a frame.

In the representational novel, the plot is part of the wider framework of background, place, and time. Raskolnikov moves in a familiar city, as does the protagonist of *Notes from Underground*. So, despite the cold horror of Raskolnikov's act, despite the effort of the man from underground to be a "characterless creature," there is a certain familiarity which allows us to overcome the numbing horror of both situations. The plot has an analogue: there are friends, family, subplots, changes in outlook (e.g., redemption through remorse, for Raskolnikov).

In the barren landscape of Kafka, Beckett, Ionesco, or any other "hermetic" novelist, however, we are given no assurance as to the existence of an outer world. Thus the separation of the author from the protagonist parallels the separation of plot and protagonist from their frame. And accordingly we may say that it is the frame, rather than the plot, that is our entrance card to the world of fiction. The frame brings with it a tone (as in Swift), a guide (as in Fielding or Thackeray), a system of values (Austen, George Eliot, Dickens).

Complexity, in the traditional sense, involves a conflict of values. The complexity of *Don Quixote* is in the protagonist's adherence to one system while existing in another. A similar complexity is a source of delight to the reader of *Tristram Shandy* experiencing the duality of the narrated moment and the moment of narration. On a deeper level, the praxis of the representational novel is the display of various and varying normative systems. The orchestration of the totality is what gives meaning to actions, reasons, causes, and outcomes.

As we all know, the absence of framework is widespread throughout the arts in this century — and it has all been done to a fatiguing degree. There are *objets trouvées* and even *retrouvées* in galleries, mountains wrapped in nylon, conceptual art, environmental art — all of which reject the notion of a framework. Yet our personal hold on our social reality has not really been shaken. The theater of the absurd can unsettle us. Is it because we are faced (as is claimed) with "pure" unadulterated experience? Is it because the inner consciousness of the person (whether protagonist or audience) is the place of our disconnection from the other person? We come to the frontiers of our experience. And literature — perhaps more than other art-forms — is, in its greatness, a reminder of our (what I call) interisolation, our mutual separatedness from one another in our togetherness. We know how to walk the inner paths of our consciousness, never sharing it, yet reporting about it in a limited way. Kafka, Beckett, Joyce, Woolf, the governess in Henry James's 'The Turn of the Screw,' reveal these inner paths. They lead to the threshold of chambers which are floorless, ceilingless. The grasp of the self by the self is, as Hegel saw, paradoxical; it is a process which cannot be tacked onto a frame, and which is, in its *Urform*, nameless. This has led to the de-concretization of the novel, along with its de-realization and de-generalization. Yet with this loss there is a gain. The opening of these new inner territories has brought awe, even fear, at the timeless power of art as shaper of man's consciousness.

Kafka, Joyce, Beckett, and Camus do not reject the value-laden world of the representational novel. It seems to me that in only some of their works is a mode of experience portrayed that does not allow for the presence of values — i.e., a mode of experience beyond all moral codes, rather than a protest against them. I am fully aware that certain novels, such as those of Sartre, while falling within the scope of the representational mode, do emphasize the situation of absurdity, anxiety, and nausea as representing the state of man in a world devoid of certainties, codes, and systems. I would rather address myself to Kafka, however, and say that the presence of other values, in a code no one can decipher, is what lies at the root of the supposedly "nonnormative" trend. And the moment the Kafka protagonist, somewhere in the course of the story, somewhere along his life, crosses into this other value-system, he is caught between two worlds he has no keys to.

I do not see this as an Existential theme with a capital E, but rather with a small e. The sense of separation I spoke of, between author, reader, and protagonist, together with the accompanying sense of loss, does not belong exclusively to a pessimistic outlook regarding man's fate. Rather, it is part of the uniquely modern conception — whether in literature or in the other arts — wherein the subject and object are separated from their common cultural frame. Thus the self leaves the milieu it was born to, and goes on an excursion beyond the laws of the here-and-now. And it is for this reason — *not*, that is, because of an ontology of anxiety and the absurd — that there are shattered the unity we had heretofore experienced, together with the three-way closeness between author, reader, and protagonist.

Not only are we, as readers, separated from the protagonists in many contemporary works, we lose them somewhere and we never find them again. They keep meandering, within our range of observation — but we do not know what we observe. Usually, at the end of the traditional representational novel, there is an evaluation or reevaluation on the part of the reader. In the "hermetic" (or obscure) examples of contemporary literature, there is no such process; as readers, we sense that the more we read the less we know. Moreover, there is a cumulative force in the traditional representational novel, a force which is even essential to it; but in the obscure contemporary work no such force is manifested. Why is it, then, that we know less and less? Why is the knowledge we have accumulated so short-lived that it does not serve the novel as a totality?

Many answers might be suggested. For one, perhaps the given novel simply does not aim at a totality. Perhaps this novel does not seek to resolve the tension between its own stasis and dynamics — so that it engulfs itself and yet reveals itself before our eyes. The idea of resolution, so fundamental to the representational novel, is by no means a positive element so far as the contemporary novel goes. This might contribute some discomfort, perhaps even anxiety, to the reader of the contemporary novel. The absence of resolution goes with the absence of reevaluation: resolutions often contain the moral material that goes into reevaluation. In the "hermetic" work we are often left in the dark as to both, since we are without guidance.

Not only is the contemporary protagonist separated from a social framework, he also moves through spheres devoid of analogy. This generates further problems of comprehension and interpretation. Indeed, since there is often no way of deciding on the meaning of what is going on, we lack the starting point for interpretation. Is the landscape through which Joseph K. moves a *place* at all? The place in which Estragon and Vladimir are marooned and are waiting for Godot — is this a "place" susceptible of description, location? It has been said that in these works we find ourselves in the world of dream, hallucination, inner consciousness, even raw experience. Whatever world it is, we may ask whether it is a metaphor connecting the objective reality with another reality that is indefinable and unidentifiable. Is it an interpretation of our own reality, in such a way as to connect it with an individualistic reality beyond the ethical? The "suspension" of the ethical, here, goes hand-in-hand with the neutrality of the absurd — and this goes with the incomprehensible and the ambivalent.

The basic question, therefore, is whether, for example, Kafka's world is essentially ethical or is altogether beyond good and evil. The answer to that question will enable us to decide as to the possibilities of interpretation. That is, is he actually commenting on an existing state of affairs, or is he beyond every such framework of evaluation and thus beyond all analogy and all comprehension?

Much has been said, in contemporary criticism, about the hero and antihero, as cultural symbols and phenomena. The hero in literature is an individual who exercises his potential. This can have a "tragic" outcome, but in different ways and with different meanings — as we can see if we contrast Adam, the ur-model of the hero, with Oedipus and Antigone. What they have in sommon is the fact of a framework, sometimes containing only one command (directly issued by God) or an entire array of inherited values, implicit and expressed. In contemporary literature we have a new sort of protagonist, the private hero of the private world. Instead of undergoing one climactic retribution, he finds himself in a continuous process of losing out. He may undergo a physical waning, but more characteristically he is losing erstwhile certainties, such as the element of causality so fundamental to our comprehension.

Further, the Kafka protagonist is a person with little or no past. Such a deficiency is usually remedied in the unfolding of a novel: the protagonist reveals himself in a way that gives a cumulative weight to his character. The Kafka protagonist, however, remains without a past, and (lacking a cumulative identity) he also has no present. He exists in what we might call an effort to overcome space — i.e., an effort to escape the outer spatial world and enter inner consciousness. This inner consciousness is translated, metaphorically, into fables which take place in a certain space, as objective correlatives to the inner consciousness striving for pure experience. But since there is this metaphorical connection between the inner and outer worlds, and since the world of inner consciousness exists in only a temporal dimension, we can say that the attempt to escape from the outer spatial framework is in essence an attempt to escape from the internal consciousness of time. Kafka's protagonist significantly falls asleep at rather important moments in the course of his experience. He is fatigued and suffers loss of concentration. He is in this way the diametrical opposite to the protagonist in the representational mode (e.g., Galsworthy) who grows and swells before our eyes, gaining complexity, reliability, and life.

The more fundamental problem — to which I alluded — is the problem of how to read and understand this "hermetic" literature, when it leaves us no opening into its inner meaning, leaves us wondering if it is indeed intended to have any, but certainly provides us with none of the usual keys with which to open the door of interpretation. "Hermetic" prose obviously does not have that instrumental function which Paul Valery said prose has: to give information, transfer meaning, and convince of its truth.

What can account for this change? One of the theses I have been arguing for is that the new problem of intelligibility and interpretation — as a problem in the relation between author and reader — has been reflected in the author's changed relationship to his protagonist. As to the change in the figure of the protagonist himself, we may summarize this if we begin with the prototype of the public hero and go from there:

(1) Don Quixote lives in two constellations. He believes he belongs to the heroic world of knight-errantry, while moving in a low realistic landscape. This tension, between the seeming and the actual, is tragic. When he discovers the reality, he turns his back and dies.

(2) The protagonist in the realistic, representational mode is a private hero in a public setting (middle class). Regardless of the nature of his existence, or the outcome of his actions, the social-normative framework exerts a major force on his existence. So far as the realistic/representational novel and its protagonist are concerned, therefore, we can say that without that framework there is no novel.

(3) The protagonist in contemporary "hermetic" literature is a private hero in an altogether private world. That world has no analogue in any experience the reader can bring to his encounter with the novel. Can we say that Kafka's Castle is a castle, and that is that? Can we say that ambivalence is a necessary thematic or structural element in such literature, and refrain from attaching any meaning not accessible in the work itself? Can we presume to take one step farther, and say that literature is implacably moving beyond the mimetic function? Can we say that there are, here, signs of a major change in aesthetic conception? Can we go on to say that we are coming to see literature not as a "temporal" but rather as a "spatial" art, producing an artifact to be perceived as *there*? Lastly, can we separate literature itself from the fabric of meaning, and relate to it as a pure object?

All these questions would have been otiose if the author — in this case Kafka — has been "present" in his work, had been there to guide us and provide a certain degree of clarity — i.e., if we could perceive his own attitude as to what happens in the novel. Yet in many of Kafka's stories we are deprived of the author's guidance, as well as of the protagonist's stream of consciousness. Where, then, do we stand? We cannot look to the genre to give us a clue as to what takes place. The form of narration limits the protagonist as it does the reader — and we know no more about what is happening to Joseph K. than he himself does.

Tone, point of view, identifiable genre — all are keys to comprehension and interpretation. Yet, as I indicated, we must read Kafka's stories as fairy tales — and fairy tales have no author, and therefore no author's viewpoint. Nor could we discern, as in the representational/realistic novel, a deterministic element that would help us understand what goes on. Not only do we know little, if anything, about the protagonist in "hermetic" literature, the little we think we know can just as readily be negated in the story.

In conclusion I should like to suggest that it would be best, in criticism of "hermetic" literature, to reject such traditional polarities as hero versus antihero, middle-class norms versus the negation of such norms, positive versus negative views of man, etc. None of these dichotomies, so useful in former criticism, is now quite apt. If, as I have argued, the contemporary novelist has made interpretation difficult (if not impossible), let us remember that he too is alone.

PART II

THE HUMAN SPIRIT ON THE REBOUND

DAVID L. NORTON

NATURE AND PERSONAL DESTINY: A TURNING POINT IN THE ENTERPRISE OF HUMAN SELF-RESPONSIBILITY

My theme is the problematic nature of the human being, and concerns one of its most important implications. My thesis is that if persons in meaningful numbers are to respond constructively to the precarious adventure in which the human being is involved, then this adventure must be recognized as intrinsically social as well as individual, and impossible of success until appropriate social responsibility is acknowledged and exercised. Where individuality itself is conceived asocially or "atomically," as by classical liberalism and by the Sartre of *Being and Nothingness*, it renders the notion of support or assistance to the individual in his fundamental responsibility of self-actualization as a contradiction in terms. I take the result of this to be conclusive as an historical lesson, which I will term the "three percent law." It is that where persons receive no support or help, the hazardous and arduous task of individuation will be undertaken by fewer than three percent of human beings, going unrecognized, disregarded, or shirked by the majority. This historical lesson has in the end brought about the profound discouragement with individualism of many of our noblest individuals. I think here, for example, of Montaigne, and of John Stuart Mill, who late in life turned toward socialism,[1] against the proud and challenging adventure he had proclaimed in the third chapter of his *On Liberty*. I think of Jacob Burckhardt, who late in his life wrote to a friend, "You know, as far as individualism is concerned, I hardly believe in it any more, but I don't say so, it gives them [his enemies] so much pleasure."[2] And I think of Kierkegaard, who prefaced some of his later books with the remark that he expected no more than three or four readers to understand him.

The problem with *moral* individualism, i.e., the individualism which claims the superior worth of lives lived in terms of self-responsibility — is that too few lives have been lived in this fashion to count. Its argument that all lives *can* be so lived and *ought* to be so lived is decisively countered by the reply, "But except for isolated instances they won't be." And this counter-argument has in the past been fully affirmed by individualists themselves, leading them either to eventual despair, or to proud but hopeless defiance as solitary voices crying in the wilderness. As a voice of individualism, Nietzsche saw that he came too early, and announced that henceforth he would write for readers one hundred years in the future. My contention is that individualism will always come too early, until individualists themselves recognize their social responsibility for creating conditions which nurture and assist the process of individuation. And in this paper I want to be specific, for I believe that enough is known today about individuation to enable us to see exactly why such is the case, and to pinpoint that precise juncture in the lives of persons when, if they are left without social sanction and support, they will succumb to the workings of the "three-percent law" and betray their own individuality. If, as Shakespeare says, "self-love is not so vile a sin as self-neglecting[3] — if, indeed, as Greek individualism maintained, self-fulfillment as the actualization by individuals of their potential worth is the first responsibility of persons — then it is time to insist that none of us can be content with recognizing this responsibility in himself, for we have an entailed responsibility to help to create the conditions under which others in increasing numbers will awaken to and accept for themselves their corresponding responsibilities.

I argue here for the abandonment of classical liberal individualism in favor of what I term eudaimonistic individualism in acknowledgment of its origin in ancient Greece. Classical liberalism employs a conception of individuality which is "atomic," nondevelopmental, and unproblematic. It is at bottom merely numerical individuality, consisting in the fact that two things are unalterably two and not one. Against this conception, eudaimonism conceives individuation not as unalterable *fait accompli*, but as the outcome of a process, a process which may or may not come about and is therefore problematic, and it understands true individuality to be inherently social. To be a person is to be a being which is social in the beginning and social in the end, but emphatically these two socialities are not the same. The sociality with which the life of the person begins is an imposed sociality which regulates the lives of persons, and must do so in virtue of the status of persons in the first stage of their lives as dependents. But the sociality in the end is voluntary sociality which requires nothing in the way of the sacrifice of true individuality, for its principle is not the essential sameness of persons "at bottom," but is instead the principle of the ideal complementarity of perfected differences.

To begin at the beginning, I will say something about "man's place in nature," i.e., about the distinctive kind of being that the human being is. It is a problematic being. And it is this recognition that grounds a profound psychology, or philosophical anthropology, or social or political theory. In our time the spokesmen for this recognition are a minority, but nevertheless an eloquent one. Among them Jean-Paul Sartre is notable for his pronouncements, first, that the human being "is what it is not, and is not what it is," and second, that it is "an individual venture," and a "project."[4] In a similar vein, Ortega y Gasset identifies the human being as in essence *drama*, in the sense that "we do not know what is going to happen." For, Ortega says, "Being man signifies precisely being always on the point of not being man, being a living problem, an absolute and hazardous adventure."[5] Recently, Roberto Mangabiera Unger, in *Knowledge and Politics*, identifies personhood as a "predicament rather than . . . a substance."[6] And if British philosophers seemed to be constitutionally impervious to this understanding, it is encouraging to find that no less than Michael Oakeshott has given this recognition the central place in his new and important book, *On Human Conduct*. For Oakeshott, human conduct is "an engagement in which Subjects continuously create and recreate themselves as the finite persons they wish to be." And he notes that "this condition may be greeted by various mixtures of revulsion, anxiety, and confidence; it is both gratifying and burdensome. The sort of self-fulfillment it promises is partnered by a notorious risk of self-estrangement or self-destruction."[7]

The origin of this recognition of man's distinctiveness lies with the ancient Greeks, and I return to them here in the belief that their conception in some ways remains better than our best. If you will recall with me the Greek creation myth, you will remember that the Creator had at his disposal raw matter and many immaterial and perfect forms or ideas. Having compounded creatures from these elements he assigned to Epimetheus the task of distributing advantages to the creatures to fit them for survival in the world. Thus the elephant received his size and his trunk, the tiger his fangs and claws, etc. But the absent-minded Epimetheus had emptied his bag of attributes and was left with nothing to give when he arrived at man. To rectify man's helplessness, brother Prometheus stole from the gods their most precious possession and delivered it to man, the gift of fire, by which to survive and perhaps to prosper in the world. Not literal fire, of course, but spiritual fire, the passion for ideals and the fervor to live by them. This fire is love, in the Greek meaning of Eros. Eros is an ingenious compensation for the oversight of Epimetheus. For thanks to his oversight what man uniquely lacks is that *necessity* by which all other creatures are forever just what they are, the metaphysical necessity of essentialism. But this very lack is also human freedom, namely, the freedom in the absence of a predetermined nature, to try to make of oneself what one chooses to become. This predicament cannot be undone (for the Creator has withdrawn, and cannot be summoned to answer complaints). But Eros affords an ingenious solution, for it represents a new kind of necessity. It is a necessity which does not antedate freedom and preclude it, but stands as an option to be chosen by freedom, which therefore presupposes it. It is, in a word, moral necessity. Man is thus equipped for his unique place in nature as an experiment with respect to which he is at the same time the experimenter.

In Greek terms, to be a person is to be an indeterminacy with an opportunity of self-determination by moral necessity. This recognition finds undying expression upon the stage of Greek tragic theater in the person of the moral hero. The great question posed upon the tragic stage is, amid warring powerful forces in the universe, does man count? And the sagacious answer worked out there is, "Man counts if he chooses to count." He stands or falls by what he freely makes or does not make of himself.

Likewise the *locus classicus* of the conception of the human individual as a problematic entity lies in ancient Greece, this time in Plato's famous depiction (*Phaedrus*) of the soul as a chariot, charioteer, and two horses — one struggling to rise aloft while the other tries to plunge below. This represents the personhood of the individual as equivocal, an argument with itself, a problem to itself. Here is the priority which Greek humanism ascribes to man. It does not contend for man as the supreme being in the universe, metaphysically or morally. It contends for man's priority as his own first problem, a problem whose solution rests with the initiative, the "spirit" of man. Furthermore, *each individual* is such a problem, and in the sense that he as an individual is *his own* first problem. Here is classical individualism, anticipating Sartre's pronouncement that the human "being is an individual venture." There can be no collective solution by which the problem which each individual is to himself is solved *for* him. There is only a cumulative solution, resulting from the additive successes of individuals with the problem that is the self of each. And despite some evidence of uniformitarianism in Plato, and a great deal of it in Aristotle, there was in the spirit of Hellas considerable support for qualitative individuality. As Socrates says (*Phaedrus*, 242d), "each person chooses from the ranks of beauty according to his own character."

It is the eudaimonistic conception of personhood that is required for a viable individualism today. It is qualitative, developmental, and not at war with, but harmonious with, the social, when the latter is rightly conceived. Accordingly, to be a person is to be a qualitatively unique and irreplaceable potential excellence, with responsibility for self-discovery (of one's daimon or unique potential excellence) and progressive self-actualization. It is this conception which underlies the great Greek imperatives, "Know Thyself" (*gnothi seauton*, inscribed on the temple of Apollo at Delphi), and "Become what you are" (Pindar's phrasing). There were, I think, two mistakes in the Greek understanding, neither of which vitiates the eudaimonistic conception of the individual, but both of which require rectification. The first is the "intellectualism" of the philosophers, Plato, Aristotle, and (perhaps) Socrates. The second is the elitism by which the Greeks excluded whole classes of persons from personhood in the eudaimonistic sense — slaves, women, *barbaroi*, and perhaps tradesmen and craftsmen. These deficiencies — representing parochialisms — have been rectified in modernity, notably by existentialism and its precursors, Kierkegaard and Nietzsche. It was Nietzsche in *Birth of Tragedy*, and Kierkegaard in *The Concept of Irony* who put to rest the sometime-Greek supposition that the destiny of all individuals alike was fulfillment as pure speculative intellect. So far as slaves are concerned they are, from Nietzsche's and Kierkegaard's point of view, self-chosen dependents who are to be encouraged to assume self-responsibility. But on the eudaimonistic personhood of women Niertzsche and Kierkegaard are, at best, equivocal. This work remained for later existentialists to do, and if they have not done it, it remains for the rest of us to do.

I am now going to center upon the problem I mentioned at the beginning, the problem of the sociality of true individuals treating it from a developmental standpoint, as eudaimonism demands.

On the eudaimonistic conception, we do not begin life as solutions to the problem we are, namely, as fulfilled and worthy individuals. We begin life as merely the potentiality for such, with the inborn responsibility to become such in actuality, and as the outcome of a course of development. As the Greeks like to put it, compliance with the first great imperative, "Know Thyself," is not to be sought until the time "when the young man's beard begins to grow." Very well, prior to that time we are dependent creatures. In this condition we are not — the Greeks recognized — self-responsible, and, as Aristotle says, the whole of the *Nichomachean Ethics* must not be thought to be applicable to children. Or, as Nietzsche put it, "He who cannot obey himself is commanded."[8] But what requires recognition here, and what escapes the common simplistic misunderstanding, is that dependence is, by the developmental nature of individuality, a provisional dependence, hence authority over the individual (social authority) is provisional authority.

The recognition that individuality is a development is the recognition that one does not begin life as a self-responsible individual, but as a dependent creature with the potentiality to become a self-responsible individual. Such an individuality is not unconditional. It has conditions, foremost of which is the condition that it be provided for, and properly provided for, in its initial period of latency or dependence. In short, eudaimonism recognizes that in the first stage of their lives, as dependent creatures, persons are social products.

Denial of this truth is the crucial failing of classical liberal individualism. By its nondevelopmental conception of the individual, persons are (numerical) individuals unalterably from beginning to end, and are intrinsically asocial or "atomic," which means that social relations among individuals are so-called "external relations" only, and do not affect the true natures of their terms. On this view society is not natural but conventional, the product of a "social contract." And within this scheme childhood is assimilated by the concept of "tacit consent." The presence of children in a given society is taken to express their "tacit choice" of the terms of that society and thereby their participation in the "social contract."

In the last hundred years the classical liberal conception of "atomic" and unalterable individuality has been convincingly refuted by the rise of what I will term the "sociological perspective." The sociological perspective has demonstrated that persons begin life as products of their culture. Children are dependents, not merely in the sense that they are not physically self-sufficient, but in the deeper, "subjective" sense that their concepts, feelings, and judgments are internalized from sources external to themselves, namely, from their parents as agents of the wider community and exponents of a distinct culture. What they thus receive is the "common sense" of their social milieu, so named because it is shared alike by all persons, and what eccentricities may obtain are considered to be nonnormative and either disregardable or else standing as "errors" to be corrected.

But with this great truth in hand, the "sociological perspective" commits precisely the same formal fallacy that it condemns in "social contract" individualism. It mistakes a part of the truth for the whole. According to classical liberalism, all sociality is voluntary and conventional. According to the sociological perspective, all sociality is natural and involuntary. The sociological perspective correctly apprehends that personhood in the beginning is entirely a social product. But, dazzled by this half-truth, it shuts its eyes and proclaims personhood to be an involuntary social product from beginning to end, and thereby it falls into grievous error.

What distinguishes eudaimonism from both classical liberalism and the sociological perspective is its recognition of individuality as a development. Eudaimonistic individualism affirms that personhood is inherently social in the beginning and it is inherently social in the end. But emphatically the two socialities are not the same in principle, and should not be the same in fact. For what happens between is the actualization within the person of his individuality, his unique potentiality. And this happens just as the young man's beard begins to grow (and correspondingly in the case of women). The individuation we speak of is, of course, normative individuation, and not the epiphenomenal differentiation which characterizes both dependents and self-responsible adults. The fact that one has a mole on his cheek of just such a color and in just such a position is nonnormative; but the fact that one prefers a rural setting more than an urban or suburban one, or that one prefers a certain kind of activity to all others, is not epiphenomenal and nonnormative. It is an indication of the kind of excellence which it is innately one's responsibility to actualize.

Prior to the first manifestation of one's individuality one is a dependent and a participant in the mode of dependency upon one kind of sociality, for which I have adopted the name "antecendent sociality."[9] This sociality is not chosen but inherited with one's birth, and one is not responsible for it, but conversely: it is responsible for one's formation as the person one is. And since individuation in this situation is mere latency and has yet to appear, the principle of this form of sociality is uniformitarianism, namely, the supposition that "at bottom" all persons are alike. Antecedent sociality is entirely correct in its supposition that there is a sense in which all persons are alike, but it is parochial in its supposition that *sociality as such* rests on a likeness which is the mere numerical replication of an identical human nature, without which pure chaotic anarchism would obtain. The emergence of individuality "when the young man's beard begins to grow" is the forecast of a new form of sociality, a sociality which is not imposed upon persons but created by them, not as a convention, but as the expression of the social nature of true individuals. I term this "consequent sociality."[10] It is a sociality *for which* individuals are responsible and not conversely, and it is a sociality which asks no least sacrifice by individuals of their distinctiveness but, on the contrary, calls for the perfection by individuals of their individuality. It is a form of sociality whose inherence in the true individual is confirmed even in those true individuals who, historically, have appeared to the world as the most resolutely antisocial. For examples, I will choose two famously antisocial philosophers, Arthur Schopenhauer and George Santayana. Santayana was a recluse who went to great lengths to avoid interpersonal contact of any sort, while gently and distantly patronizing the human race in his writings. On the other hand, Schopenhauer caustically scalded his fellow human beings, and openly proclaimed that the only worthwhile companion he ever found in life was himself. But out of his solitude, Santayana revealed himself in *Dialogues In Limbo*, and precisely the same self-revelation appears in the informal, first-person writings of Schopenhauer. Both men longed for companionship, but the companionship of persons no less accomplished at individuation than themselves. They did not find it in the world, but such was their longing that they invented it, as attested by the writings mentioned. These writings are their dialogues, respectively, with their chosen company — a handful of historical persons who were their equals — now long dead, but resurrected for purposes of companionship by Santayana and Schopenhauer.

What these "antisocial" individuals attest to is that the most pristine individuality is not "atomic" but inherently social. It is social, however, in a sense which refuses to compromise its own individuality. It is, therefore, social on a principle which does not compromise individuation but presupposes it. It is this sociality that I have termed "consequent sociality." What Schopenhauer and Santayana (with other true individuals before them) condemned, is but one form of sociality, namely, uniformitarian sociality — the sociality of mere numbers, the sociality of the "herd," the "mob," and the "mass." Both men found little else available and were uncompromising in their refusal of it. But their resolute uncompromise is our beacon light, pointing to the possibility of the second form of sociality, the sociality of true individuals. I attach great importance to such exemplary individuals as Santayana and Schopenhauer for the following reason. By their intransigence they tell us that for true individuals, the sociality which is at the expense of individuality is unsatisfactory. By their fantasy — as exemplified in Santayana's *Dialogues in Limbo* — they tell us of the intrinsic sociality of true individuals. And by their peopling of their fantasy with historical figures they tell us that the unavailability to them of the society for which they longed was a contingent fact of their lives, and not an inevitable and unalterable one. Following this up, my contention is that the historical fact of the scarcity of lives of genuine integrity, of true individuality, in any given time and place, represents a *collective* failure for which we are all responsible. And because "ought" implies "can," I am therefore called upon to suggest the means by which to ensure that true individuals shall not be so few as Schopenhauer and Santayana found them to be. Once again: that true individuals are so few as to find themselves in their experience alone, as Schopenhauer and Santayana did, marks the failure of individualism. There can be nothing but failure until such time as persons, having extricated themselves from the uniform sociality of their dependence by the discovery within themselves of something unique, something for which only each alone can speak, are assured of finding others who are living likewise as those individuals. The developmental release from antecedent sociality must in short be the opportunity, not of solitude and alienation, but of consequent sociality. This must be so if normative individuation is to be anything more than "a voice crying in the wilderness."

To recognize that, as Sartre says, "being is an individual venture" it is not sufficient, I am arguing, to leave individuals to their own resources. The result of this "devil take the hindmost" individualism is that individuality will not only be isolated, a voice crying in the wilderness, but will be contaminated by resentment or discouragement. What legislates against this simplistic approach is the recognition that individuation — the solution to the problematic nature of the human being — is a development, and a problematic one which today Michael Oakeshott calls a "practice"[11] and which in Hellenic Greece was termed a *destiny* as distinguished from a mechanical *fate*. As a development for which the individual is himself responsible, individuation has conditions. By the doctrine of "individualism," individuals themselves bear first responsibility for providing to themselves the necessary conditions of their own individuation. But the lesson of developmentalism is that individuation has *some* conditions which individuals cannot reasonably be held to be responsible for providing to themselves.

I said at the outset that we could identify a critical point at which supports are needed, if true individuality is to be a viable human prospect, and "individualism" is to be anything more than an elitist *taste*, like the preference for Mercedes-Benz or the couture of Ives St. Laurent. That critical point is identified by Abraham Maslow as the point of emergence of what he figuratively terms the "tender growth tip" of individual personhood. It first appears at the onset of adolescence, but because on the eudaimonistic conception of the human being, to be (in this mode) is to grow, the point in question is a *moving point*, just as the plant (except for its periods of dormancy) has perpetually (so long as it lives) a "tender growth tip." But the crucial point is the first appearance of the tender growth tip, its original manifestation in adolescence.

The first intimation of one's unique individuality marks the exchange of childhood for the stage of life we term adolescence. It appears as the reactive feeling of being misrepresented by those very persons who were authoritative with respect to oneself previously, namely, one's parents. It is, in a figure of speech, an "inner voice" which finds its first expression as an event which marks the onset of adolescence. And it is recognized by us as the first occasion on which what our parents said in our behalf, as spokesmen for us, authorized as such by our dependence as children, is misrepresentative. It first appears, then, as negation, somewhat analogous to the Hindu description of Atman-Brahman as "Neti, Neti," — "nothing of that sort, nothing of that sort." It is this negation which Camus termed the *rebellion* which is necessary to autonomy. But as Camus saw, this negation is likewise and simultaneously an affirmation.[12] It is an intuition of a "something which I, alone, can do." But it first appears without content, as the individual's small conviction of his own irreplaceable worth.

The treachery is that this small conviction as it first appears is wholly unequipped to withstand the drubbing it takes from the world, and from which all too often it never recovers. At its first appearance it is buffeted by alarms and commotion, and trampled beneath the scurrying crowd. Propped upright it is conscripted to his cause or that where roll call is "by the numbers," truth is established and preached, and responsibility is collective, with the individual's share being determined by arithmetic apportionment. What remains is a merely numerical individuation, the individuation that characterizes two otherwise indistinguishable things as being irreducibly two and not one.

But on the eudaimonistic conception, individuality is not merely numerical; it is qualitative; and the work of maximizing human value can only be done by increasing the numbers of persons who hear their inner voice, learn to understand its murmurings, and strive to strengthen it and live in truth to it, thereby actualizing their personal worth as objective worth in the world which is, in principle, of worth not to themselves merely, but to all other persons.

We have identified the point in the development of persons where the task of individuation requires support, if *individualism* is ever to become a viable enterprise rather than an idle utopian preference. For if individuality is a powerful force in the end, it is weak and tentative in the beginning. Through long exercise of self-discipline the integrity of the mature individual possesses the tensile strength of moral necessity, the inner imperative, "I must." Here is what the Greeks deemed moral heroism, and portrayed upon the tragic stage in the person of the tragic hero as the answer to the great question, "Does man count?" But individuality in the beginning is a tender "growth tip." It is timorous and untested, and in this condition it is no match for the juggernaut that is the world.

In the spirit of eudaimonism, I propose that the problematicity of the human being is the very opportunity for the manifestation of an as yet scarcely suspected human dignity and human worth. Such manifestation depends upon the recognition by human beings that each of us is his own first problem, the solution to which lies within us individually, as our own unique potential worth, and the requirement this lays upon us of self-responsibility and self-truth. This picture has been grasped in all of its profundity by a handful of moral heroes in history, and this handful of true individuals teach by example, they show us what can be done. But the real work has not been undertaken — the work without which the human experiment shall not be measured as other than a failure. It is the work of universalizing the recognition of individual self-responsibility and of all that follows from it. My argument here is that we now know enough about the rigors, the rewards, and the impediments to this work to be able to see that the developmental nature of individuation demands social supports, and that the crucial place for such supports is in adolescence, at the inception of individuation itself. By its nature adolescence is, as everyone recognizes, a restless time, but it is restless, not out of perversity as we commonly suppose, but because it is charged *intrinsically*, by what it essentially is, with the task of exploration. Its restlessness signals its intrinsic requirement, first, for authentic, firsthand acquaintance with the opportunities the world affords in terms of alternative life-styles. But second, it requires exploration in the interest of self-knowledge. And while this interest also requires exploratory enactment of alternatives in the world — for self-knowledge is not gained by introspection, but *a posteriori*, by acting in the world and reflecting on the results — it is served not by inculcation, but by elicitation.

If the problematicity of the human being is to generate a meaningful response in more than a handful of persons, certain conditions must obtain. The first of these, in order of importance, is that adolescence be no longer regarded as a temporary aberration in an otherwise sensible life, but the crucial stage of development which it is, a stage with its own intrinsic values, virtues, and obligations, fulfillment of which are prerequisite to qualitative individuation, and thus to the maximization of human worth.

NOTES

[1] 'Chapters on Socialism,' reprinted from the *Fortnightly Review* (1879), under the title *Socialism — John Stuart Mill*, ed. W. D. P. Bliss (Linden, Mass.: Social Science Library Series, 1891).
[2] Cited in Steven Lukes, *Individualism* (New York: Harper & Row, 1973), p. 25, n.
[3] *Henry V*, act 2, scene 4.

[4] Jean-Paul Sartre, *Being and Nothingness, An Essay on Phenomenological Ontology*, trans. Hazel E. Barnes (New York: Philosophical Library, 1956), pp. 67, 619.
[5] Jose Ortega y Gasset, *Man and People*, trans. Willard R. Trask (New York: W. W. Norton, 1957), p. 25.
[6] Roberto Mangabeira Unger, *Knowledge and Politics* (New York: Free Press, 1975), p. 196.
[7] Michael Oakeshott, *On Human Conduct* (Oxford: Oxford University Press, Clarendon Press, 1975), pp. 260, 236.
[8] Friedrich Nietzsche, *Thus Spoke Zarathustra*, in *The Portable Nietzsche*, trans. Walter Kaufmann (New York: Viking Press, 1954), p. 226.
[9] David L. Norton, *Personal Destinies: A Philosophy of Ethical Individualism* (Princeton: Princeton University Press, 1976). The idea is developed in chaps. 8–10.
[10] Ibid.
[11] Oakeshott, *On Human Conduct*, e.g., pp. 14–15.
[12] Albert Camus, *The Rebel*, trans. Anthony Bower (New York: Alfred A. Knopf, 1954), e.g., pp. 14–22.

VEDA COBB-STEVENS

AMOR FATI AND THE WILL TO POWER IN NIETZSCHE

Philosophy finds its impetus and its ground in the activity of questioning. For Nietzsche, this statement formulates a principle demanding more than a merely cognitive recognition. The degree of an individual's sensitivity to prereflective suppositions is an index of his or her "intellectual conscience." But most human beings, he claims, are so alienated from any sense of the wonder and strangeness of reality that they are unable to comprehend the real meaning of interrogation.[1] The awakening of consciousness to questioning, however, is not without dangers of its own. An inquiry, modest in its inception, oftens expands profusely in the very course of its elaboration. As a consequence, the inquirer becomes possessed with a fervor for resolution. Nietzsche warns that " ... an unrestrained thirst for knowledge for its own sake barbarizes men just as much as a hatred for knowledge."[2] Overturning every triviality and questioning simply for the sake of questioning can result in a dispersion and loss of self not unlike that alienation which does not question at all. "Philosophical activity," he says, "is ever on the scent of those things that are most worth knowing."[3] The route which Nietzsche takes toward this most valuable knowledge involves a critique of previous religion and philosophy and of the *décadence* of nineteenth-century culture. But even though it might appear that Nietzsche's entire enterprise is imbued with negativity, he is, in the last analysis, an advocate of affirmation. He says of himself, "I contradict as no one has contradicted before, and nevertheless I am the reverse of a negative spirit."[4]

I. *AMOR FATI*: EXISTENTIAL AFFIRMATION OF THE ETERNAL RECURRENCE

We may take an initial step toward understanding this paradox of affirmation/ negation in Nietzsche's thought by a reflection on his statements concerning the developmental logic of Western culture. The fundamental axiom upon which rest all our philosophy, religion, and science can be discerned in " ... a faith that is thousands of years old, that Christian faith which was also the faith of Plato, that God is the truth, that truth is divine."[5] It was the unquestioning avowal of the primacy of truth which led to that world-historical irony which Nietzsche saw exhibited in the culture of his contemporaries, viz., that the highest values devalue themselves.[6] A religion such as Christianity which is built upon this affirmation of truth must inevitably, Nietzsche thought, culminate in its own destruction. Under the banner of "truth at any price," the belief in the Christian God becomes unbelievable.[7] At first, the culture is not fully cognizant of its act of deicide. The radical consequences of this "monstrous logic of terror" are hidden by those "shadows" of God, traditional morality, for example, which continue to linger and veil the emptiness. There are those who, like Schopenhauer, renounce God and call themselves pessimists while yet clinging to the objective distinction between good and evil.[8]

Others, Nietzsche observes, forego both religion and the hope of an objective foundation for moral values, but still adhere to the standards of scientific exactitude. They continue to base their existence upon " ... a belief in the unity and perpetuity of scientific work, so that the individual may work at any part, however small, confident that his work will not be in vain."[9] But, ultimately, science too undergoes self-destruction. "It is not the victory of science that distinguishes our nineteenth century, but the victory of scientific method over science."[10] Science's piety, the faith in the value of truth, leads to the subversion of its own rigor.

One idol, however, continues to hold dominion: that of intelligibility as such. This ideal of intelligibility, grounded in (the belief in) the law of non-contradiction and interlacing language and the functions of consciousness, is the last shadow to be effaced. In *Twilight of the Idols*, Nietzsche says, "I fear we are not getting rid of God because we still believe in grammar."[11] Grammar, he says elsewhere,[12] is the metaphysics of the people. The truth which one must finally confront is that " ... there is no intelligible world."[13] Being is a myth and becoming is the ultimate character of reality. Moreover, it is a becoming which appears terrible in its complete indifference to anything human: "The total character of the world ... is in all eternity chaos – in the sense not of a lack of necessity but of a lack of order, arrangement, form, beauty, wisdom, and whatever other names there are for aesthetic anthropomorphisms."[14] We must come to realize that the categories of reason (aim, unity, being) are not true of reality, that there is nothing in the nature of what is which could correspond to the meaning of these concepts. In brief, there is no "true world," everything is false.[15] If the categories most highly esteemed are found not to apply to reality, then reality appears meaningless. The question "Why?" receives no answer, unless it be the cry, "In vain!" Such a response is heard from the soothsayer, " ... the proclaimer of the great weariness who taught, 'All is the same, nothing is worthwhile, the world is without meaning, knowledge strangles.'"[16] This is the lament of one who is a nihilist in possibly the most extreme sense, "... a man who judges of the world as it is that it ought *not* to be, and of the world as it ought to be that it does not exist."[17]

Here we have the ultimate culmination of metaphysical values: these values devaluate themselves. From an unquestioning faith in the value of truth, all "truth" has been rendered false. According to Nietzsche, the devolution of a civilization takes place on its own impetus: "For some time now, our whole European culture has been moving as toward a catastrophe, with a tortured tension that is growing from decade to decade: restlessly, violently, headlong, like a river that wants to reach the end, that no longer reflects, that is afraid to reflect."[18] The nihilism of the soothsayer is the inevitable consequence of the explosion of this tension. Nietzsche holds, "Extreme positions are not succeeded by moderate ones but by extreme positions of the opposite kind."[19] Once belief in God and a moral order becomes impossible, existence appears immoral and meaningless, appears as false. The type of nihilism which denies that there is truth shares a characteristic with the kind of nihilism which affirms truth, viz., in both cases truth remains the canon of value by which reality is judged.

When Nietzsche maintains that the proclamation of the soothsayer need not be taken as the final verdict upon reality, we can detect the beginnings of his own critical activity, that negative philosophy which, as we noted earlier, ends in affirmation. In order to guarantee that the nihilism of meaninglessness and the cry of "In vain!" represent only a "pathological transitional stage," Nietzsche must undermine the positive valuation placed on truth and all the categories of reason. This is his revaluation of all values: "Against the value of that which remains eternally the same ... the values of the briefest and most transient, the seductive flash of gold on the belly of the serpent *vita*."[20] The estimation placed on unchangingness, eternity and being must be overturned. "Truth is stood on its head" not only in the sense that it finally renders itself false, but more significantly, in the sense that its *worth* must be reassessed. The world of flux, which for millenia has been "mummified by the Egyptianism of philosophers," must become the object of a renewed affirmation. Thus we see, in a preliminary way at least, how Nietzsche can "contradict as no one has contradicted before, and nevertheless remain the reverse of a negative spirit."

However, the affirmative moment of Nietzsche's reflections must be explored more fully, if we are to discern the existential significance of his docrine of *amor fati*, a doctrine which is, in fact, merely another expression for the revaluation of becoming. When being is seen to be an illusory ideal, when all unchanging "truth" is shown to be false, only the immanent world of becoming remains. It is this world which must be accepted, which must be *loved*. Yet this act of acceptance is not merely a facile affirmation, the right to which is easily won. For according to Nietzsche, the world of becoming does not involve progressive transition but rather circular repetition. Becoming repeats itself infinitely. Any given moment is a "gateway" opening onto an eternal lane leading backward and also onto an eternal lane leading forward. Everything that can happen has happened before and will happen again, an infinite number of times.[21] The thought of the eternal recurrence of all things in endless, aimless cycles is something which Nietzsche called "the greatest weight."[22] We could say of the eternal recurrence what Nietzsche said of the complementary doctrine that denies an intelligible world, viz., that it is " . . . at once fruitful and terrible, facing the world with that Janus-face possessed by all great knowledge."[23]

One reason for the difficulty of affirming the eternal recurrence lies in the fact that such a view entails that reality could yield no ultimate novelty.[24] As Arthur Danto points out, " . . . this might cause some consternation: one might recall Mill's sadness at the fact that there is only a finite number of musical combinations so that all the musical possibilities would some day be exhausted."[25] More seriously, however, the doctrine of the eternal recurrence can be the source not just of mere consternation, but of anguish and nausea. When one reflects not merely on *how* things recur or on what, in general, can recur, but rather on what, in particular, must recur, then the doctrine appears as the greatest weight. Zarathustra says:

"Eternally recurs the man of whom you are weary, the small man" — thus yawned my sadness and dragged its feet and could not go to sleep. . . . Naked I had once seen both, the greatest man and the smallest man: all-too-similar to each other, even the greatest, all-too-human. All-too-small, the greatest! — that was my disgust with man. And the eternal recurrence even of the smallest — that was my disgust with all existence.[26]

Yet this condition of despair need be only a "pathological transition stage." The ideal of the eternal recurrence is, Nietzsche holds, a "doctrine which sifts men."[27] Personal strength is measured by the extent to which an individual can live in a world without meaning and, ultimately, *love* such a reality. Nietzsche asserts: "My formula for greatness in man is *amor fati*: that a man should wish to have nothing altered, either in the future, the past or for all eternity. Not only must he endure necessity and on no account conceal it

... but he must *love* it."[28] This affirmation of becoming, the great yea-saying to life, is unconditional and transcends any negative moments of pessimism or nihilism. Thus we have the real test of an individual's strength. With the momentous collapse of the intelligible world, the affirmation of total unintelligibility could be made only by a truly free spirit, one who " ... would take leave of all faith and every wish for certainty, being practiced in maintaining himself on insubstantial ropes and possibilities and dancing even near abysses."[29] This free spirit par excellence is the *Übermensch*, " ... the man who is able to affirm eternal recurrence, the man who experiences eternal recurrence as his own inner being."[30]

This, then, is Nietzsche's view of a cosmos shorn of meaning (indeed, no longer a cosmos) and of a person's authentic relation to it. Several questions should be raised regarding this account. We ought to ask, first of all, whether such a yea-saying individual is totally bereft of any faith or security, as Nietzsche would have us believe. Certainly, traditional religion, metaphysics, science, and morality have been renounced. Yet perhaps the most crucial question facing a human being, that of death, has been answered. Furthermore, it has been answered in such a way as to provide infinite security. A courage *is* needed to face the "most paralyzing thought" that all that is objectionable in reality will be repeated *ad infinitum*. But ultimately, this courage is one which " ... slays even death itself, for it says, 'Was *that* life? Well, then, once more!' "[31] Like the Dionysian mysteries of the Greeks, the *amor fati* which Nietzsche urges is " ... a triumphant yes to life beyond death and change."[32] In one of the aphorisms from *Beyond Good and Evil*, he declares: "One should part from life as Odysseus parted from Nausicaa – blessing it rather than in love with it."[33] By blessing, or *loving* life, we accept it as it is and make no demands upon it. But like any lover who loves out of weakness, those who are *in love* with life are prisoners of their own captivation. Such individuals who grasp and clutch at life as a support for their own debility, are, in Nietzsche's eyes, contemptible. But is it not the case that the love of fate which Nietzsche extols can be free from tenacity only because death is no longer something to be feared? In *The Will to Power* we read:

A certain emperor always bore in mind the transitoriness of all things so as not to take them too seriously and to live at peace among them. To me, on the contrary, everything seems far too valuable to be so fleeting: I seek an eternity for everything: ought one to pour the most precious salves and wines into the sea? My consolation is that everything that has been is eternal: the sea will cast it up again.[34]

When this promise of eternity is contrasted with the bleak horizon which faced Freud, we wonder if the supposedly most paralyzing thought is really the ultimate occasion for a test of strength. Freud also upheld a doctrine of *amor fati*, which he expressed as "making friends with the necessity of dying."[35] Yet for Freud, dying is not one moment among others; it is not the merely arbitrary designation of a point on a circle which will come around again and again. A life is linear and its first termination is its last. To love a fate which receives and annihilates one's very being, never to cast it up again, might demand a strength greater than that exacted by the thought of the eternal recurrence.[36]

Furthermore, in light of Kierkegaard's minute reflections upon eminent moments of faith, we could perhaps question Nietzsche's condemnation of religion as a source of hypocrisy and a bulwark for weakness. To be sure, a religious ideal, like any other, can serve as the basis for self-deception and *bad* faith. Kierkegaard, whose primary concern was the task of becoming a Christian within Christendom, realized these dangers as clearly as Nietzsche. For Kierkegaard, a genuine religious experience is very far removed from the easy security of herdlike assent, demanding a faith which, like Nietzsche's *amor fati*, must devaluate universal moral standards.

Thus it may be seriously doubted whether the affirmation of becoming commended in Nietzsche's doctrine of *amor fati* is the only valid, or even the most valid, existential act which a human being can perform. Certainly, to justify any alternative as existentially authentic, a supporting ontology distinct from Nietzsche's would have to be articulated. However, the question to which I would like to address the following sections of this paper is whether or not Nietzsche's ontology can ground the possibility of that existential affirmation so dramatically portrayed in his writings.

II. NIETZSCHE AS TRADITIONAL METAPHYSICIAN

We have already seen that Nietzsche denies the existence not only of a transcendent God, but that he also refuses to grant the status of reality to any of God's shadows. Morality, science, even language itself, are structural falsifications that impose an order upon a fundamentally chaotic becoming. All the hypostases which Nietzsche designates as "metaphysical" (in the wide sense of the term)[37] have their origin in the human ego. One must be careful, however, to specify the exact sense in which the human subject (considered as a self or soul-substance) forms the basis of all other reifications. It is not that the subject is the only stable entity amid a world of flux which, from its fortress of self-identity, generates a world of beings in its own image. Rather, "it is only from the *conception* 'ego' that there follows, derivatively, the concept 'being.'"[38] Just as the outer world is mere phenomena, so too is the inner. Consciousness simplifies the subtly multiple world of instinct, just as it structures the outer world of sense. "The whole world of 'inner experience' rests on the fact that a cause for an excitement of the nerve centers is sought and imagined — and that only a cause thus discovered enters consciousness...."[39] Thus the exultant dance of affirmation performed by the *Übermensch* is the expression, not of a conscious willing, but of an overflowing instinctual energy: "Everything good is *instinct* — and consequently, easy, necessary, free. Effort is an objection, the *god* is typically distinguished from the hero (in my language *light* feet are the first attribute of divininty)."[40] We noted earlier that Nietzsche views the doctrine of eternal recurrence as one which "sifts men, driving the weak to decisions, the strong as well." By contrast, in the passages now under consideration the language employed is quite inimical to the positive valuation of conscious decision. Further, in *The Gay Science*, Nietzsche says:

I have learned to distinguish the cause of acting from the cause of acting in a particular way, in a particular direction, with a particular goal. The first kind of cause is a quantum of damned-up energy that is waiting to be used up somehow, for something, while the second kind is, compared to this energy, something quite insignificant, for the most part a little accident in accordance with which this quantum discharges itself in one particular way — a match versus a ton of powder.[41]

Thus if I, on the one hand, am unable to overcome despair at the thought of eternal recurrence, whereas you, on the other hand, begin to dance for joy, it may not be said that I have failed in my human task (and might conceivably have succeeded), whereas you have succeeded (and might conceivably have failed): "To expect that strength will not manifest itself as strength, as the desire to overcome, to appropriate, to have enemies, obstacles, and triumphs, is every bit as absurd as to expect that weakness will manifest itself as strength."[42] Hence, it would seem that the possibility of existential authenticity, which depends upon a certain degree of personal authonomy, becomes at this point radically compromised.

We now begin to wonder whether the prevalent characterization of Nietzsche as a philosopher of human existence can, in spite of numerous supporting passages, finally be upheld. Heidegger, certainly, views Nietzsche's philosophy as much more than a diagnosis of modern cultural alienation. According to his interpretation of the significance of Zarathustra, Nietzsche appears as a *metaphysical* thinker. In Nietzsche's affirmation of the eternity of the recurring cycles of becoming, he must find himself squarely within the tradition that he thought he had transcended. Heidegger comments:

The *Yes* to time is the will that would have transcience abide, would not have it degraded to nihility. But how can transcience abide? Only in such a way that, as transcience, it does not just constantly pass, but always comes to be. It would abide only in such a way that transcience and what ceases to be return as the self same in its coming. But this recurrence is abiding only if it is eternal.[43]

Heidegger also sees Nietzsche as following in the tradition of subject-ism, initiatied in modern philosophy by Descartes and expanded by Leibniz, for whom the term "subject" included not only the human self, but all beings. Moreover, for Leibniz, a being was considered in its dynamic aspect, its *appetitus*.[44] Thus Nietzsche's doctrine of the Will to Power, which Heidegger takes to be his account of the Being of beings, is merely the culmination and completion of a path prepared before him.[45] Furthermore, Heidegger claims, Nietzsche's doctrine of the Will to Power represents the essence of becoming, while that of the eternal recurrence represents its existence.[46] This constitutes a further concession on Nietzsche's part to the Western metaphysical tradition.

Setting aside the question of the validity of Heidegger's views on the history of Western metaphysics, we can nonetheless find his claim that Nietzsche continues that tradition, to be heuristically fruitful for an interpretation of Nietzsche's philosophy. In addition to the notions of eternity, essence, etc., which Heidegger discovers, we may also discern in Nietzsche's writings an implicit endorsement of the affinity between "microcosm" and "macrocosm" as well as the distinction between appearance and reality.

In our delineation of Nietzsche's views on the ontology of human nature, we found that he opposes the phenomenal world of consciousness to the unconscious workings of physiological instinct. "For it is our energy which disposes of us," he claims, "and the wretched spiritual game of goals and intentions and motives is only a foreground – even though weak eyes may take them for the matter itself."[47] These last words are, of course, suggestive of the notorious thing-in-itself, that *horrendum pudendum* of the metaphysicans. The consciousness which simplifies not only the indeterminate flux of the outer world, but also the infinite modulations of the inner, must itself retain the status of mere superficiality. But Nietzsche thinks that one can reach even beyond the depth of instinct. Underneath the discrete entities posited by consciousness and underneath the fluctuating streams of instinct, there lies a single unity. In *Beyond Good and Evil*, he speculates that one might succeed " ... in explaining our entire instinctive life as the development and ramification of *one* basic form of the will — namely, of the will to power. ... "[48] Thus it would seem that Nietzsche sponsors a kind of "soul-substance" of his own, except that it is not an unchanging, self-identical, and thinglike substrate, but rather a dynamic and energetic force seeking its own discharge in appropriation.

This notion of appropriation is central to Nietzsche's conception of Will to Power. Moreover, it is through this notion that we are able to see how he extends the explanatory scope of his basic conception over ever-widening domains. Many passages throughout *Beyond Good and Evil* indicate that the ultimate nature of all organic existence, animal as well as vegetable, can be discerned in this one principle. But perhaps the most felicitous passage is found in *The Will to Power*: "Increasing 'dissimulation' proportionate to the rising order of rank of creatures. It seems to be lacking in the inorganic world — power against power, quite crudely — cunning begins in the organic world; plants are masters of it."[49] The idea of a Will to Power present in plants so appealed to Nietzsche that he began to construe human nature according to a botanical model and to speak of "the plant, man." In part 9 of *Beyond Good and Evil*, he praises the noble individual as " ... comparable to those sun-seeking vines of Java ... that so long and so often enclasp an oak tree with their tendrils until eventually, high above it but supported by it, they can unfold their crowns in the open light and display their happiness."[50]

Ultimately, this same Will is seen as underlying the whole world of nature, inorganic as well as organic. " 'Thinking' in primitive conditions (pre-organic) is the crystallization of forms, as in the case of the crystal. — In *our* thought, the essential feature is fitting new material into old schemas, . . . *making* equal what is new."[51] With this final extension of the Will to Power, Nietzsche seems to have fallen within the metaphysical perspective of Schopenhauer which he repudiated in his preface to *The Birth of Tragedy*. The concluding sentence of *Beyond Good and Evil*, section 36, is expressed in terms reminiscent of the "multiplicity of phenomena" and the "one will": "The world viewed from the inside, the world defined and determined according to its 'intelligible character' — it would be the 'will to power' and nothing else."[52] At this point, Heidegger would emphasize Nietzsche's subject-ism and his doctrine of the Will to Power as the essence of becoming. The passage can be interpreted also, it seems to me, under the category of appearance/reality. All diversity in the apparent world is resolved into the underlying unity of this Will.

This dynamic, ever-changing Will makes false all our claims about a stable and univocal reality. As Danto notes, "because [Nietzsche] wanted to say that all our beliefs are false, he was constrained to introduce a world for them to be *false about*; and this *had* to be a world without distinctions, a blind, empty, structureless thereness."[53] Such a view of reality has repercussions, as Heidegger saw, upon the notion of truth.

If truth according to Nietzsche is a kind of error, then its essence lies in the manner of thinking which always and necessarily falsifies the real, in so far as every act of representation causes the unexposed "becoming" to be still and sets up something that does not correspond (i.e., something incorrect) with what has been established in contradistinction to the fluent "becoming," thereby establishing something erroneous as allegedly real. In Nietzsche's defining of truth as incorrectness of thinking, there lies the concession to thinking of the traditional essence of truth as the correctness of making as assertion (*logos*).[54]

In passages where Nietzsche speaks of the "character of the world in a state of becoming as incapable of formulation, as "false," as "self-contradictory,"[55] his implicit capitulation to the standard of *adequatio* is evident. Just as the anguish of the soothsayer at the meaningless of reality is conditioned by his positive valuation of meaning, so too Nietzsche's account of falsity is simply the complement of the correspondence view of truth.

Nietzsche's doctrine of the Will to Power involves not only the traditional distinction between appearance and reality along with a "correspondence" view of truth, but also the notion of an affinity between "microcosm" and "macrocosm." The relation of the individual human being to being as a whole (i.e., to the dynamic Will of becoming) simply expresses a variation upon a predominant motif of pre-Socratic philosophy. The ontological structure of a human being is a duplication, a miniaturized mirror image of the structure of reality itself. The human ideal is, according to Nietzsche, that of " . . . a spirit who plays innocently (that is to say, involuntarily, out of his superabundance of power) with everything that has hitherto been called holy, good, inviolable, divine . . . "[56] So too he can say of reality: "The origin of all good things is thousandfold: all good, prankish things leap into existence from sheer joy: how could one expect them to do that only once?"[57] The same fundamental energy which is the source of the exuberance of the *Übermensch* is the origin of the eternal dice game of becoming itself.

In all these ways in which Nietzsche may be seen to adhere to previous metaphysical categories, the same presupposition is found. It is the implicit endorsement of what may be termed a "vertical ontology." Although this spatial metaphor is perhaps the most appropriate regarding the affinity betweem "microcosm" and "macrocosm," it is meant, as metaphor, to transcend any literal relationship to extension and to designate fundamentally any ontology which posits some principle of difference. In this sense, Plato's distinction between an idea and its instances, Aristotle's principles of form and matter, and Kant's phenomenon and noumenon, are all representative of ontological verticality. The distinction between the undifferentiated Will to Power and its multiple manifestations constitutes, in Nietzsche's philosophy, the principle of difference. Heidegger, then, would seem to be correct in designating Nietzsche as a metaphysican who attempts to think the Being of beings as the Will to Power.

Heidegger sees his own task, of course, as the overcoming of all such ontologies. While positing a principle of difference between *Sein* and *Seiendes*, these ontologies, he claims, have concentrated upon articulating the elements of this relation and have forgotten the relation *as* relation.[58] Nonetheless, Heidegger himself may be placed within the development of Western metaphysics to the extent that he assumes that there *is* a difference between Being and beings.[59] There are passages in Nietzsche's writings, it seems to me, which attempt to overcome precisely this assumption.

III. NIETZSCHE AS RADICAL ONTOLOGIST

In the preceding sections of this paper, Nietzsche was found, first of all, to be an advocate of existential authenticity, one who demanded that a universe in which all events recurred in endless, aimless cycles be joyously affirmed. An examination of his ontology of the Will to Power, however, led us to question whether the individual could retain enough autonomy to affirm, in any meaningful sense, this *amor fati*. Since the individual is only an appearance underneath which lies the Will to Power, any specifically *human act* of affirmation seemed to be compromised by the dynamism of a general ontological principle. However, an even more radical level of the ontology of the Will to Power can be discerned in Nietzsche's writings, one which calls into question not only the individual, but the ontological inquiry itself.

In the final entry to *The Will to Power*, we read:

And do you know what "the world" is to me? Shall I show it to you in my mirror? This world: a monster of energy, without beginning, without end; a firm, iron magnitude of force that does not grow bigger or smaller, that does not expend itself but only transforms itself; as a whole, of unalterable size, a household without expenses or losses, but likewise without increase or income; enclosed by "nothingness" as by a boundary; not something blurry or wasted, not something endlessly extended, but set in definite space as a definite force, and not a space that might be "empty" here or there, but rather as force throughout, as a play of forces and waves of forces, at the same time one and many, increasing here and at the same time decreasing there; a sea of forces flowing and rushing together, eternally changing, eternally flooding back, with tremendous years of recurrence, with an ebb and a flood of its forms; out of the simplest forms striving toward the most complex, out of the stillest, most rigid, coldest forms toward the hottest, most turbulent, most self-contradictory, and then again returning home to the simple out of this abundance, out of the play of contradictions back to the joy of concord, still affirming itself in this uniformity of its courses and its years, blessing itself as that which must return eternally, as a becoming that knows no satiety, no disgust, no weariness: this, my *Dionysian* world of the eternally self-creating, the eternally self-destroying, this mystery world of the twofold voluptuous delight, my "beyond good and evil," without goal, unless a ring feels good will toward itself — do you want a *name* for this world? A *solution* for all its riddles? A *light* for you, too, you best-concealed, strongest, most intrepid, most midnightly men? — *This world is the will to power — and nothing besides*! And you yourselves are also this will to power — and nothing besides![60]

The last words of this passage demand a new reflection. *"The world is will to power — and nothing besides*! And you yourselves are also this will to power — and nothing besides!" It might appear at first glance that these words express again the "microcosm — macrocosm" distinction which was previously discussed. That Nietzsche has here transcended the categories of traditional metaphysics may be shown by considering additional supporting passages and then returning to the this one.

One of the central notions in the preceding account of Nietzsche as metaphysician was the distinction between appearance and reality: the Will to Power as opposed to its various manifestations. Moreover, this distinction was maintained on two levels. The inner world of consciousness as well as the diverse beings of nature were seen to be masks of the one Will to Power. These differences themselves conceal another differentiation which Nietzsche, according to the viewpoint which we are now considering, seeks to overcome, viz., the distinction between inner and outer, between a human being and the rest of reality, between interpretation and text. For Nietzsche claims that there is no text; there are only interpretations. "Against positivism, which halts at phenomena — 'There are only *facts*' — I would say: 'No, facts is precisely what there is not, only interpretations.' "[61] Likewise, there is no one who interprets. "One may not ask: 'Who then interprets?' for the interpretation itself is a form of the will to power, exists (but not as a 'being' but as a process, a becoming) as an affect."[62] Hence, there is no distinction between the Will to Power and its manifestations.

The Will to Power *is* its manifestations. The lightning is not something which flashes, but is merely the flashing itself.[63]

What then becomes of the *doctrine* of the Will to Power as ontology? Nietzsche says: "Supposing that this also is only interpretation — and you will be eager enough to make this objection? — well, so much the better."[64] We are now in a position to grasp more fully the significance of the long paragraph from *The Will to Power*. It might have seemed that it was an account distinct from and supposedly corresponding to reality. Yet we now notice that Nietzsche opens the passage with words which confirm the passage just quoted from *Beyond Good and Evil*. He asks: "Do you know what 'the world' is to me? Shall I show it to you in my mirror?" The theory which speaks of the flux of becoming is simply an aspect of that becoming itself, one more mirror in a world of mirrors. One might say of such a reality that it is " . . . a jugglery in which the juggler is part of the whirl he manages to keep aloft through some miraculuous feat of lighthandedness."[65] From this viewpoint, which is no longer a viewpoint, meaning itself becomes meaningless, there being nothing of which it would be the meaning. If there are *only* interpretations, then there are *no* interpretations. In this kind of ontology where there is no principle of difference, all of reality is "horizontalized." The stillness *is* the dancing, the oneness *is* the multiplicity of interpretations. There is no Will to Power behind interpretations, as reality is behind appearance. Rather, the Will to Power *is* the interpretations, and becoming is "at the same time one and many."

Is Nietzsche unique in attempting to think this thought which destroys thought *as* thought? Whether intentionally or not, the words at the end of *The Will to Power* echo the formulation of the Secret Doctrine in Plato's *Theaetetus* (156a): "The universe is *kinesis* and nothing else besides." This doctrine in its most radical form destroys all distinctions, just as does the present account of the Will to Power. The destruction of difference with the consequent collapse of discourse, however, is precisely the reason given by Socrates for the rejection of such an ontology (181d ff.). It would seem that for Nietzsche, on the contrary, this is a reason for its affirmation.

Furthermore, both the ontology of universal *kinesis* as well as that of the Will to Power exhibit an essential affinity to Parmenides' Way of Truth. The two former doctrines express an ontology of dynamism, of movement, whereas the latter considers reality as absolute stability. But these ontologies, though different in content, are formally the same, i.e., to the question which is posed to reality, they all give a simple and unconditional answer: reality is such-and-such and nothing else besides. These are what might be termed "ontologies of identity." The exclusive existence of movement, just insofar as it is movement, would involve a pure ontological self-coincidence, just as would the exclusive existence of rest or unchangingness.

Yet perhaps such ontologies exhibit more than a merely formal affinity. The world which is mirrored in the final section of *The Will to Power* is without beginning, without end; as a whole, of unalterable size; enclosed by "nothingness" as by a boundary. In strikingly similar terms, Parmenides says:

One way is left to be spoken of, that it *is*; and on this way are full many signs that what *is* is uncreated and imperishable, for it is entire, immovable and without end. It *was* not in the past, nor *shall* it be, since it *is* now, all at once, one continuous; for what creation wilt thou seek for it? How and whence did it grow?[66]

From the viewpoint of the whole and of eternity, Nietzsche's "sea of forces" is just as immovable as Parmenides' sphere, "a firm, iron magnitude of force that does not grow."

Thus it would appear that Nietzsche is not alone in trying to think this thought-destroying thought. However, in the *Theaetetus*, the Secret Doctrine in its radical form is finally rejected. Parmenides, furthermore, although claiming that there is only one path to be spoken of, depicts also the route of mortals which he calls the Way of Seeming. But for Nietzsche, the Way of Truth *is* the Way of Seeming.

Have we here discerned the real significance of that philosophical questioning which is "ever on the scent of these things that are most worth knowing"? The very surface of the world is its depth. But how are we to understand such an answer which, as just another aspect of that surface, does not possess sufficient verticality to *be* an answer? And even more importantly, how are we to understand such questioning? Even if it should be the case that there is no questioner, in any ordinary human sense, but only the endless surging of the ocean of Power, must not reality contain its own principle of difference, if it would question itself at all? If reality *is* self-coincident, how could it be sufficiently concealed from itself such that a question would be demanded? Heraclitus asked: "How could anyone hide from that which never sets?" (fr. 16). Nonetheless, he thought that most human beings succeeded in doing so, just as Nietzsche saw the "herd" as lacking in intellectual conscience. But if we are trying to think Nietzsche's thought, we must not fall back into the ordinary human categories of the natural attitude. We are considering the possibility, given an ontology of identity, of reality's questioning itself. But in such an ontology, who are "we"? Who (or what) is asking these questions? *Are* they really questions?

The essential thrust of an ontology of identity is to overthrow that unquestioned assumption which has characterized metaphysics from Thales to Heidegger (and even a certain level of Nietzsche's thought): the assumption that there is an ontological difference. Is this a necessary assumption? More radically, how could it *not* be?

In Plato's *Sophist* (253e–254b), the Eleatic Stranger remarks that both the philosopher and the sophist are equally difficult to discern. The sophist who "takes refuge in the darkness of Not-being" cannot be perceived because of the obscurity of the region which he inhabits. The philosopher "whose thoughts dwell upon the nature of reality" is difficult to see because his realm is so bright. Nietzsche is, as we have seen, a thinker who would "stand Plato on his head," having the Sophist inhabit the region of light, letting the philosopher dwell in Dionysian darkness. Why is *Nietzsche* so difficult to see? Was he too dark or too bright?

NOTES

[1] Friedrich Nietzsche, *The Gay Science*, trans. Walter Kaufmann (New York: Random House Vintage Books, 1974), p. 76 (II, 36–37); hereafter cited as *GS*. In general, any number in parentheses refers to the volume and page number(s) of Nietzsche, *Werke in Drei Banden*, ed. Karl Schlechta (Munich: Hanser, 1959–1961).

[2] Friedrich Nietzsche, *Philosophy in the Tragic Age of the Greeks*, trans. Marianne Cowan (Chicago: Henry Regnery, 1962), pp. 30f. (III, 355); hereafter cited as *PTAG*.

[3] *PTAG*, p. 43 (III, 364). This view of the 'most valuable knowledge," or philosophic wisdom, can be found in Heraclitus, a philosopher whom Nietzsche greatly admired. Heraclitus, too, maintained that "much learning does not teach one to have intelligence" (fragment 40). The transition from the conception of wisdom as quantitative multiplicity implicit in the Homeric language (e.g., *poluphron*) to the notion of depth or profundity expressed the language of Heraclitus (*bathuphron*) is quite parallel to Nietzsche's juxtaposition of the fragmentary inquiries of nineteenth-century *Wissenschaft* with that Dionysian *Weisheit* which permeates his writings. (For an account of the former transition, cf. Bruno Snell, *The Discovery of Mind* (Oxford: Basil Blackwell, 1953), p. 18.) However, with any statement concerning either Heraclitus or Nietzsche, extreme caution must be exercised. Heraclitus also said: "Men who love wisdom must be inquirers into many things indeed" (fragment 35). And we find Nietzsche claiming in *The Will to Power*: "Inertia needs unity (monism); plurality of interpretations is a sign of strength." One of the principal tasks of this paper will be to demonstrate that this tension between unity and multiplicity can, at least under one interpretation of Nietzsche's philosophy, be viewed as representing complementary aspects of one philosophical position. (Cf. part III, "Nietzsche as Radical Ontologist.")

[4] Friedrich Nietzsche, *Ecce Homo*, trans. Clifton P. Fadiman, in *The Philosophy of Nietzsche* (New York: Random House, 1954), p. 924 (II, 1152); hereafter cited as *EH*.

[5] *GS*, p. 283 (II, 208).
[6] Friedrich Nietzsche, *The Will to Power*, ed. with commentary by Walter Kaufmann, trans. Walter Kaufmann and R. J. Hollingdale (New York: Random House Vintage Books, 1967), p. 9; hereafter cited as *WP*. Copyright © Random House, Inc.
[7] *GS*, p. 279 (II, 205-6).
[8] Friedrich Nietzsche, *Beyond Good and Evil*, trans. Walter Kaufmann (New York: Random House Vintage Books, 1966), p. 99 (II, 644); hereafter cited as *BGE*.
[9] *WP*, p. 325.
[10] *WP*, p. 261.
[11] Friedrich Nietzsche, *Twilight of the Idols*, trans. R. J. Hollingdale (Harmondsworth: Penguin Books, 1968), p. 38 (II, 960); hereafter cited as *TwI*.
[12] *GS*, p. 300 (II, 222).
[13] *EH*, p. 884 (II, 1123). Nietzsche sometimes means by the term "metaphysics" (or "intelligible world" or "true world") a transcendent, unchanging realm as posited by either Platonism or Christianity. "The law of contradiction provided the schema: the true world, to which one seeks the way, cannot contradict itself, cannot change, cannot become, has no beginning or end" (*WP*, p. 315). At other times, he uses "metaphysics" in a wider sense to include all hypostases, even those effected upon the world in which we live: "The 'A' of logic is, like the atom, a reconstruction of the thing – If we do not grasp this but make of logic a criterion of true being, we are on the way to positing as realities all those hypostases: substance, attribute, object, subject, action, etc.; that is, to conceiving a metaphysical world, that is a 'real world' . . . " (*WP*, p. 279).
[14] *GS*, p. 168 (II, 115-16).
[15] *WP*, p. 13.
[16] Friedrich Nietzsche, *Thus Spoke Zarathustra*, trans. Walter Kaufmann (New York: Viking Press, 1966), p. 241 (II, 480f.); hereafter cited as *Z*.
[17] *WP*, p. 318.
[18] *WP*, p. 3 (preface, 2).
[19] *WP*, p. 35.
[20] *WP*, p. 310.
[21] *Z*, p. 158 (II, 408-9).
[22] *GS*, p. 273 (II, 202).
[23] *EH*, p. 885 (II, 1123).
[24] *WP*, p. 546.
[25] Arthur C. Danto, *Nietzsche as Philosopher* (New York: Macmillan, 1965), p. 210; hereafter Danto.
[26] *Z*, p. 219 (II, 465).
[27] *WP*, p. 39.
[28] *EH*, p. 853 (II, 1098).
[29] *GS*, p. 290 (II, 213).
[30] Joan Stambaugh, *Nietzsche's Thought of the Eternal Return* (Baltimore: Johns Hopkins University Press, 1972), p. 88.
[31] *Z*, p. 157 (II, 408).
[32] *TwI*, p. 109 (II, 1031).
[33] *BGE*, p. 83 (II, 629).
[34] *WP*, pp. 547f.
[35] Sigmund Freud, 'The Theme of the Three Caskets,' in *Character and Culture* (New York: Collier, 1963), p. 68.

36 Bernd Magnus (*Heidegger's Metahistory of Philosophy: Amor Fati, Being and Truth* [The Hague: Martinus Nijhoff, 1970], p. 30) expresses a thought along these lines, without, however, any reference to Freud: "If I were persuaded that beyond my temporal existence nothingness awaited me, this thought too might be a 'greatest stress,' in so far as I could ask myself the question: do you want this once and *once only?*"
37 Cf. n. 13 above.
38 *TwI*, p. 38 (II, 959f.); italics mine.
39 *WP*, pp. 265f.
40 *TwI*, p. 48 (II, 972).
41 *GS*, p. 315 (II, 233f.).
42 Friedrich Nietzsche, *The Birth of Tragedy and the Genealogy of Morals*, trans. Francis Golffing (Garden City: Doubleday, 1956), p. 178 (II, 789); hereafter cited as *GM*.
43 Martin Heidegger, 'Who is Nietzsche's Zarathustra?' trans. Bernd Magnus, *Review of Metaphysics*, 20 (1967), 424.
44 Martin Heidegger, *Holzwege* (Frankfurt: Vittorio Klostermann, 1950), pp. 121, 212; hereafter cited as *HW*.
45 *HW*, p. 213.
46 *HW*, p. 219.
47 *WP*, p. 518.
48 *BGE*, p. 48 (II, 601).
49 *WP*, p. 292.
50 *BGE*, p. 202 (II, 728).
51 *WP*, p. 273.
52 *BGE*, p. 48 (II, 601).
53 Danto, p. 96.
54 Martin Heidegger, *Platons Lehre von der Wahrheit* (Frankfurt: Vittorrio Klostermann, 1947), pp. 45f. (Quotation translated in Magnus' *Metahistory*, p. 91.) More recently, P. Christopher Smith has argued quite convincingly that there are certain passages in Nietzsche's writings (e.g., the preface to *The Gay Science*) which evince an awareness of truth as the process of dis-closure very much in the vein of Heidegger's notion of *alētheia*. Cf. his article, 'Pre-scientific Truth and *Alētheia*: A Reconsideration of Heidegger's Relationship to Nietzsche,' in *Proceedings of the Heidegger Circle* (New Orleans, 1977).
55 *WP*, p. 280.
56 *EH*, pp. 895f. (II, 1131).
57 *Z*, p. 174 (II, 423).
58 Martin Heidegger, *Identity and Difference*, trans. Joan Stambaugh (New York: Harper and Row, 1969). Stambaugh's comment in her introduction is quite appropriate: "For Heidegger thinking is concerned with Being in regard to its difference from beings. Heidegger doesn't ask about Being as the ground of beings; he goes from what is yet unthought, from the difference between Being and beings *as difference* (the ontological difference), to that which is thought, the *oblivion* of that difference. ... Instead of progressing toward an all-inclusive totality, thinking for Heidegger attempts to move *forward* by the step back into the realm of the essence of truth which has never yet come to light. This step back allows Being as difference to come before thinking without being its object" (p. 16).
59 Although Heidegger's view of the "ontological difference" is in one sense quite

distinct from the principle of difference in Plato, Aristotle, Kant, etc. (cf. Stambaugh's comment in the previous note), it is just precisely insofar as his philosophy admits of difference at all, that it is sufficiently close to the tradition to be classed with previous views in opposition to Nietzsche's "radical ontology."

[60] *WP*, pp. 549–50.
[61] *WP*, p. 267.
[62] *WP*, p. 302.
[63] *GM*, pp. 178f. (II, 789f.).
[64] *BGE*, pp. 30f. (II, 586).
[65] Danto, p. 101.
[66] G. S. Kirk and J. E. Raven, *The Presocratic Philosophers* (Cambridge: Cambridge University Press, 1971), p. 273.

STEPHEN L. WEBER

JORGE LUIS BORGES – LOVER OF LABYRINTHS: A HEIDEGGERIAN CRITIQUE

Borges is, above all else, a lover of labyrinths, a contemporary Daedalus, constructing an intricate world in which reality gives way to unreality, and unreality, in turn, becomes real. His is a world in which all paths turn back upon themselves; there is no escape:

> They knew it, the fervent pupils of Pythagoras:
> that stars and men revolve in a cycle; . . .
> this writing hand will be born from the same
> womb; and bitter armies will contrive their doom.
>
> (The philologist Nietzsche made this very point) . . . .
>
> It returns, the concave dark of Anaxagoras; in my human
> flesh, eternity keeps recurring, and an endless poem,
> remembered or still in the writing . . . .
>
> They knew it, the fervent pupils of Pythagoras . . . .[1]

For Borges, as for Nietzsche,

All things themselves are dancing . . . . Everything goes – everything comes back – eternally rolls the wheel of being. Everything dies – everything blossoms again – eternally runs the year of being . . . . In every now, being begins; round every here rolls the sphere there. The center is everywhere. Bent is the path of eternity.[2]

Borges does not offer us a very optimistic picture of the world and of our condition in it. The labyrinth is not simply the mosaic dance floor in front of the Cretan palace on which dancers performed the spring dance of the partridge. More than that, the world itself is a mosaic and we are all dancers, caught in the labyrinth. We come from nowhere and we go nowhere. Our life is an empty dance reflected in an infinite hall of mirrors. Unlike Theseus, we have no thread to guide us through the labyrinth. We cannot identify the real, we cannot distinguish the genuine from the illusory, there is no good or bad or right or wrong, nothing to fight for, nothing to fight against: "One destiny is no better than another."[3] No justice rewards our struggles. There is no good to triumph in the end. There is no end. It is all a dance on an infinite mosaic floor beneath a spinning mirrored sphere.

It is my intention to explore Borges' presentation and discussion of the labyrinth within which we find ourselves, to juxtapose his discussion of the poet's response to the labyrinth with that of Martin Heidegger and, finally, to argue that Heidegger's view of the poet is superior, maintaining that, in the end, Borges' love of labyrinths, which is perhaps his greatest strength, is also his ultimate weakness.

We begin to appreciate the complexity of Borges' work when we understand that the labyrinth is indeed more than a mosaic dancing floor, more than a collection of bewildering, twisting underground passageways. Borges' labyrinth is nothing less than language itself. Language itself is inextricable, bewildering, a perplexity, a maze.

The inability of language to grasp the genuine is suggested in 'A Yellow Rose':

> Then came the revelation. Marino *saw* the rose as Adam might have seen it in paradise. And he sensed that it existed in its eternity and not in his words, and that we may make mention or allusion of a thing but never express it at all; and that the tall proud tomes that cast a golden penumbra in an angle of the drawing room were not — as he had dreamed in his vanity — a mirror of the world, but simply one more thing added to the universe.[4]

In 'Parable of the Palace' he again offers his melancholy recognition of the inevitable defeat of all poets: "The poet was the emperor's slave and died as such; his composition fell into oblivion because it merited oblivion and his descendants still seek, and will not find, the word for the universe."[5] All language ultimately fails; it does not deliver us from itself. Language is a closed labyrinth within which, and by means of which, we dance.

But still there is no reason to despair just because language reveals itself to be a hall of mirrors; it remains a limited, finite labyrinth — we may still our voice, lay down our pen, and step off the dance floor at any moment into a silent, but nonetheless real, world. Or can we? Borges is not so naive. Not only is language a human construction, more importantly, the world we imagine to be "out there," beyond the dance floor, is itself only an interpretation, a human construction. The world, the room within which you are sitting, is an interpretation, a poem. There is no world but that which shines in and through the mirrors of language. The labyrinth is, then, complete and inescapable. Which is to say, there is no stepping off the dance floor.

Borges urges us to extend Norwood Hansen's famous maxim that "all perception is theory-laden" (meaning that there is no "pure" inductive base from which science may proceed) to the more general assertion that "all perception is language-laden." There is no world but that which shines in and through the mirrors of language. The labyrinth is, then, complete and inescapable. It is the English Department — more than the Physics Department — which is the broker of the world. They are the dance masters, and there is no way in which we may sit out the dance.

In his essay 'Kafka and his Precursors,' Borges notes that the voice of Kafka may be recognized in texts from diverse literatures and periods, among them Zeno's famous paradox against movement, the parables of Kierkegaard, and Browning's poem 'Fears and Scruples.' Reflecting on these "precursors," each of which resembles Kafka, but none of which resemble each other, Borges observes:

In each of these texts we find Kafka's idiosyncrasy to a greater or lesser degree, but if Kafka had never written a line, we would not perceive this quality; in other words, it would not exist. The poem "Fears and Scruples" by Browning foretells Kafka's work, but our reading of Kafka perceptibly sharpens and deflects our reading of the poem. Browning did not read it as we do now .... The fact is that every writer creates his own precursors. His work modifies our conception of the past, as it will modify the future.[6]

Note the clear recognition that without the labyrinth of language which Kafka constructs in his work, we would not *perceive* this quality. But more than that, note the future recognition that if we did not perceive it, "it would not exist." Like Daedalus, Kafka constructs a labyrinth which "modifies our conception of the past as it will modify the future."

In an essay entitled 'Avatars of the Tortoise,' Borges considers this theme more directly. The tortoise to which the title refers is the tortoise of Zeno's second paradox. The swift Achilles cannot overtake the slow moving tortoise who has a head start of ten meters: "Achilles runs those ten meters; the tortoise one. Achilles runs that meter; the tortoise runs a decimeter. Achilles runs the decimeter; the tortoise runs a centimeter. Achilles runs the centimeter; the tortoise a millimeter. Fleet-footed Achilles the millimeter; the tortoise a tenth of a millimeter. And so on to infinity."[7] Among the avatars or embodiments of this paradox, Borges notes the third man argument of Aristotle (if A and B are related by virtue of C, then $(A + B)$ and C must be related by D, and so forth *ad infinitum*), Agrippa, the Skeptic — who "denies that anything can be proven, since every proof requires a previous proof"[8] — and Sextus Empiricus — who argues that "definitions are in vain, since one will have to define each of the words used and then define the definition."[9]

From these, and others, Borges is led to suggest that "the vertiginous *regresses in infinitum* is perhaps applicable to all subjects."[10] Looking at Zeno's paradox and its heirs, he concludes:

> Let us admit what all idealists admit: the hallucinatory nature of the world. Let us do what no idealist has done: seek unrealities which confirm that nature . . . . The greatest magician (Novalis has memorably written) would be the one who would cast over himself a spell so complete that he would take his own phantasmagorias as autonomous appearances. Would not this be our case? I conjecture that this is so. We . . . have dreamt the world. We have dreamt it as firm, mysterious, visible, ubiquitous in space and durable in time; but in its architecture we have allowed tenuous and eternal crevices of unreason which tell us it is false.[11]

In the latter collection entitled *A Personal Anthology*, of which Borges says "My preferences have dictated this book. I should like to be judged by it . . . ,"[12] the theme continues. In a poem entitled 'The Moon,' Borges tells of one man's heroic project. He conceived a plan of "ciphering the universe in one book." To this end he:

> built his high and mighty manuscript,
> shaping and declaiming the final line.
>
> But when about to praise his luck,
> He lifted up his eyes, and saw
> A burnished disc upon the air; startled,
> He realized he'd left out the moon.

Reflecting on this tale, Borges concludes:

> Though contrived, this little story might well exemplify the mischief that involves us all who take on the job of turning real life into words. Always the essential thing gets lost. That's one rule holds true of every inspiration.[13]

It is, of course, apparent to even the most untutored of readers, as it is to Borges himself, that these are not new ideas. A sense of the frailty of human understanding dates back at least to the pre-Socratics. (One recalls the sophist Gorgias who claimed: "Nothing exists. Even if anything does exist it is inapprehensible by man. Even if anything is apprehensible, yet of a surety it is inexpressible.")[14] Indeed, an awareness of the dilemma which all human beings face as agents trying to know the world, which itself is altered by their very knowing it, has been shared by most modern thinkers since Kant. Nonetheless, for much of the twentieth century, Western thought has fallen back into a naive realism accompanied by a rather crude empiricism. It is only within the last twenty years that the Western mind has begun to work its way back to the insights to which Borges has devoted so much of his life's work. Only lately have we seen again the labyrinth. In the late 1950s, philosophers of science began rejecting empiricism's claim of direct access to the real, and suggesting instead that human knowledge is far more complex and circular. One may see this development in the work of Norwood Hansen,[15] who maintains that observations are not "pure," that there are no "brute facts," that reality does not "lie there" for the taking. On the contrary, all observation is "theory-laden," which is to say that theory does not follow upon observation so much as it precedes (and hence influences) observation. In the words of Borges, "Our eyes see what they are accustomed to see."[16] Thomas Kuhn agrees, arguing that historically relative paradigms, "ways of seeing the world," structure our experience, and that there are no criteria for adjudicating the claims of alternative paradigms. One might extend the list of scientific critics of empiricism to include still others, such as Stephen Toulmin, Michael Polanyi, or Paul Feyerabend, but the above should suffice to make our point. If these scholars are correct, the foundations of empiricism must lie in ruin. We can no longer naively assume direct access to reality. We have lost the thread; there is no escape from the labyrinth.

Of all the critics of naive empiricism, perhaps the most eloquent is Martin Heidegger. Like Borges, Heidegger concludes that, "to go straight to the things cannot be carried out."[17] It is assumed by many, observes Heidegger, that science has an immediate grounding in things which affords it "a direct transition and entrance to them starting out from everyday representations." And yet, "there always remains the possibility that we only exchange subjective pictures of things with one another, which may not thereby become any truer because we have exchanged them continually."[18] Recognizing that "things stand in different truths," that the "description of things and their interdependence corresponds to what we call 'the natural conception of the world,'" and that "what is 'natural' is not 'natural' at all," which is to say not "self-evident for any given ever-existing man," it follows that "the natural is always historical," and hence that "the question 'what is a thing?' is a historical question."[19]

Given the historicity of our knowledge, and the consequent limitations of all science and objectivity, we are driven to ask with Heidegger, "Is science the measure of knowledge, or is there a knowledge in which the ground and limit of science ... are determined. [If so] is this genuine knowledge necessary for a historical people, or is it dispensable or replaceable by something else?"[20]

We need go no further to establish our claim that Borges and Heidegger have much in common. They share a humility which frees them from the presumption that human knowledge can grasp the real. Moreover, they hold a common conviction that the world is not what we see it to be. What is it then which separates Borges and Heidegger, and how am I to substantiate the claim I made earlier that Heidegger's view of the poet is superior, and that in the end Borges' love of labyrinths is not a strength but a weakness? And why is it that I wish to second the self-judgment which Borges makes at the conclusion of his prologue to *A Personal Anthology*, "I know now that my gods grant me no more than allusion or mention"? [21]

For Borges, the poet is the caller of the partridge dance, the master of ceremonies in the hall of mirrors. For Heidegger, on the contrary, the poet is one who names the "holy," i.e., that which lies beyond the labyrinth. In his commentary on Hölderlin's poem "Homecoming" (which title refers precisely to the human struggle to escape from the labyrinth), Heidegger identifies the poet's task as that of letting "the heavenly share itself out." [22] It is the task of the poet to help the holy illumine the spirit of men "so that their nature may be open to what is genuine in their fields, towns and houses." [23] To do this, the poet must come forth, must struggle out of the labyrinth, to stand in the presence of the genuine. For Heidegger, mankind is being's co-respondent, but we are often forgetful, losing our way in the labyrinth. It is the poet's task to call us forth, out of the labyrinth. The poet seeks to come home, to emerge from the labyrinth into the light of the genuine. Here we see more clearly the distinction between Borges and Heidegger. Borges is a skeptic. He embraces that skepticism, fundamental to all sophists, which maintains that wisdom is the acceptance of our ignorance. Heidegger, on the contrary, is a philosopher — one who recognizes, but does not accept, his ignorance, and hence who is sustained by a "love of wisdom." Heidegger and Borges are one in their joint recognition that we humans are not wise. The difference between them lies in the fact that Borges accepts the limitations of the human condition while Heidegger does not.

The resignation, characteristic of Borges, which is the subject of my criticism, is apparent in "Inferno I, 32" in which Borges tells the story of a leopard locked in a cage, to which God spoke the following words:

> You live and will die in this prison, so that a man I know of may see you a certain number of times and not forget you and put your figure and your symbol in a poem which has its precise place in the scheme of the universe. You suffer captivity, but you will have given a word to the poem.

Borges goes on to tell us that the leopard understood his fate and even accepted it: "And yet, when he awoke, he felt merely an obscure resignation, a gallant ignorance, for the machinery of the world is overly complex for the simplicity of a wild beast." Borges then draws a melancholy parallel:

> Years later, Dante lay dying in Ravenna, as little justified and as much alone as any other man. In a dream, God revealed to him the secret purpose of his life and labor.... Tradition holds that on awakening he felt he had received and then lost something infinite, something he could not recuperate, or even glimpse, for the machinery of the world is overly complex for the simplicity of men.[24]

We may further distinguish Borges and Heidegger in reference to Nietzsche. Both quote Nietzsche and cite his influence in their work, but Heidegger is a true student of Nietzsche and Borges is not. We began our essay recalling, as does Borges himself, Nietzsche's eternal return: "All things themselves are dancing [think here of the labyrinth] ... everything goes, everything comes back, eternally rolls the wheel of being ...."[25]

This is the lament of the defeated Zarathustra of book 3; the Zarathustra who has succumbed to the spirit of gravity (reason); the Zarathustra who has fallen from the tightrope into the labyrinth and has abandoned hope of the *Übermensch*, resigning himself instead to the eternal passages of the labyrinth. It is this Zarathustra of book 3 who concludes, as does Borges, "I love you, O eternity."[26] But there are four books to Zarathustra; the story does not end with book 3, with the defeat of Zarathustra. There is the joyful Zarathustra of book 4, the Zarathustra who reasserts the struggle for the overman and who, in our terms, continues the struggle to emerge from the labyrinth.

Heidegger's refusal to resign himself to the inevitability of falsehood is suggested in his discussion of the poet's "care." For Heidegger, as for Borges, "that which remains (that which stands forth) is established by the poet."[27] But it is not mere whim or caprice that guides the poet's hand, not the idle desire to simply add one more mirror to the hall of mirrors. Rather, for Heidegger there is a clear recognition (which I do not detect in Borges) that the poet's mirrors reflect something beyond themselves: "That which supports and dominates the existent in its entirety [the labyrinth] must become manifest. Being must be opened out, so that the existent may appear."[28] It is the task of Heidegger's poet to stand between that which lies beyond the labyrinth and the labyrinth, between Being and its expression in existents: " ... in this Between it is decided who man is and where he is settling his existence."[29] Hence, for Heidegger, the poet's task is monumental: it is to bring the labyrinth closer to the genuine.

Precisely for this reason, Heidegger knows a "care" of which Borges is not cognizant. Heidegger's poet, like Borges', must ultimately become a mirror — he cannot be the window (the transparency that would allow escape from the labyrinth), that he wishes to be. But he dedicates his "mirror" to reflecting the genuine, and not merely to reflecting the reflections. And precisely because of this dedication, he fears that the dweller in the labyrinth may mistake the mirror for that of which it imperfectly speaks. Heidegger is concerned to preserve not the truth of illusion, as is Borges, but the illusion of "truth."

In conclusion, we must ask, is the poet the dance master, singing for the partridge dancers beneath the mirrored sphere? Or is he or she struggling to bring the labyrinth closer to the genuine? It is the question posed earlier by Heidegger: "Is this genuine knowledge necessary ... or is it dispensable?" For Heidegger it is necessary; for Borges, it is dispensable. Borges' poet is a dancer, delighting in the illusions of the hall of mirrors. Heidegger's poet is a marcher, dedicated to escape from the labyrinth.

Heidegger's poet is one who loves truth, who would escape from the labyrinth. Borges, on the contrary, loves illusion and delights in compounding it. He is indeed a lover of labyrinths, a dancer in the hall of mirrors.

NOTES

[1] Jorge Luis Borges, 'The Cyclical Night,' trans. Alastair Reid, in *A Personal Anthology*, ed. Anthony Kerrigan (New York, 1967), pp. 155–56.
[2] Friedrich Nietzsche, *Thus Spoke Zarathustra*, in *The Portable Nietzsche*, trans. and ed. Walter Kaufmann (New York, 1954), pp. 329–30.
[3] Jorge Luis Borges, 'Biography of Tadeo Isidoro Cruz (1829–1874),' trans. Anthony Kerrigan, in *A Personal Anthology*, p. 164.
[4] Jorge Luis Borges, 'A Yellow Rose,' trans. Anthony Kerrigan, in *A Personal Anthology*, p. 83.
[5] Jorge Luis Borges, 'Parable of the Palace,' trans. Carmen Feldman Alvarez del Olmo, in *A Personal Anthology*, p. 88.
[6] Jorge Luis Borges, 'Kafka and His Precursors,' trans. James E. Irby, in *Labyrinths*, ed. Donald Yates and James E. Irby (New York, 1962), p. 201.
[7] Jorge Luis Borge, 'Avatars of the Tortoise,' trans. James E. Irby, in *Labyrinths*, pp. 202–3.
[8] Ibid., pp. 204–5.
[9] Ibid., p. 205.
[10] Ibid., p. 207.
[11] Ibid., p. 208.
[12] Jorge Luis Borges, 'Prologue,' in *A Personal Anthology*, p. ix.
[13] Jorge Luis Borges, 'The Moon,' trans. Edwin Honig, in *A Personal Anthology*, p. 196.

[14] Milton C. Nahm, *Selections from Early Greek Philosophy* (New York, 1964), p. 233.
[15] See, for instance, *Patterns of Discovery* (Cambridge, 1965).
[16] Jorge Luis Borges, 'The Modesty of History,' trans. Anthony Kerrigan, in *A Personal Anthology*, p. 179.
[17] Martin Heidegger, *What is a Thing?*, trans. W. B. Barton and Vera Deutsch (Chicago, 1967), p. 27.
[18] Ibid., p. 12.
[19] Ibid., pp. 39, 43.
[20] Ibid., p. 10.
[21] Jorge Luis Borges, 'Prologue,' in *A Personal Anthology*, p. x.
[22] Martin Heidegger, 'Remembrance of the Poet,' trans. Douglas Scott, in *Existence and Being* (Chicago, 1949), p. 241.
[23] Ibid., p. 252.
[24] Jorge Luis Borges, 'Inferno I, 32,' trans. Anthony Kerrigan, in *A Personal Anthology*, p. 80.
[25] *Zarathustra*, pp. 329–30.
[26] Ibid., p. 340.
[27] Martin Heidegger, 'Hölderlin and the Essence of Poetry,' trans. Douglas Scott, in *Existence and Being*, p. 280.
[28] Ibid., p. 281.
[29] Ibid., pp. 288–89.

JAMES B. SIPPLE

LAUGHTER IN THE CATHEDRAL: RELIGIOUS AFFIRMATION AND RIDICULE IN THE WRITINGS OF D. H. LAWRENCE

I

The greatness of D. H. Lawrence as a writer in the classic-modern period, and that which, as well, makes him a monumental figure for the distinctly religious imagination, is his combination of the modern critical and skeptical intelligence with the consistent reassertion of the *Mysterium Tremendum et Fascinans* of the Creative Life from which flows all meaning and being.

George Ford in his exceptional study of Lawrence, *Double Measure*, anticipates this insight by noting that the "recurring tactic" of the Brangwen women in Lawrence's novel, *The Rainbow*, is to "laugh in church"; that is, "to let the light of critical intelligence flood in upon dark religious mysteries."[1] Ford's attention notices as well that the role of Kate Leslie in *The Plumed Serpent* offers a striking example of the same phenomenon. He says, "Only rarely has *The Plumed Serpent* been given a fair reading in this regard, one which pays adequate need to the ironic juxtaposition of the religious solemnities with Kate's mocking intelligence, her sense of absurdity, her capacity to laugh in church."[2] Ford's intention is to illustrate the dual nature (the "Double Measure") pervasive in the fiction of Lawrence. Lawrence's greatness, according to Ford, is a result of his ability to demonstrate "Mankind has been moved, is moved, and will be moved by two forces."[3] From the vista of literary criticism this insight underlines the perceptions of Coleridge, I. A. Richards, and Cleanth Brooks, to cite a few in a brilliant pantheon, whose words are echoed in the argument of R. W. B. Lewis in *The American Adam* when he says, "All genuine fiction is, by nature, ironic. For fiction ... dramatizes the interplay of compelling opposites."[4]

At first glance then, since the figures in Lawrence's novels who "laugh in church" are generally women (the list is long: Anna and Ursula in *The Rainbow*, Harriett Somers in *Kangaroo*, Kate Leslie in *The Plumed Serpent*) it would seem well to conclude that we are essentially being made aware of the projection onto the novels of the relationship between Lawrence and his wife Frieda. The relationship has been described as a "very religious man married to a very irreligious woman." Since this case is so easy to make (Are not all "heroes" in Lawrence' novels actually incarnations of the author?

Are not all female counterparts Frieda in disguise?), it probably accounts for the reason no one has considered "Laughter in the Cathedral" as a paradigm for a critical understanding of Lawrence's work as a whole. George Ford has seen it and then collapsed his insight into a larger consideration.

From the point of view of Christian theology, a significant study of thematic motifs in Lawrence occurred in *D. H. Lawrence and Human Existence* by Martin Jarrett-Kerr (1951). This work appeared during the cultural vogue of existentialism and imposes an agenda upon the works of Lawrence that bend them a bit to echo the Christian existentialist vocabulary. Jarrett-Kerr ignores completely the radicalizing of the traditional dualisms inherent in the existentialist perspective (nature–history, God–World, mind–body) which Lawrence expends so much energy to overcome. Lawrence is related to existentialism, but it is not a positive relation as Jarrett-Kerr believes. That is, throughout Lawrence's work one is haunted by the "intrinsic otherness" of each character. Such "otherness" creates a gulf of estrangement between the characters over which Lawrence seeks to construct a mystical bridge. But it is rarely successful. I find a great deal of skepticism in Lawrence's work (similar to that of Jean-Paul Sartre in *Being and Nothingness*) in regard to any hope of sustained fulfillment in interpersonal relationships. Jarrett-Kerr's enthusiasm was helped along, no doubt, by F. R. Leavis who contended for the placement of Lawrence within the "Christian camp." The timing of the publication of *D. H. Lawrence and Human Existence* (shortly after the court trial concerning the publication of the unexpurgated version of *Lady Chatterley's Lover* in England) had a great deal to do with the tone of this book: it is special pleading for the acceptance of the works of Lawrence by "Church people."

There is a hint, however, of the subject of this paper in Jarrett-Kerr's first section:

The famous scene from the same novel, *Women in Love*, in the Pompadour Cafe is another instance. Gudrun and Gerald overhear some "arty" young men and women laughing about a letter one of them has received from Birkin. They are all a bit drunk, and one of them, Halliday, reads out the letter mockingly as if preaching a sermon. Gudrun, furious at the insult to Birkin, slips up to them and asks to see the letter. They give it to her unsuspecting, and she strides out of the cafe with it before they can stop her. Now the interesting thing about this is that, according to Mr. Murry, this scene actually took place. Some young men (and it is known that Philip Heseltine – "Peter Warlock" – understood Halliday to refer to himself) were overheard by Katharine Mansfield joking about Lawrence's volume of poetry, *Amores*, and she did what Gudrun did. But in the novel Lawrence loads the dice against himself even more heavily by turning the book of poems into a letter from Birkin; and, moreover, a letter which reads like an exaggerated version of Lawrence's own mystical philosophy of the time. "There is a phase in every race when the desire for destruction over comes every other desire.... It is a desire for the reduction-process in itself, a reducing back to the origin, a return along the Flux of Corruption to the original rudimentary conditions of being.... Surely there will come an end in us to this desire − for the constant going apart − this passion of putting asunder − everything − ourselves ... reducing the two great elements of male and female from their highly complex unity − reducing the old ideas, going back to the savages for our sensations − always seeking to lose ourselves in some ultimate black sensation, mindless and infinite − burning only with destructive fires, raging with the hope of being burnt out utterly...." This is the letter which is read, punctuated with hiccups, and interspersed with giggling comments, in the tone of a clergymen reading the lessons. Lawrence has here produced a perfect self-parody; and a writer who can do that, without losing control of the novel, knows what he is doing.[5]

Jarrett-Kerr perceives clearly affirmation and ridicule outside the aforementioned circle of affirmation (male) and ridicule (female), but uses it no further than to dismiss the charge that Lawrence was an "obstinate, temperamental, predatory, emotionalist" with no capacity for self-criticism or artistic control of his material. That is, Jarrett-Kerr seizes the insight into a basic movement in the tensive counterpoise of the novel, but applies it in a primarily defensive context rather than proceeding with it as a positive position to account for his enthusiasm for Lawrence's writing from a Christian point of view.

The importance for our consideration of Jarrett-Kerr's discovery is that *despite* the frequent juxtaposition of "very religious men with very irreligious women" (which suggests the embattled Lawrence−Frieda relationship) the essential dualism of the religious and the critical intelligence is joined within the character of the author himself. In the recent biography (1975) of Lawrence's early years by Philip Callow entitled *Son and Lover*, the mounting dichotomy within Lawrence is the split between his saturation in nonconformist Christianity and his "questioning of religion."[6] In his late teens, he was "becoming impatient with religion as he delved into such writers as Darwin, T. H. Huxley, and Haeckel."[7] Jessie Chambers, the "Miriam" of *Sons and Lovers*, once questioned him about the addition of the figure of Annable the gamekeeper in the second version of his first novel, *The White Peacock*. "He has to be there," he told her, "Don't you see why? He makes a sort of balance. Otherwise it's too much one thing, too much me." According to Philip Callow, "Jessie did not see at all. Lawrence was kicking out at religion and embracing materialism for want of something better, and Jessie thought Annable's presence somehow symbolized this dilemma."[8] As Lawrence's Christian background became abridged by the plentitude of doubt swirling in his mind, his characteristic pose was to laugh, mock, and jeer at what he considered the ineptitude of the chapel services that he had so faithfully attended for so many years, to "laugh in church."

The counterpoise of affirmation and ridicule is the fundamental texture of the best Laurentian novels. As essential movement and tension it is the linguistic window into the complex matrix of the larger canvas upon which Lawrence is fashioning his vision. When the various dimensions of the Laurentian corpus are seen in their interrelationship, these aspects become apparent:

(1) We cannot speak of affirmation and ridicule as the essential dialectical tension in which Lawrence is working without seeing this tension in relation to the struggle going on within Lawrence about the meaning of power. In the novel *Women in Love* he distinguishes between a German phrase from Nietzsche, *Wille zur Macht* (*Will to Power*) and a French phrase *Volonte de Pouvoir* which he translates for us as "will to ability." Mark Schorer, in the introduction to the Grove Press edition of *Lady Chatterley's Lover*, offers an explanation of the difference between the two phrases by quoting Fromm's book *Escape from Freedom*:

> The word "power" has a twofold meaning. One is the possession of power over somebody, the ability to dominate him; the other meaning is the possession of power to do something, to be able, to be potent. The latter meaning has nothing to do with domination: it expresses mastery in a sense of ability. If we speak of powerlessness we have this meaning in mind; we do not think of a person who is not able to dominate others, but of a person who is not able to do what he wants. Thus power can mean one of two things, *domination or potency*. Far from being identical, these two qualities are mutually exclusive. Impotence, using the term not only with regard to the sexual sphere but to all spheres of human potentialities, results in the sadistic striving for domination; to the extent to which an individual is potent; that is, able to realize his potentialities on the basis of freedom and integrity of his self, he does not need to dominate and is lacking the lust for power. Power, in the sense of domination, is the perversion of potency, just as sexual sadism is the perversion of sexual love.

As we continue we will not only be dealing with Lawrence's own confusion over the meaning of power, but with the *impotence* of the Christianity that existed in Lawrence's mind. It is this which is the object of ridicule within the Christian circle. At the same time, Lawrence as a religious writer could only respond with a *counterreligion* which again would be ridiculed and the process would go forward. The ambiguity in Lawrence's own character (and in our own as well) is the mixture of will to potency and will to domination. As Rollo May puts it, "Power is the birthright of every human being. It is the source of his self-esteem and the root of his conviction that he is interpersonally significant."[9] What Lawrence knew all too well was that a basic source of his own rage came at a deeper level than "religious questioning." The source was rather in his own need to feel impowered; a need frustrated first by a father who ridiculed him into oversensitivity, a mother intent upon molding him to her own model, a marriage filled with strife in which ridicule of his philosophizing was a chief instrument of inserting the sliver of scorn.

We have then a most remarkable situation. The author who struggles against a sickly body (Lawrence died of tuberculosis at age forty-five, from an advanced stage of an illness that had never left him since he first became ill at age fifteen) which contains a tortured soul, filled with self-doubts and doubts in regard to the Ultimate, was as the one who railed against a sickly and puny and impotent Christianity.

In the course of this paper, I will illustrate just how the tensive quality of Lawrence's work accounts for our aesthetic distinctions between his best work (a heightened sense of affirmation and ridicule) and his less valuable writings in which the tension has been collapsed. For the moment we can point to the salient nature of this turn in the discussion. We look again at Rollo May's angle of vision in *Power and Innocence*: "There is one way, however, of confronting one's powerlessness by making it a seeming virtue. This is the conscious divesting on the part of an individual of his power.... I call this innocence ... to be free from guilt or sin, guileless, pure; and in actions it means 'without evil influence or effect, or not arising from evil intention.'"[10]

It is this quality in Lawrence's final works that give them that "closed in" tonality. This innocence is that nurtured in the school of religious apocalypse; that is, through the lens in which power and innocence (the Lion and the Lamb) cease to be opposed, but are the perfect simile (in which innocence *is* power, and power *is* innocence). The vision is Utopian; it does not see that the vision itself can become evil; it does not see the creative – destructive tension in all activity.

(2) An extension of the discussion of power is the analysis of the ontological structure of Being itself. In the ontological structure there is an essential unity of form-creating and form-destroying strength. The twentieth-century theologian Paul Tillich in one of his early essays entitled 'The Demonic' writes that this unity

> must always be retained: the unity of form-creating and form-destroying strength. That is true of the demon who determines the great destiny which disrupts all forms of existence; it is true of the demon who drives the personality beyond the limits of its allotted form to creations and destructions it cannot grasp as its own. Where the destructive quality is lacking, one can speak of outstanding power, or genius, of creative force,

not of demonry. And *vice versa*, where destruction is evidenced without creative form, it is fitting to speak of deficiencies, flaws, declines or the like, but not of demonry.[11]

At another point in his writings, Tillich speaks of the polar forces that break apart, and operate in a destructive fashion; these forces he calls the "impulse for power and the impulse for Eros." They are identical to Lawrence's distinction between the *Wille zur Macht* and *volonte de pouvoir*, the contrast between dominance and potency. Against the background of Lawrence's interest in theosophy, Lawrence would have been attracted to Tillich's understanding of the demonic which belies the influence of the great German mystic Jacob Boehme (in the terms "Ground" and "Abyss") and the romantic values of Schelling:

> The possibility of the demonic is based on the fact, first, that the sacred is at the same time the absolute support and the absolute claim, the depth and the form, the ground and the abyss; and that second, in the creature these elements may separate. The creature desires to draw up into itself the inexhaustibility of the divine depth, to have it for its own. By this means the creative potency becomes destructive. For creation occurs where the abyss and the form are united in one entity, and destruction occurs where the abyss arises and dominates and breaks the form.[12]

This analysis of the ontological dimension will be fundamental to our conclusions about the relative value of Lawrence's work. To proceed against this background will be to follow the lead of Jarrett-Kerr when he affirms that literary criticism must be "axiological and ontological" as well as aesthetic. This procedure will enter the space suggested by Stephen Miko in his book *Toward Women in Love* when he proposed that the direction of Lawrence criticism must be ontological. Miko himself, however, provides no framework to understand how this might be worked out. My suggestion is that this structure is provided by the analysis of Being in the theology of Paul Tillich. George Ford catches a glimpse of the significance of this suggestion in a footnote in *Double Measure* when he names Tillich's major influences: "The historian of ideas can demonstrate how Lawrence's awareness of double rhythms may derive from a long tradition evident in such writers as Paracelsus, Schelling, and Hegel, which in England became especially prominent in the 19th century."[13] We also agree with Miko in his contention that Rupert Birkin, the major male character in *Women in Love*, is Lawrence's most fully realized character. For any who may have seen the Ken Russell film of *Women in Love* (which the *New Yorker* magazine suggests is a "purple pastiche"), they will remember one redeeming feature (apart from the splendid performance by Glenda Jackson as Gudrun) was the spirited combination of affirmation and ridicule in the Birkin character as played by Alan Bates. In addition, this analysis of the ontological dimension will demonstrate the reason why, for all their virtues, *The Man Who Died* and *Lady Chatterley's Lover*, remain among Lawrence's lesser accomplishments. These works represent a flowering of innocence that will move considerably beyond the charm of *authentic* innocence present throughout Lawrence's writings (use of diminutives, the celebration of the awe and wonder of the natural landscape, the freshness and beauty of life in its essential character) to an apocalyptic vision which becomes, for Lawrence at the end of his life, "dreaming innocence", a vision divested of its roots in actual experience.

II

Rudolph Otto's discussion of the Holy is the starting point for a more detailed examination of our subject. The experience of the Holy is described by Otto in the Latin words, "Mysterium tremendum et fascinans." The word *mysterium* (mystery) points to the inexhaustible nature, the infinite ground, of the experience itself. A mystery is that experience which never ceases to be mysterious no matter how deeply one penetrates into its nature and *will not* cease to be mysterious in the light of future research and new information. The word *tremendum* points to the element of "numinous dread," the "daunting and repelling" moment of the numinous, the "awe," the terrible otherness of the God that would consume us, burn us to ashes, annihilate us utterly if we were to *directly* encounter it. The "burning bush" episode of Moses is a classic example. Before the terrible majesty of the Holy God, the Center of all the power there is, human beings can only have oblique experiences (gazing upon the back of God rather than the face and the eyes of God) for to encounter *directly* the gaze of God would destroy the human being as such. Another classic example is the experience related in Isaiah 6:

In the year that King Uzziah died I saw the Lord sitting upon a throne, high and lifted up; and his train filled the temple. Above him stood the seraphim; each had six wings: with two he covered his face, and with two he covered his feet, and with two he flew. And one called to another and said:

"Holy, holy, holy is the Lord of hosts;
the whole earth is full of his glory."

And the foundations of the thresholds shook at the voice of him who called, and the house was filled with smoke. And I said: "Woe is me! For I am lost; for I am a man of unclean lips, and I dwell in the midst of a people of unclean lips; for my eyes have seen the King, the lord of hosts!"

Then flew one of the seraphim to me, having in his hand a burning coal which he had taken with tongs from the altar. And he touched my mouth, and said: "Behold, this has touched your lips; your guilt is taken away, and your sin forgiven". And I heard the voice of the Lord saying, "Whom shall I send, and who will go for us?" Then I said, "Here I am! Send me."

In both examples, there is a strenuous commitment demanded. We have then, in these examples, the experience of the Holy offered as a story of vocational origin. That is, the strenuous commitment needed for the journey of mission and proclamation is rooted in the *a priori* of the encounter with the Holy. This encounter includes an expression of the Divine Will which directly addresses the individual in the encounter as *personal* communication. This occurs again in the conversion of St. Paul. St. Paul is addressed directly by a visionary manifestation of the Risen Jesus. We will note later in our attention to the Laurentian texts, that it is this *interpersonal* element in the Holy that Lawrence will seek to obliterate. He will seek to erase the notion that God and man are linked in any relationship of mutual vulnerability. The Isaiah illustration is explicit in its addition of another element in Otto's discussion, the dimension of Grace. According to Otto, the *tremendum* element carries with it "ideas of justice, moral will, and the exclusion of what is opposed to morality," but the *fascinans*, the attracting and alluring element in the numinous, refers to the experience of forgivenness, new being, love, mercy, Grace. Theologian Paul Tillich understood the encounter with the Holy, especially as an experience of Grace.

Sometimes at that moment a wave of light breaks into our darkness, and it is as though a voice were saying: You are accepted. *You are accepted*, accepted by that which is greater than you, and the name of which you do not know. Do not ask for the name now; perhaps you will find it later. Do not try to do anything now; perhaps later you will do much. Do not seek for anything; do not intend anything. *Simply accept the fact that you are accepted*![14]

As we shall see presently "the Holy" in Lawrence will not only be impersonal, but will be conceptualized as the "Quick" (*elan vital*, life force) in the midst of life, rather than the God of love of Christian faith. The challenge he offers to the Christian community is nevertheless considerable: Is the Christian God of love only loving? An easygoing benevolence who "rolls around heaven all day"? Is he not the God of life (imagined as the power of being)? Is not the Negative a dimension of the life of the one Holy God?

Lawrence struggled throughout his career with the meaning of the word "love": the long discourses between Ursula and Rupert in *Women in Love* and the dialogue throughout that excellent story of male—female conflict, 'The Captain's Doll,' are only a couple of many instances where his fiction is the threshing floor for his struggle for definition. In 1921, he wrote to his friends Earl and Achsah Brewster that "the word love for me has gone pop." Lionel Trilling notes in a 1958 article in *Encounter* that for Lawrence "love was anathema". One wonders if the recent recasting of Harry T. Moore's biography of Lawrence, *The Intelligent Heart*, should wisely have been retitled, *The Priest of Love*. The abiding value of Lawrence's theological perceptions as Jarrett-Kerr has noted, "was his sense of the ISness rather than the OUGHTness of religion. ... It is important to see that he believed in his dark gods not because they 'worked,' but because they were true."[15]

The paradigm of our study is the visit of Will and Anna to Lincoln Cathedral (chapter 7) in the novel, *The Rainbow*. As they walk to the cathedral we read that Will's "heart lept." "It was a sign in the heaven [the cathedral], it was the Spirit hovering like a dove, like an eagle over the earth." He turned his glowing, ecstatic face to her, his mouth opened with a strange ecstatic grin: "'There she is,' he said." The couple is not yet in the cathedral, but a great deal has already been packed into the anticipation of the encounter with the Holy that Will is presently going to experience (even the word "ecstatic" has already been used twice in one sentence). Already Lawrence is loading his language with theological concepts by describing the cathedral as "the Spirit hovering like a dove, like an eagle." Isn't "the Spirit" a dove only in the Christian tradition? In this short sentence we meet what Lawrence meant when he described his religious perspective as the "religion of the Holy Ghost." In a letter written to Eleanor Farjeon on 7 October 1915 (during the year *The Rainbow* was taken out of publication and banned by the Home Office), Lawrence explains "the dove and the eagle"; it has to do with the "anathema of love":

Dear Eleanor: I mean the "of" to be there. When Christ said, "the blasphemy against the Father should be forgiven, and the blasphemy against the Son, but not the blasphemy against the Holy Ghost," he meant, surely, that which is absolute and timeless, the supreme *relation* between the Father and the Son, not a relation of love, which is specific and relative, but an absolute relation, of opposition and attraction both, this should not be blasphemed. (That is, the encounter with the Holy in its "wholeness," tremendum and fascinans) (and now the "anathema of love"). And it seems to me a *blasphemy* to say that the Holy Spirit is Love. In the Old Testament it is an Eagle: in the New it is a Dove. Christ insists on the Dove: but in his supreme moments he includes the Eagle.

Can you not see that if the relation between Father and Son, in the Christian theology, were only *love*, then how could they even feel love unless they were separate and different, and if they are divinely different, does this not imply that they are divine opposites, and hence the relation *implied* is that of eternal opposition, the relation *stated* is eternal attraction, love? . . .

P. S. Christ himself is always going against the Holy Spirit. He must *insist* on the love, because it has been overlooked. But insistence on the one is not to be interpreted as negation of the other. In his purest moments, Christ knew that the Holy Spirit was both love and hate – not one only.[16]

Implicit in this quotation is the tension between affirmation and ridicule ("it seems to me a *blasphemy* to say that the Holy Spirit is love"). Lawrence's revulsion against what Nietzsche called the "pseudo-morality of pity" will eventually take shape in the absurd posture of believing the Jesus of the New Testament is best characterized as a "corpse" (dead, cold flesh without "Quick" even in life). In *The Plumed Serpent*, Quetzelcoatl will be conceptualized as the "Lord of the Two Ways" (the "two ways" discussed in the letter) as a protest against the Christianity of the "pseudomorality of pity" symbolized in the lifeless figure of the crucifix that Lawrence saw throughout the churches of Mexico.

As "meaning *against* meaning," the ecstatic behavior of Will is contradicted by the skepticism of Anna:

The "she" irritated her.
Why "she"? It was "it."
What was the cathedral, a big building, a thing of the past, obsolete, to excite him to such a pitch?

In this pregnant paragraph is a summary of volumes of phrases characteristic for the modern period: " 'thingification' (It was 'it'); The Cathedral considered as 'dead matter,' rather than an organic construction whose meaning grew from within; 'Man dwells in a sanctuary where there is no presence' (just 'a big building')." The organic growth from within is denied as well as the implied "subjectivity." The cathedral is "dead matter" surrounding meaningless space; it is not "the perfect womb": "All constructions of reality have become a debris". (The cathedral is "a thing of the past, obsolete".) Religion in the modern period is for many the superstitious residuum of a more primitive stage in human development. Retreating before the advancement of enlightenment, religion and its trappings became "obsolete" remains to excite only the antiquarian interest to whom the cathedral is a museum of antiquities.

How well Lawrence knows the modern skeptical intelligence! As Will pushes open the door he beholds "the perfect womb." And his "transport" begins. At first Anna is caught up in the mystery; "She too was overcome with wonder and awe." Then follows some of the best prose Lawrence (or anyone) was ever to write, soaring in its repetitive intensity, "Away from time, always outside of time." The interior of the cathedral, "Spanned round with the rainbow, the jewelled gloom folded music upon silence, light upon darkness, fecundity upon death, as a seed folds leaf upon leaf and silence upon the root and the flower, hushing up the secret of all between its parts, the death out of which it feel, the life into which it has dropped, the immortality it involves, and the death it will embrace again" (*The Rainbow*, p. 189).

In this mystic prose is what I. A. Richards would call the "Cross connexion after cross connexion" which provide "vital" imaginative writing. Against the mystical transport is the experience of Anna. Beyond the wonder and awe, "she remembered that the open sky was no blue vault, no dark dome hung with many twinkling lamps, but a space where stars were wheeling in freedom, with freedom above them always higher"; "even in the dazed swoon of the cathedral, she claimed another right. The altar was barren, its lights gone out. God burned no more in that bush. It was dead matter lying there." It is there, contemplating her feelings, that Anna finds her representative: the mocking and leering figures of the gargoyles. "She caught sight of the wicked, odd little faces carved in stone, and she stood before them arrested" (i.e., arrested at worship at what Lawrence believes is a countershrine):

These sly little faces peeped out of the grand tide of the cathedral like something that knew better. They knew quite well, these little imps that retorted on man's own illusion, that the Cathedral was not absolute. They winked and leered, giving suggestion of the many things that had been left out of the great concept of the church. "However much there is inside here, there's a good deal they haven't got in," the little faces mocked.

The couple of pages in *The Rainbow* which follow illustrate how deep the tension of affirmation and ridicule is within Lawrence and how much the tension is played out against the backdrop of struggle for power: Anna points out to Will the "adorable faces" of the gargoyles; then follows an oft-quoted line from this novel, "This was the voice of the serpent in his Eden." We are led to suppose that the mocking other (evil) has appeared and in (her) voice is the persistent jeering that dissolves the affirmation of the cathedral. Its expression is an uncontrollable laughter. (She "laughed with a pouf! of laughter") ("She laughed with malicious triumph"). In such simple juxtaposition is the ground for the belief that Anna is Frieda and Will is Lawrence and we are receiving some kind of report of a visit to Lincoln Cathedral which resulted in one of their typical rows over the question of power between the "very religious man" and the "very irreligious woman." Such juxtaposition is suggested by the opposition to religious affirmation implied in the "serpent in his Eden" image and by such apparent rehearsal of the constant violence going on between the author and his wife in everyday life outside the novel. (It could probably be proved that exactly the same words passed between Lawrence and Frieda on such an occasion.) Lawrence, however, is the kind of writer who merits close reading. What may be easily explained upon first reading becomes complex and unusually subtle upon subsequent readings. This is why Lawrence, often the hero of the college undergraduate, is essentially beyond the scope of any but the most sophisticated and intense student. *Women in Love*, despite its frequent use in upper-level English classes, remains one of the most difficult novels in the English language. Our present scene in *The Rainbow* is a scene of such complexity. To give it full value, these aspects are important:

(1) The "serpent in his Eden" can only be appreciated if we look beyond "Frieda" or "Woman" as the referent of this phrase. Beyond the embattled marital situation with its built-in power struggle (both in his own life and that portrayed between Will and Anna in the novel) is the tension of affirmation and ridicule that Lawrence has adopted as the artistic method for his novel. It is for this reason that he wrote to Edward Garnett not to look for "the old stable ego" in this novel. The real struggle is the analogical struggle between cultural forces: the struggle is between the inclusive God-mystery, deeper than consciousness, in conflict with the critical intelligence that surfaces in action in the pose of a jeering interrogator. Dietrich Bonhoeffer in his study of Genesis 1–3 entitled *Creation and Fall* interprets the role of the Serpent in the Garden of Eden in the original narrative as an interrogator. Having the "modern query" in mind, Bonhoeffer sees the Serpent as the originator of the first "religious discussion"; that is, the Serpent is the agent of the distancing of human being from the Divine Reality which makes the Reality a subject of critical analysis about which one may have opinions, for or against, agree or disagree, as about a dispassionate and abstract subject. Prior to the appearance of the Serpent, Adam and Eve live in unconscious relation – even the other one is not objectified – from the Center. There is a hint here already why *Lady Chatterley's Lover* will represent the Eden in which the voice of the Serpent (Clifford) is inconsequential and life is lived to what is – according to the story – its full, because it exists from (and in) the Center itself. In *The Rainbow*, the denouement is in the apparent contradiction to Will's "transport," given in this report of his inner thoughts following the laughter

of Anna: "His mouth was full of ash, his soul was furious [as we would expect]. He hated her for having destroyed another of his vital illusions. Soon he would be stark, stark, without one place wherein to stand, without one belief in which to rest. [All this sounds like the aftermath of "two people" at odds with each other in a bitter encounter about fundamental issues; then comes the *contradiction*.] *Yet somehow in him he responded more deeply to the sly little face that knew better, than he had done before to the perfect surge of his cathedral*" (italics my own).

The cleft between religious and modern sensibilities is within the author himself! Lawrence is more of a "modern" than he is usually given credit to be! What we can learn here is an essential lesson in reading Lawrence: do not mistake his occasional harrangue for a doctrinaire capsule of the author's "philosophy." His "polyanalytics" as he called them is his dialectical method of "meaning against meaning" for which the discursive paragraph pitted against the jeering and critical response is his fundamental method. However, *both* the solemn doctrine and the jeers are Lawrence! "He" can only be evaluated on the strength of his novels as composed in their wholeness, not in the truth or falsity of this or that opinion contained within them which is identified in the popular imagination as the author's "doctrine." The struggle in *The Rainbow* as indicated in the apparent contradiction of these italicized words comes close to *our* experience, and that is just the point. *The modern dilemma is that mystical wholeness is not sustained by the glaring estrangement of our actual experience.* It is this revelation that Will is opened to in the counterpoise of affirmation and ridicule.

(2) In the second place, there is an important point to be made in our discussion about the relationship of laughter to the experience of the Holy. Rudolph Otto uses the phrase "creature feeling" to express what he calls the "nothingness" that sweeps over one in the situation of being "overwhelmed" in the mystical Encounter. The same words "nothingness" and "overwhelmed" appear in Lawrence's novel written just prior to *The Rainbow, Sons and Lovers*; he uses Otto's words to describe the desire that passes between Paul Morel and Clara Dawes:

to know their own nothingness, to know the tremendous living flood which carried them always, gave them rest within themselves. If so great a magnificent power could overwhelm them, identify them altogether with itself, so that they knew they were only grains in the tremendous heave that lifted every grass blade its little height, and every tree, and living thing, then why fret about themselves? They could let themselves be carried by life, and they felt a sort of peace each in the other. (*Sons and Lovers*, p. 436)

Despite Lawrence's rejection of "the old stable ego" (apparent even before he wrote to Garnett about *The Rainbow* in this quotation from his previous novel), *laughter has an essential function to perform in the protection of the ego from annihilation* in the face of "so magnificent a power" which "could overwhelm them." *Laughter is the moment of self-awareness in the hypnotic swoon of the Holy.* For clergy, laughter is, in fact, a common experience: a giggle that bubbles up in the midst of the most solemn liturgy on the most solemn of days, before the High Altar, while elevating the Host before Almighty God with the deepest seriousness. This giggle is an automatic response of the ego as it grasps an instant of self-awareness, in which the whole exercise — kneeling, the wearing of elaborate vestments, resonant intonations — seems absurd and ridiculous (i.e., an object of *ridicule*). This insight will stay with Lawrence to the end, though it is dubious that he ever understood it directly. It must be thought of as an insight of which he may have had an intuitive glimmering. In *Lady Chatterley's Lover*, the insight is expressed in one of the few instances of humor in that humorless novel in Connie's "snirt" of laughter before the solemnities of the "Resurrection of the Body" and again in Lawrence's humorous awareness of just how silly, objectively viewed, an act of sexual intercourse between a man and a woman actually is in its absurd postures and contorted discomfort. This suggests that Reinhold Niebuhr was not quite on target when he suggested in volume 1 of *The Nature and Destiny of Man* that "The flight of the self into the other and the escape into oblivion are recurring themes in D. H. Lawrence's analysis of sex. . . . It will be noted that the motif of escape into subconscious nature is more dominant than the sense of loss of the other. . . . Sometimes Lawrence explicitly identifies the sex impulse with the longing for death." Lawrence is saved from going this far because of the "snirt," "the pouf! of laughter," which hold him back from the sirens of the abyss and allow him to maintain artistic control of his material as well as to keep, in the face of declining health, a relative grip on his sanity.

III

In the concluding section, I want to sketch the affirmation–ridicule tension in application to the major novels of Lawrence (and one of the tales) as a method of determining their relative value as significant literature. As a "preview of coming attractions," I offer this outline in brief form:

1. *Sons and Lovers*

The major weakness of this novel is the ambivalent character of Paul Morel.

Paul is not clear about the actual nature of the tension of his life and this accounts for the confusion and ambivalence. On the one hand, he has not yet fully abandoned his saturation in Christianity, and, on the other, not fully developed the "dark mysteries" which he will pose as an alternative. Paul's mother and Miriam are essentially the same person which indicates that the spirit–flesh dichotomy which the author proposes to us in the novel is not the actual problem at issue. The issue is a struggle for power (in the sense of dominance) between the characters, but preeminently between Paul and his mother (even Morel's rage, so bitter at times one has to put the book aside and rest for a moment, is, of course, being communicated to us, the readers, by the author, Morel-Lawrence).

2. *The Rainbow and Women in Love*

The novels are the best Lawrence was ever to write. The reason for this is the development of the affirmation–ridicule tension. In the characters of Ursula Brangwen (*The Rainbow*) and Rupert Birkin (*Women in Love*) the alternations of these contradictory forces are brilliantly unified.

3. *Naked Innocence: The Plumed Serpent, The Man Who Died, and Lady Chatterley's Lover*

In *The Plumed Serpent* the tension exists in provisional fashion in the major character of Kate Leslie, but she is essentially confused and therefore does not embody the contradictory forces with consistency. This novel is Lawrence's least spontaneous; the best example of the modern critical intelligence in Lawrence ("White Consciousness") at work to present a carefully worked out mosaic to which the characters are made to fit.

In *The Man Who Died* and *Lady Chatterley's Lover*, we have two examples of "last words" written by a man facing the finality of death. The tension of affirmation and ridicule is essentially abandoned and it is this which gives to these works their testamental, humorless, doctrinal tone.

In all three — *The Plumed Serpent, The Man Who Died*, and *Lady Chatterley's Lover* — the symbol of the testament is "naked innocence." In *Lady Chatterley's Lover* it is the "naked innocence" (almost) of the Garden of Eden itself.

4. *Sons and Lovers*

In the final lines of *Sons and Lovers*, Paul Morel *refuses* to give in to darkness,

"He would not take that direction, to the darkness, to follow her" (to follow his mother in death). There is a confusion here about the meaning of "darkness"; in fact it is *the* dilemma at this point in Lawrence's development. The "darkness" of these lines is different from the rich, fecund darkness of the Creative Mystery. In an early scene, Mrs. Morel, while pregnant with Paul, is "thrust" out into the yard by Morel and (we are told) enters the energy of the Life-Force, teeming around her:

> She became aware of something about her. With an effort she roused herself to see what it was that penetrated her consciousness. The tall white lilies were reeling in the moonlight, and the air was charged with their perfume, as with a presence. Mrs. Morel gasped slightly in fear. She touched the big, pallid flowers on their petals, then shivered. They seemed to be stretching in the moonlight. She put her hand into one white bin: the gold scarcely showed on her fingers by moonlight. She bent down to look at the binful of yellow pollen; but it only appeared dusky. Then she drank a deep draught of the scent. It almost made her dizzy.
>
> Mrs. Morel leaned on the garden gate, looking out, and she lost herself awhile. She did not know what she thought. Except for a slight feeling of sickness, and her consciousness in the child, her self melted out like scent into the shiny, pale air. After a time the child, too, melted with her in the mixing-pot of moonlight, and she rested with the hills and lilies and houses, all swum together in a kind of swoon. (p. 24)

In this passage you can feel the early intuitions of Lawrence into the surging vitality of the Mystery, not unlike the experience of Will Brangwen in Lincoln Cathedral. Mrs. Morel enters an experience of transport, a "swoon" with the same hypnotic power as the interior of the cathedral. It is well to bear in mind, however, that this "swoon," despite Lawrence's attempt to envelope Mrs. Morel in a mysterious "something about her," is more likely an accurate report of the feelings of a woman following a bout of violence (an instance of how Lawrence is the unchallenged master when it comes to a male writing through a female perspective) and, therefore, is more directly related to the drama of struggle for power than the author intended. A hint of the affirmation— ridicule tension (more like the clergyman's giggle cited above) is given after Mrs. Morel returns to the house following the "swoon": "As she fastened her broach at the mirror, she smiled faintly to see her face smeared with the yellow dust of lilies."

Returning to the final lines of the novel, it is clear there that Paul is intentionally going toward "the city's gold phosphorescence." Biblical scholars identify city culture with the Tower of Babel, the center of human culture-(power), and this is, indeed, the direction that the author's quest will take him. The "city's gold phosphorescence" is the direction the author will go to clarify his essential tension between affirmation and ridicule as it will appear in his next novel, *The Rainbow*. By contrast, at this point in his career the "darkness" is confused between life force and death. *They are not identical* (contrary to the opinions of Nathan A. Scott, Jr. and Reinhold Niebuhr).[17] Darkness as sheer oblivion is best expressed in a poem of the period (from *Look, We Have Come Through!*):

> In the darkness we are all gone, we are gone with the trees
> And the restless river, – we are lost and gone with all these ...
> Not sleep, which is grey with dreams,
> nor death, which quivers with birth,
> but heavy, sealing darkness, silence, all immovable. ...

The basic tension in *Sons and Lovers* is a struggle for dominance. A paradigm of this struggle is the feelings expressed in the chapter, 'The Test on Miriam':

His [Paul's] heart was hard against her. She [Miriam] sat full of bitterness. She had known – oh, well she had known! All the time he was away from her she had summed him up, seen his littleness, his meanness, and his folly. Even she had guarded her soul against him. She was not overthrown, not prostrated, not even much hurt. She had known. Only why, as he sat there, had he still this strange dominance over her? His very movements fascinated her as if she were hypnotised by him. Yet he was despicable, false, inconsistent, and mean. Why this bondage for her? Why was it the movement of his arm stirred her as nothing else in the world could? Why was she fastened to him? Why, even now, if he looked at her and commanded her, would she have to obey? She would obey him in his trifling commands. But once he was obeyed, then, she had him in her power, she knew, to lead him where she would. She was sure of herself! (298–99)

This is a portion of the larger vicious circle of "bondage" in which all of the major characters are trapped: Morel goes away, only to inevitably return; Mrs. Morel continues her grip upon Paul, even in the face of his agency in her "euthanasia," following her death – character is entwined with character without escape and without rest.

What all depressed personalities find out, sooner or later, is that they must break out of the depression through an act of anger; what persons in a "vicious circle" must do is to "break the circle." It is against the background of such therapeutic destruction that I regard the destruction of "Arabella" the doll in chapter 4 as the central symbolic act of the novel, that act necessary before the "movement to the city" of the final lines:

So long as Annie wept for the doll, he sat helpless with misery. Her grief wore itself out. She forgave her brother [Paul] – he was so much upset. But a day or two afterwards she was shocked. "Let's make a sacrifice of Arabella," he said. "Let's burn her."
She was horrified, yet rather fascinated. She wanted to see what the boy would do.

He made an altar of bricks, pulled some of the shavings out of Arabella's body, put the waxen fragments into the hollow face, poured out a little paraffin, and set the whole thing alight. He watched with wicked satisfaction the drops of wax melt off the broken forehead of Arabella, and drop like sweat into the flame. So long as the stupid big doll burned he rejoiced in silence. At the end he poked among the embers with a stick, fished out the arms and legs, all blackened, and smashed them under stones.

"That's the sacrifice of Missis Arabella," he said. "An' I'm glad there's nothing left of her."

5. *The Rainbow and Women in Love*

In *The Rainbow* we have Lawrence's continuing movement toward the "city's gold phosphorescence." The "fall" from the Edenic blood-intimacy of the Marsh farm portrayed at the beginning of *The Rainbow* is into knowledge. The major protagonist in this movement is the "new Eve" in the third generation of Brangwens, Ursula. *The Rainbow* is Ursula's story, and the family chronicle is essentially background for the emergence of her character. The highly charged exchange by her parents after the visit to Lincoln Cathedral will now be the model for her relationships and inner musings as the parents have become "drowsed" by the process of procreation and childrearing.

Ursula's questions concerning herself are emblematic of her development as the "new Eve": "How to act, that was the question? Whither to go, how to become onself? One was not oneself, one was merely a half-stated question. How to become oneself, how to know the question and the answer of oneself" (*The Rainbow*, p. 267). It is through the process of "affirmation and ridicule" that an answer will be given to this set of questions. The symbol is the "watch" of the moon:

Something was looking at her. Some powerful glowing sight was looking right into her, not upon her, but right at her. Out of the great distance, and yet imminent, the powerful, overwhelming watch was kept upon her.... She sighed in pain. Oh, for the coolness and entire liberty and brightness of the moon. Oh, for the cold liberty to be oneself.... If she could just get away to the clean free moonlight. (*The Rainbow*, p. 300)

Drenched in moonlight (consciousness), the quest for the Truth and for her own essential being is launched.

Immediately upon asking herself the set of fundamental questions, she had turned to a consideration of the Christianity that she had inherited. The cleft between the "visions" of the Gospel and "weekday meaning" seemed insurmountable:

Nor could one turn the other cheek. Theresa slapped Ursula on the face. Ursula in a mood of Christian humility silently presented the other side of her face. Which Theresa, in exasperation at the challenge, also hit. Whereupon Ursula, with boiling heart, went meekly away.

But anger, and deep writhing shame tortured her, so she was not easy till she had quarrelled with Theresa and had almost shaken her sister's head off.

"That'll teach *you*," she said, grimly. And she went away, unChristian but clean. (*The Rainbow*, p. 268)

Then follows a delightful segment in which Ursula asserts herself ("in her pride"), but "in the end there returned the poignant yearning from the Sunday world":

The passion rose in her for Christ, for the gathering under the wings of security and warmth. But how did it apply to the weekday world? What could it mean, but that Christ should clasp her to his breast, as a mother clasps her child? And, oh, for Christ, for him who could hold her there and lose her there. Oh, for the breast of man, where she should have refuge and bliss forever! All her senses quivered with passionate yearning. (*The Rainbow*, p. 269)

Transmuted into physical desire for a man ("the weekday world"), the passion led to an occasion of ridicule:

All the time she walked in a confused heat of religious yearning. She wanted Jesus to love her deliciously, to take her sensuous offering, to give her sensuous response. For weeks she went in a muse of enjoyment.

And all the time she knew underneath that she was playing false, accepting the passion of Jesus for her own physical satisfaction. But she was in such a daze, such a tangle. How could she get free?

She hated herself, she wanted to trample on herself, destroy herself. How could one become free? She hated religion because it lent itself to her confusion. She abused everything. She wanted to become hard, indifferent, brutally callous to everything but just the immediate need, the immediate satisfaction. To have a yearning towards Jesus, only that she might use him to pander to his own soft sensation, use him as a means of re-acting upon herself, maddened her in the end.

There was then no Jesus, no sentimentality. *With all the bitter hatred of helplessness she hated sentimentality.* (*The Rainbow*, pp. 270–71; italics my own)

The italicized sentence above brings to mind our discussion sketched earlier in the paper, relating power and our general theme of affirmation and ridicule. Without the inclusion of the importance of power in Lawrence's writing, we would have had a less intense and nervous prose to observe, but because we have asserted power as intrinsic to the ontological analysis (the structure of life itself carries the impulse for power and the impulse for Eros), now *ridicule* becomes the correct word, rather than criticism. *Ridicule is an act of violence brought on by helplessness. Ridicule* is the voice of "modernism" *fighting back* out of its own malaise. Rollo May devoted his entire study, *Power and Innocence*, to this single insight: sources of violence lie in *impotence*, not in power, and (further, and relevant for our study) that innocence is ultimately powerless. Later, under the influence of Winifred Inger, Ursula's quest for knowledge will be advanced as she gains intellectual distance from the Christianity of her youth, but impowerment is very much in the picture:

The motive of fear in religion is base [she had learned from humanist Winifred], and must be left to the ancient worshippers of power, worshippers of Moloch. [And here comes a bit of Lawrence's satire.] We do not worship power, in our enlightened souls. Power is degenerated to money and Napoleonic stupidity.

Ursula could not help dreaming of Moloch. Her God as not mild and gentle, neither Lamb nor Dove. He was the Lion and the Eagle. (*The Rainbow*, p. 322)

Ursula is on a quest. Her next step (perhaps because the "sweet reasonableness" of Winifred's ideas) is to move into the "magic" of academia. Her belief is that knowledge will widen the circle of her being, to allow her to move beyond the pettiness and smallness of her previous life. Lawrence's ideas (cf. 'The Education of the People', one of Lawrence's essays) are echoed in her vision of the "sense of retreat and mystery" represented by the college spirit (at least as it exists at this time in her own mind). "Her soul flew straight back to the medieval times, when the monks of God held the learning of men and imparted it within the shadow of religion" (*The Rainbow*, p. 406). In time, the magic declined. Then near the end of her college career comes the key counterpoise of affirmation and ridicule which will answer the fundamental questions that Ursula asked at the beginning of her quest. Here is the enabling dialectic which results in the creation of the "new Eve": "She had on her slide some special stuff come up from London that day ... as she focused the light on her field ... she was fretting over a conversation she had had a few days ago with Dr. Frankstone, who was a woman doctor of physics in the college." Note how both Winifred and Dr. Frankstone are "modern thinkers" similar to Anna in the cathedral scene. All, of course, are women. Ursula, the genuine "new Eve," will be the discoverer of a new religious insight and is to be seen as the "compelling opposite" of the other women as well as all the men in the novel:

"No, really," Dr. Frankstone had said, "I don't see why we should attribute some special mystery to life — do you? We don't understand it as we understand electricity, even, but that doesn't warrant our saying it is something special, something different in kind and distinct from everything else in the universe — do you think it does? May it not be that life consists in a complexity of physical and chemical activities, of the same order as the activities we already know in science? I don't see, really, why we should imagine there is a special order of life, and life alone." (*The Rainbow*, p. 416)

But what was the nature of the phenomena that Ursula saw "within the field of light, under her microscope"? "She saw it move – she saw the bright mist of its ciliary activity, she saw the gleam of its nucleus.... What then was its will? If it was a combination of forces, physical and chemical, what held these forces unified, and for what purpose were they unified? ... What was its intention? To be itself? Was its purpose just mechanical and limited to itself? It intended to be itself. But what self? Suddenly in her mind the world gleamed strangely, with an intense light, like the nucleus of the creature under the microscope. Suddenly she had passed away into an intensely-gleaming light of knowledge. ... It was a consummation, a being infinite. Self was a oneness with the infinite. To be oneself was a supreme, gleaming triumph of infinity" (*The Rainbow*, pp. 416–17). "The new world" which begins through the "new Eve" is founded on the discovery of self, not as finite and limited, but as *infinite spirit*. To be oneself is to participate in a universe of intentionality and purpose and expectation and freedom and hope!

This is the background for one of the most significant paragraphs in the history of literature and a great tribute to the "feminizing" power of this remarkable novel:

And the rainbow stood on the earth. She knew that the sordid people who crept hard-scaled and separate on the face of the world's corruption were living still, that the rainbow was arched in their blood and would quiver to life in their spirit, that they would cast off their horny covering of disintegration, that new clean, naked bodies would issue to a new germination, to a new growth, rising to the light and the wind and the clean rain of heaven. She saw in the rainbow the earth's new architecture, the old, brittle corruption of houses and factories swept away, the world built up in a living fabric of Truth, fitting to the over-arching heaven. (*The Rainbow*, p. 467)

In *Women in Love*, the journey to the "city's gold phosphorescence is completed and we encounter Lawrence's most public novel. *Women in Love* can be used as a guidebook to the circle of Lady Ottoline Morrell – most of the visitors to Garsington Manor (Breadalby in the novel) are present as the guests of Hermione Roddice. Ursula has regressed in character from her role in *The Rainbow*. In this novel she is the commonsensical companion of Rupert Birkin, who replaces her as the major protagonist. Ursula is generally portrayed speaking in the words of Dorothy, the advocate of "love," in the late chapters of *The Rainbow*. Rupert is the spokesman for "love is anathema" in his pose of anti-intellectual intellectual. In *Women in Love*, Birkin is the one who is on a quest. The quest is pursued in the midst of the "putrefaction" of culture analogous to the cultural crisis of the west in the years following 1914. The collapse of civilization is a major motif. The novel is analogical; it is about World War I without ever mentioning it.

The tone is decisively set by the counterpoint of affirmation and ridicule:

> There was a sense of violation in the air, as if too much was said, the unforgivable. Yet Ursula was concerned now only with solving her own problems, in the light of his words. She was pale and abstracted.
> "But do you really *want* sensuality?" she asked, puzzled. Birkin looked at her, and became intent in his explanation. "Yes," he said, "that and nothing else at this point. It is a fulfillment – the great dark knowledge you can't have in your head – the dark involuntary being. It is death to oneself – but it is the coming into being of another."
> "But how? How can you have knowledge not in your head?", she asked, quite unable to interpret his phrases. "In the blood," he answered; "when the mind and the known world is drowned in darkness – everything must go – there must be a deluge. Then you find yourself a palpable body of darkness, a demon – ."
> "But why should I be a demon – ?", she asked. "*Woman wailing for her demon lover*" – he quoted – "why, I don't know."
> Hermione roused herself as from a death-annihilation. "He is such a *dreadful* satanist, isn't he?" she drawled to Ursula, that ended with a shrill little laugh of *pure ridicule*. The two women were jeering at him, jeering him into nothingness. (*Women in Love*, pp. 46–47; italics for *pure ridicule* my own).

Hermione, the one with the shrill little laugh of pure ridicule, is portrayed earlier (p. 24) as having "the face of an almost demoniacal ecstatic." One of the weaknesses of this novel is the monodimensional character of Hermione. She seems to have no redeeming characteristics. Nevertheless, viewed in the light of her "pure ridicule" we get some explanation why this is so: she is the embodiment of the demonic expressed through the psychology of helplessness. The violence manifest in smashing Rupert's head with the ball of lapis lazuli is a direct result of her impotence in his presence and before his rantings. Lawrence describes well the experience of Hermione's helplessness: "Her whole mind was a chaos, darkness breaking in upon it, and herself struggling to gain control of her will, as a swimmer struggles with the swirling water ... the terrible tension grew stronger and stronger, it was the most fearful agony, like being walled up" (*Women in Love*, p. 118). Similar to the "sacrifice of Missis Arabella" in *Sons and Lovers*, the prison of "being walled up" must be destroyed. As I observe this experience again and again in the fiction of Lawrence, I think of a depressed young woman I know who described her condition as "suddenly feeling like a glass jar has been shoved down over you." It was not surprising when one of her attempts at self-affirmation was to put her first through a glass windowpane.

The tensions of *Women in Love* are the best example in Lawrence's fiction of the *contradictions* that make for great literature. He explores the total landscape of human thought and action. And it is this totality *in extremis*, as it were, that leads him to his fascination with the demonic that gives this novel its chilling tone. It is not the "silver river of life" (like the "rainbow which stood on the earth") which can sustain the weight of modern experience. In the cultural crisis, the silver river is matched by the "other river":

> "It seethes and seethes, a river of darkness," he said, "putting forth lilies and snakes, and the ignis fatuus, and rolling all the time onward. That's what we never take into account – that it rolls onwards."
> "What does?"
> "The other river, the black river. We always consider the silver river of life, rolling on and quickening all the world to a brightness, on and on to heaven, flowing like a bright, eternal sea, a heaven of angels thronging. But the other is our real reality – "
> "But what other? I don't see any other," said Ursula.
> "It is your reality, nevertheless," he said; "that dark river of dissolution. You see it rolls in us just as the other rolls – the black river of corruption . . . . " (*Women in Love*, pp. 195–96)

Through Lawrence's vision of the black and silver rivers which roll "within us" (another example of how the whole cosmic drama is enacted by the self) we catch a glimmering of why the dialectic of affirmation and ridicule is about to run down. The optimism about self and world present at the end of *The Rainbow* has been smashed by the horror of the war which is raging as *Women in Love* is being composed. The demonic has been unleashed in all its destructive potential, and it is destruction (rather than fresh new creation) that is on Lawrence's mind as he writes this novel.

It is this creative–destructive tension which makes this the greatest novel of his career, but the dialectic is carried on with such ferocity (affirmation and *ridicule*) that at the end one is only left with the painful longing for rest and peace. The affirmation of the creative mystery is present, but it is combined with a hatred of humanity which abhors the notion that God and man would be linked in mutual vulnerability. The impersonal character of the God-mystery reaches a fever pitch in Birkin's musings to "best leave it all to the vast, creative, non-human mystery" in the final pages of the novel.

At the level of history, Lawrence perceived the cultural crisis: the romantic ethos of "splendid singleness" was not enough. (Though later, in *Aaron's Rod*, as an act of despair, it is "singleness of being" which will *have* to be enough.) In *Women in Love* a remnant community is on his mind because he needs connection with human beings. His sanity depends upon it. In the same instant he knows that such community cannot be built on the old terms ("love your neighbor as yourself") because that ignores the reality of power. His metaphor of "balanced stars" is a valiant attempt to provide a foundation for such community as it grows out of the fundamental human community of man and woman in perfect polarity. The novels written at the end of his life demonstrate that the quest for wholeness in actual "weekday" relationship was never fulfilled. The journey from the "city's gold phosphorescence" will lead first to a confusion of history and apocalypse (*The Plumed Serpent*) and finally to his "testamental" works, *The Man Who Died* and *Lady Chatterley's Lover*.

6. *Naked Innocence: The Plumed Serpent, The Man Who Died, and Lady Chatterley's Lover*

An outstanding characteristic of these three works is their decided lack of humor. *The Plumed Serpent*, published in 1926, is the earliest of the three works and at first glance would seem to stand the best chance of reviving the affirmation—ridicule tension, but the dialectic is crushed under the heavy foot of Lawrence's intentional scheme. The scheme is built upon the contrast, familiar to readers of Lawrence, between the modern world of mechanical forms and superficial human relationships (the "meeting and mingling" of *Women in Love*) and an apocalyptic vision of organic forms and human communion through the affirmation of instinctual being. The drama of the novel is played out upon the stage of religious sensibility, and it is precisely the confusion between history and apocalyptic vision that leads to a kind of clumsiness in Lawrence's deliberate forcing of character and situation. It is a religious drama, the conflicting symbolic personae are Christ, depicted as "dead matter" like that of the stone of the altar in *The Rainbow*, and Quetzalcoatl, the Lord of the Two Ways, the restorer of organic life to dead old Mexico. The novel would have had a chance if Kate's "mocking intelligence, her sense of absurdity" could have been sustained (like Rupert Birkin's "mocking intelligence and sense of absurdity" persists throughout *Women in Love*). The *scheme* tells us that Kate is to reach a fulfillment as she moves from disillusionment to rebirth. It is a credit to the sharpness of Lawrence's perception that Kate knows she is unfulfilled as the novel ends, despite all that we have been told (according to the scheme) that fulfillment is inevitable. This perception is clear in Kate's inner thoughts in the final scene: "*What a fraud I am! I know all the time it is I who don't altogether want them. I want myself to myself. But I can fool them so they won't find out."*

Her greatness as a character is precisely in such candid honesty. Early in the novel she is the agent of the tension between affirmation and ridicule:

> "One is driven, at last, back to the far distance, to look for God," said Ramon uneasily. "I rather hate this search-for-God business, and religiousity," said Kate.
> "I know!" he said with a laugh. "I've suffered from would-be-cocksure religion myself." "And you can't really 'find God'!" she said. "It's a sort of sentimentalism, and creeping back into old, hollow shells."

But Kate is on a quest. The quest takes her from the prose of modern skepticism (actually the focus of genuine interest in her as a fictional character that resonates with the tension of actual experience) to an initiate in a religious community in which she is part of a pantheon of those who (we are told) have become one with the gods. In a confusion of utopian innocence with historical ("weekday") event, she becomes momentarily engulfed in a "sexual swoon" which is pondered with a humorless solemnity:

> Strange how naive he was! He was not like Ramon, rather ponderous and deliberate in his ceremonials. Cipriano in his own little deeds to-night with her, was naive like a child. She could hardly look at that bud of light which he said meant their united lives, without a catch in her heart. It burned so soft and round, and he had such an implicit, childish satisfaction in its symbol. It all give him a certain wild, childish joy. The strange convulsions like flames of joy and gratification went over his face!
> "Ah, God!" she thought. "There are more ways than one of becoming a little child."
> . . .
> And she pressed him to her breast, convulsively. His innermost flame was always virginal, it was always the first time. And it made her again always a virgin girl. She could feel their two flames flowing together. How else, she said to herself, is one to begin again, save by refinding one's virginity? And when one finds one's virginity, one realizes one is among the gods. He is of the gods, and so am I. Why should I judge him!
> So, when she thought of him and his soldiers, tales of swift cruelty she had heard of him: when she remembered his stabbing the three helpless peons, she thought: Why should I judge him? He is of the gods. And when he comes to me he lays his pure, quick flame to mine, and every time I am a young girl again, and every time he takes the flower of my virginity, and I his. It leaves me insouciant like a young girl. What do I care if he kills people? His flame is young and clean. He is Huitzilopochtli, and I am Malintzi. What do I care, what Cipriano Viedma does or doesn't do? Or even what Kate Leslie does or doesn't do! (*The Plumed Serpent*, pp. 430–31)

The distinction between apocalyptic vision and the "weekday" world is an essential and crucial one in these final works. Apocalyptic vision means the projection of a postulated world in which the "weekday" world ("mammon," estrangement, self-importance, mechanization, the prostitution of nature) is transcended and surpassed by the vision of the way-it-ought-to-be ("the democracy of touch," "the resurrection of the body," becoming "born again"). The symbol of this postulated world is the state of "naked innocence," which is beyond good and evil. In this vision the human manifestations of genuine and vital relations are essentially impersonal. "Getting to know you," "meeting and mingling" is not the form of these relationships. It is a belief in "being in touch" which is beneath and beyond the perimeters of verbal and attitudinal familiarity, "prior to personality" as Middleton Murry once remarked. Lawrence's problem in *The Plumed Serpent* (as is probably obvious from the silly portion just quoted above) is that apocalypse and "weekday" world are hopelessly confused. The "scheme" so deliberately imposed on character and situation did not fit, and it is clear that the author knew it in the ambivalence of the ending. He had ceased to believe in the scheme he had designed prior to the actual interactions of the characters in living tension with each other. Such confusion is evident in the "dreaming innocence" which is blind to the obvious cruelty of blind, unbridled power. Such confusion accounts as well for the collapse within the novel of the separation between the purely religious agenda of the Quetzalcoatl movement and the turbulent political intrigues always in the background.

In *The Man Who Died* (*The Escaped Cock*) the "weekday world" is synthesized with apocalyptic vision as the stage for what is an attempt to reconstruct an interpretation of Jesus. The background is the Augustinian–Calvinist tradition in Christianity with its deep suspicion of sexuality and the sentimentalized portrait of the Jesus of Ernst Renan's "Life" which had profound impact upon the inquiring young Lawrence. *The Man Who Died* is an excellent example of how Lawrence's method of "meaning against meaning" is abandoned in his final works. In this tale, the counterpoint is against the unseen straw man of a Christianity of Lawrence's own manufacture. Any resemblance to the Jesus of the Gospels (eating and drinking, furious, in tears, acting out his message with what he did with his body) is not present. This lack of resemblance is a major flaw in the credibility of this work.

The intention of *The Man Who Died* is to harmonize the story of the Christ of the Gospels ("The Risen Lord") with the Egyptian myth of Isis and Osiris. This is an interesting combination! In the combination he brought together two contradictory traditions: the patriarchal tradition of Judaeo-Christianity with the matriarchal tradition of Egypt. And in the combination in the story it is the matriarchal that dominates (perhaps an explanation why Kate Millett did not consider this story in *Sexual Politics*). Charles Rossman in his recent article, "D. H. Lawrence and Women,"[18] notes that the word "lure" is central to the structure of the combination. "Lure" connotes more than passive magnetic attraction. It implies that the woman is the agent of her own needs and the initiator of sexual activity, a woman with an independent self to match the "aloneness" of the man who died. In their isolation, two such beings remain forever unknowable and ultimately unreachable to each other. Yet "to be" means to reach out toward the other in a gesture which fuses giving and taking."[19]

The logic of the story is the logic of the Isis–Osiris narrative. It is the priestess of Isis-in-Search who is the protagonist (after the splendid rooster has departed quickly from the scene). What we observe is her *action* and the *inner thoughts* of the man who died. The ending of the tale is consistent with this logic: against Eugene Goodheart,[20] the woman is not *abandoned* at the end of the story. Rather the man who died as the reassembled Osiris has fulfilled his destiny in regard to the "Great Mother," to fecundate her womb. Lawrence humanely saves the man who died from certain oblivion by projecting him into an adventure of escape and the Christian vision of the second coming ("I will come again," says the man who died).

The relationship between "the man" and the priestess is the best example in Lawrence's fiction of the notion (from *Women in Love*) of the male–female polarity of "balanced stars." In its mythic recasting, however, it is "dreaming innocence." The dialectic of the "impulse for power and the impulse for Eros" is missing. That is to say, the essential basis for the proposal of a relationship beyond "love," beyond "meeting and mingling," has been withdrawn in the projection of a postulated world in which one is "free from guilt or sin, guileless, pure." Without the dialectic of actual experience, the laughter is stilled and the characters are (humorless) in paradise. Because the "bodyless" Christianity of Lawrence's critique within the story is essentially of his own manufacture we have none of the tension of this issue as an actual debate within Judaeo-Christian history (e.g., as an ascetic reaction against the sexual excesses of the Roman world, or as a protest against the worship of natural vitalities enshrined as deities in Canaanite and other ancient cultures). Instead of "meaning against meaning," we witness an easy synthesis of contradictory differences. Our conclusion, therefore, is that Lawrence is offering to the reader his religious testament. The raging affirmation and ridicule of *Women in Love* called for stasis and rest. The stasis and rest is in the vision of *The Man Who Died*. Its roots are deep in Lawrence's imagination. In an essay intended as an introduction to Frederick Carter's book, *Apocalypse*, Lawrence wrote: "From early childhood I have been familiar with Apocalyptic language and Apocalyptic image: not because I spent my time reading Revelation, but because I was sent to Sunday School and to Chapel, to Band of Hope and to Christian Endeavor, and was always having the Bible read at me or to me."[21] His favorite hymn, reflected upon years later in his essay, "Hymns in a Man's Life" contained these revealing words: "Sun of my soul, Thou Saviour dear, / It is not night if Thou be near." As Emile DeLavenay has noted, "Steeped in the Bible before reaching an age in which he [Lawrence] could hope to understand its language, he was to respond above all to its poetry. His imagination was able to develop all the more freely because the meaning of the hymns and the Bible stories was so vague: his vision was of some promised land wrapped in distant haze, a land where all his dreams would come true."[22]

Lady Chatterley's Lover is a deliberate attempt on the part of Lawrence to perform what he intended when he wrote to Earl and Achsah Brewster of his goal to present "regenerate man, in the Garden of Eden." The model is idyllic: the hut is "in the center," surrounded by the wood, the larger estate, and beyond, the nearby towns, and finally, the world arena of London, Paris, etc. The only note of ridicule is directed by the humorless Mellors to a vague industrial and mechanical complex at some distance from the action of the novel. The "serpent in the garden" (Clifford) is also at some remove, and he becomes absurd in his little motorized chair when he attempts to actually move into the wood which contains the hut. Since Clifford is officially impotent, it disguises the actual psychology of the novel which is a psychology of the powerless. This is why the insight of Rollo May is so crucial in the understanding of this novel (and worth repeating again): "There is one way, however, of confronting one's powerlessness by making it a seeming virtue. This is the conscious divesting on the part of an individual of his power ... I call this innocence...."

The Midlands vernacular familiar to the readers of this novel is the language of innocence. It is direct, uncomplicated speech, distinguished by its freedom from irony and ambiguous meanings. The frequent use of biblical fragments in this novel serve the same purpose. Note how visionary sermons of our own generation such as the famous "I have a dream" sermon of the late Martin Luther King, Jr. consist largely in apocalyptic biblical fragments which are used in the service of the powerless strata of our society.

The struggle during the phases of composition give us some insight into the effort of the author to write a story within the setting of the "larger world." The novel was written three times. In the first version, Lawrence wants his characters to make an impact in the industrial malaise, and Parkin (Mellors of the third version) goes to work in a Sheffield steel mill and becomes the secretary of the local branch of the Communist party. As the novel is rewritten the life "in the center" becomes increasingly isolated from the actual world of experience, and in the third and final version, futuristic innocence is the predominant mode. That is to say, despite several attempts, the story simply did not actualize itself as a genuine history. It was "out of this world," a fragment of vision, innocent and pure, removed from the countervailing power struggle of the economic and political arena.

At the time of the writing of *Lady Chatterley's Lover*, Lawrence experienced periods of illness that would eventually result in his death. Mysteriously, the author reached a "new quality of serenity and introspective harmony" during this time.[23] The note of ridicule is gone and we are presented with pure affirmation. The analogy of this affirmation is one of the few glimmerings of paradise that human beings are given: the experience of orgasm. As Connie and Mellors make love for the fifth time, she "lets go" for the first time and overcomes her "inward anger and resistance." In this experience, a metamorphosis occurs: that which is despised is now the entry point into the wonder and aweful mystery of meaning and being. In this respect, Lawrence has progressed beyond the illumination given to Ursula peering into the microscope in *The Rainbow*. Connie's insight is that *love opens to knowledge*. Lawrence, the heretic, "in the last issues" is an ironic "little Sir echo" of a major figure (St. Augustine!) in the Christian tradition.[24]

Literary criticism has a primary task to perform in distinguishing significant texts from those of lesser value. In the course of this paper I have tried to show how this can be done in regard to some of the writings of D. H. Lawrence. In all his works Lawrence was an aristocrat of the Spirit. He is in good company: Nietzsche, Dostoyevsky, and Blake are among his companions.

NOTES

[1] George Ford, *Double Measure*, p. 120.
[2] Ibid., p. 121.
[3] Ibid., p. 118.
[4] Quoted in ibid., p. 119.
[5] Martin Jarrett-Kerr, *D. H. Lawrence and Human Existence*, p. 27. Copyright © 1961 SCM Press, London.
[6] Philip Callow, *Son and Lover*, p. 117.
[7] Ibid., p. 102.
[8] Ibid., p. 112.
[9] Rollo May, *Power and Innocence*, p. 243.

[10] Ibid., p. 48.
[11] Paul Tillich, *The Interpretation of History*, pp. 81–82.
[12] Quoted in James Luther Adams, *Paul Tillich's Philosophy of Culture, Science, and Religion*, p. 230.
[13] Ford, p. 31.
[14] Paul Tillich, *Shaking of the Foundations*, p. 161.
[15] Jarrett-Kerr, p. 144.
[16] Harry T. Moore, *The Collected Letters of D. H. Lawrence*, p. 369.
[17] In Scott's discussion in "D. H. Lawrence: Chartist of the Via Mystica" in his book *Rehearsals of Discomposure*, he argues that Lawrence's couples desire "that ravishment into a transcendent state beyond which considered from the standpoint of life is death: Lawrence's name for it is darkness." Throughout this essay Scott identifies Lawrence's romanticism with classic romantic motifs (Tristan and Isolde, Romeo and Juliet, etc.). The parallel is not quite correct for the reasons I have outlined.
[18] *D. H. Lawrence Review*, Fall 1975.
[19] Charles Rossman, "D. H. Lawrence and Women," *D. H. Lawrence Review*, Fall 1975, p. 322.
[20] Cf. *The Utopian Vision of D. H. Lawrence*.
[21] Quoted in Harry T. Moore, *The Priest of Love*, p. 20.
[22] Emile DeLavenay, *D. H. Lawrence: The Man and His Work*, p. 13.
[23] Charles Rossman, p. 316.
[24] Quoted in Joseph Sittler, *Essays on Nature and Grace*, p. 118: "Non intradit veritatem nisi caritatem" – there is no entrance to truth save by love.

SELECTED BIBLIOGRAPHY

Works by D. H. Lawrence

Aaron's Rod, New York: Viking, 1961.
Complete Poems of D. H. Lawrence, Edited by Pinto and Roberts, New York: Viking, 1971.
The Escaped Cock, Los Angeles: Black Sparrow, 1973.
John Thomas and Lady Jane, New York: Viking, 1972.
Kangaroo, New York: Viking, 1973.
Lady Catterley's Lover, New York: Grove Press, 1959.
The Later D. H. Lawrence, Edited by Tindall, New York: Knopf, 1952.
Psychoanalysis and the Unconscious; Fantasia of the Unconscious, New York: Viking, 1967.
Phoenix, Edited by McDonald, New York: Viking, 1972.
The Plumed Serpent, New York: Vintage, 1959.
The Rainbow, New York: Random House, Modern Library, 1943.
4 Short Novels, New York: Viking, 1972.
Sons and Lovers, Edited by Julian Moynahan, New York: Viking, 1968.
Women in Love, New York: Random House, Modern Library, 1951.

Biographies and Letters

Callow, Philip, *Son and Lover*, New York: Stein and Day, 1975.
Chambers, Jessie, *D. H. Lawrence: A Personal Record*, London: Cass, 1965.
DeLavenay, Emile, *D. H. Lawrence: The Man and His Work, 1885–1919*. Carbondale: Southern Illinois University Press, 1972.
Lucas, Robert, *Frieda Lawrence*, New York: Viking, 1973.
Moore, Harry T., *The Collected Letters of D. H. Lawrence*, New York: Viking, 1962.
Moore, Harry T., *The Priest of Love*, New York: Farrar Straus Giroux, 1974.

Lawrence Criticism

Andrews, W. T., *Critics on D. H. Lawrence*, Coral Gables: University of Miami Press, 1971.
Clark, L. D., *Dark Night of the Body*, Austin: University of Texas Press, 1964.
Clarke, Colin, *River of Dissolution*, New York: Barnes and Noble, 1969.
Cowan, James C., *D. H. Lawrence's American Journey*, Cleveland: Case Western Reserve Press, 1970.
Ford, George, *Double Measure*, New York: Norton, 1969.
Gilbert, Sandra M., *Acts of Attention*, Ithaca: Cornell University Press, 1972.
Heilbrun, Carolyn G., *Toward a Recognition of Androgyny*, New York: Knopf, 1973.
Jarrett-Kerr, Martin, *D. H. Lawrence and Human Existence*, London: SCM Press, 1961.
Kinwood-Weekes, Mark, *Twentieth Century Interpretations of The Rainbow*, Englewood Cliffs, N. J.: Prentice-Hall, 1971.
Miko, Stephen, *Toward Women in Love*, New Haven: Yale University Press, 1971.
Miko, Stephen, *Twentieth Century Interpretations of Women in Love*, Englewood Cliffs, N. J.: Prentice-Hall, 1969.
Nin, Anais, *D. H. Lawrence: An Unprofessional Study*, Chicago: Swallow Press, 1964.
Oates, Joyce Carol, *The Hostile Sun*, Los Angeles: Black Sparrow Press, 1973.
Spilka, Mark, *The Love Ethic of D. H. Lawrence*, Bloomington: Indiana University Press, 1971.

Theology, Etc.

Adams, James Luther, *Paul Tillich's Philosophy of Culture, Science, and Religion*, New York: Harper and Row, 1965.
Bonhoeffer, Dietrich, *Creation and Fall*, New York: Macmillan, 1964.
May, Rollo, *Love and Will*, New York: Norton, 1969.
May, Rollo, *Power and Innocence*, New York: Norton, 1972.
Niebuhr, Reinhold, *The Nature and Destiny of Man*, New York: Scribner's, 1964.
Otto, Rudolph, *The Idea of the Holy*, London: Oxford University Press, 1973.
Sittler, Joseph, *Essays on Nature and Grace*, Philadelphia: Fortress Press, 1972.
Tillich, Paul, *The Interpretation of History*, New York: Scribner's, 1936.

Articles and Dissertations

Davies, Horton, 'A re-evaluation of D. H. Lawrence.' *Religion in Life*, Summer 1969.
The D. H. Lawrence Review 8, no. 3 (Fall 1975).
Eichrodt, John Morris, 'D. H. Lawrence and the Protestant Crisis,' Ph.D. dissertation, Columbia University, 1962.
Scott, Nathan A., Jr. 'D. H. Lawrence: Chartist of the Via Mystica.' In *Rehearsals of Discomposure*, New York: Kings Crown Press, 1952.

REINHARD KUHN

THE ENIGMATIC CHILD IN LITERATURE

Gide's *Voyage of Urien* is a tale of such apparent transparency that the commentator is constantly tempted to provide simple interpretations that then turn out to be so simplistic as to miss the point. The novella is cast in the traditional form of the imaginary voyage that is the reflection of an inner odyssey. The symbols are equally traditional. So one can hardly fail to equate the algae-choked Sargasso Sea in which the boat is becalmed with the concept of ennui, or the ice-bound arctic landscape that the travelers eventually find with the notion of sterility. Nor does it require much perspicacity to see in Eric, the brutal killer of birds, the personification of the Nietzschean superman. Yet such explanations remain unsatisfactory, for they fail to sound the depths that the reader senses underneath the limpid surface of the narration.

One of the most mysterious of the seemingly clear events that occur during the course of the voyage is Urien's encounter with a young boy on a beach. After having drunk from the icy waters of a glacial spring, Urien and his companions are delighted by a sensation of beatitude. After this libation they rediscover one of the miracles of childhood, for they become capable of marveling at everything ("nous nous étonnions de toute chose"). At noon they descended toward the shore, and it is there that the meeting with the young child took place:

On the edge of the sea we met a mysterious child seated on the sand who was dreaming. He had large eyes as blue as a glacial sea; his skin shone like lilies and his hair was like a cloud tinted by the rising sun. He was trying to understand the words that he had traced on the sand. He spoke; from his lips his voice sprung forth in the same way as a matinal bird while shaking off the dew soars away; we would have gladly given him our shells, our insects, and our stones. His enchanting voice was so sweet that we would have happily given him everything we had. He smiled, and his smile was one of infinite sorrow. We would have liked to have taken him away with us to the boat, but, poring over the sand, he had resumed his tranquil meditation.[1]

It would be easy to see in this episode a prefiguration of the many other youths that appear in Gide's later works and to interpret it as an early translation into symbolic terms of the longing after "palingenesis," the expression the author himself used frequently to define his obsessive desire for rebirth. It is equally obvious (and equally unenlightening) that this is an early

A.-T. Tymieniecka (ed.), The Philosophical Reflection of Man in Literature, 245–264.
Copyright © 1982 by D. Reidel Publishing Company.

manifestation of Gide's pederastic tendencies. There is more to this incident, however, than such Freudian readings reveal.

This incident actually represents the discovery by the adult of the child as pure enigma. The confrontation with what is the essence of inscrutability has an immediate effect on the explorers. They approach the child as a god and are ready to offer to the newly discovered divinity any and all of their possessions. In approaching him, they themselves become childlike, as is indicated by the trivial and yet portentous nature of their proposed offerings (shells, insects, and stones). But the child ignores their willingness to sacrifice to him. The voyagers return alone to their vessel, but the impact of the child's presence makes itself felt long after the seekers after truth had left him. Previously, Urien and his companions had sought solace in the waters and took a sensual delight in swimming. After they leave the shore this time, however, the narrator soberly states, "This day we did not go swimming." And, indeed, they never do again. After having met the young boy, there is no further need for ablutions or for sensual pleasures.

The child as presented by Gide is an unfathomable being, a stranger with whom adults cannot communicate although they are irresistibly attracted to him. The child speaks, but Urien and his companions do not hear the words; instead they are struck by the soaring beauty of his voice. The aura of incomprehensibility is enhanced by the fact that even the child itself cannot decipher the mysterious symbols he traces in the sand, a method of transcription that suggests the evanescent nature of the signs. The child is the dreamer, living in his own world of the imagination. His existence represents a closed system, but one that exerts an extraordinary attraction on adults. His appearance is transparent, but in its inexplicability forever opaque.

The child as an enigma that no amount of speculation can resolve is a recurrent theme in literature. The examples abound; we need think only of Miles and Flora in James's *The Turn of the Screw*. They remain as inscrutable to the narrator and to the reader as to their governess. An even more striking instance is that presented by the same author in *What Maisie Knew*. The child-heroine finds herself in the center of a chaotic world of divorced parents, stepparents, and lovers whose relations to one another are constantly changing and who all compete for her affection, not out of love but in order to use her. In their attempts to maintain the fiction of her innocence, they pretend to shield her from the truth of their adulterous affairs. While the kaleidoscopic patterns of their relationships are portrayed through the eyes of Maisie, she herself, in her very stability, represents a puzzle for her elders as well as for the reader. She remains inscrutable and apparently uninvolved emotionally in the various events of which she is the focal point. Her personal reactions to what goes on around her seem transparent but are, in fact, unfathomable. Her consciousness appears as objective in its recording of experience as the lens of a camera. If Maisie knows everything, the reader is left wondering whether she understands anything. And even the extent of her knowledge remains in doubt as the novel comes to its ambiguous end:

Maisie waited a moment: then "He wasn't there" she simply said again.
Mrs. Wix was also silent a while. "he went to *her*," she finally observed.
"Oh I know!" the child replied.
Mrs. Wix gave a sidelong look. She still had room for wonder at what Maisie knew.[2]

The figure of the enigmatic child is not a modern invention. In fact, in the guise of the *puer senex* the cryptic child, old beyond its years, became a commonplace topos in the Latin literature of late antiquity and retained its popularity in the Middle Ages when it was employed as one of the clichés of hagiography. Before becoming a current theme in modern literature, it was used by Chaucer as the object of caricature in *The Canterbury Tales* (1387-1400). It is the mawkishly sentimental Prioress who recounts the martyrdom of a seven-year-old child in a tale that seems in many respects to occupy a place apart in the universe of Chaucer. Although the narrator, as portrayed in the prologue, is so sensitive that she weeps at the sight of a mouse maimed by a trap, the tale she recounts is a brutal one. She tells of a pious schoolboy who is so entranced by the *Alma redemptoris* that, with the help of an older companion, he learns the hymn by heart. He understands not a word of it, but nonetheless sings it "wel and boldely." Unfortunately for him, it is every day on the way to and from school, as he is passing through the Jewish quarter, that he prefers to intone this song of praise to the Virgin Mary. Feeling provoked and insulted, the Jews find an assassin who slits the child's throat and throws the body into a privy. Yet piteously the child continues to chant his hymn until a monk removes the cardamon seed that the Mother of God had placed on his tongue. And so at last the tormented child is allowed to find peace in death. What emerges clearly from this otherwise murky story is that the child is the incarnation of simplicity and innocence. As such, he remains from beginning to end inscrutable. He is a puzzle to his own mother who discovers the singing corpse, to the monk who buries him, to the Prioress herself who recounts his brief life, and to the reader who attempts to discover the meaning of this strange existence. The child appears on the stage briefly; his transitory presence and his disappearance are in fact inconsequential and yet seem fraught with a significance that passes our understanding.

There are other instances in literature of a somewhat different sort in which the child, while still an emblematic abstraction, does have a direct and crucial effect on the course of the narrative. This is the case in a work of the late Goethe, *The Novella*. The story is a simple one. While the Countess is out for a ride with her retinue, a conflagration breaks out in the village where a country fair is taking place. The flimsy stands go up in flames, and a tiger and a lion escape from their cage. One of the courtiers believes the life of the Countess to be in danger, and after one unsuccessful attempt manages to shoot the tiger. The owners of the menagerie, a couple and their young child, discover the corpse of the animal and are distraught. It appears that the wild beasts were perfectly tame. Their distress becomes even greater when the Count proposes to hunt down the lion as well. They plead with him, and promise that their child will bring the lion back to his cage without any danger to anyone. The Count agrees, and, as he is explaining the precautions he plans to take the child, who has not spoken a word, begins to play on his flute: "The child pursued its melody, which was not a melody at all but a series of tones without structure, and perhaps for that very reason so overwhelmingly moving; the people standing around seemed as if entranced by the movement of a songlike mode."[3] As in the case of the martyred child, his only means of expression is music. The boy proceeds to the cave in which the lion has sought refuge, alternately playing his flute and singing triumphantly. His song consists of three stanzas whose verses are constantly altered, again without apparent reason. However, from these variations there does emerge a clear theme:

> Blessed angels gladly counsel
> And accompany good children,
> In order to prevent evil will
> And to encourage beautiful deeds.[4]

The lion limps out of the cave after the child who, in a reenactment of the story of Daniel, removes a thorn from the lion's paw. And the story ends with the child still playing its flute and singing, constantly transforming the order of the verses. Within the framework of *The Novella* there is a striking contrast between the child and the young courtier, Honorio, who had slain the tiger. The latter had acted sensibly, as a young man, and could hardly be blamed for the fact that his deed was based on a false assumption. Yet he is disgraced; the last we see of him, he is looking toward the setting sun, toward the evening of his life. The innocent child, on the other hand, continues to inhabit his autonomous sphere, a tonal construct whose durability is assured by the eternal nature of music.

The insoluble conundrum posed by the presence of an uncommunicative child is not always as reassuring a symbol as it is in *The Novella*. In Blake's 'The Mental Traveller' (1803), the infant becomes a symbol of terror. The narrator-voyager of the poem had, in his travels through the land of men and women, seen "such dreadful things / As cold Earth wanderers never knew" (3–4). But of all the horrors imaginable, the most frightening is the encounter with a frowning baby:

> But when they find the frowning Babe,
> Terror strikes thro' the region wide:
> They cry "The Babe! the Babe is Born!"
> And flee away on Every side.

The presence of this child is so menacing that it causes people to panic, is destructive of the individual, and is an awesome threat even to nature.

That the apparently defenseless innocence of a quiet child can actually present a menace for the adult world is a concept by no means unique to Blake. It is also the case of the strange little being who one evening appears unexpectedly at the door of Jude and Sue in Thomas Hardy's *Jude the Obscure*. His very origins are shrouded in mystery. Born in far off Australia, he is the son of Jude's first wife, Arabella, but he does not know his father, who might be Jude himself. He is unbaptized and nameless, and is simply called Little Father Time. Hardy describes him as a *puer-senex*:

He was Age masquerading as Juvenility, and doing it so badly that his real self showed through the crevices. A ground-swell from ancient years of night seemed now and then to lift the child in this his morning-life, when his face took a back view over some great Atlantic of Time, and appeared not to care about what it saw.[5]

From the outset Hardy describes this unwanted offspring as separated from the rest of mankind. In the railroad carriage taking him to the home of his putative father, all the other travelers make themselves comfortable and eventually close their eyes. Not so the child who does not have the same confidence in the security of the train compartment as do his companions: " ... the boy remained just as before. He then seemed to be doubly awake, like an enslaved and dwarfed divinity, sitting passive and regarding his companions as if he saw their whole rounded lives rather than their immediate figures."[6] This otherworldly creature, who knows more than the others and perceives reality in a different fashion from them, is constantly on his guard.

Not only is he set apart from the adult world, but he is separated from his peers as well, in fact from all of nature. This feeling of total estrangement is depicted by Hardy when he describes the boy's lonely walk from the station to the house:

The child fell into a steady mechanical creep which had in it an impersonal quality — the movement of the wave, or of the breeze, or of the cloud. He followed his directions literally, without an inquiring gaze at anything. It could have been seen that the boy's ideas of life were different from those of the local boys. ... To him the houses, the willows, the obscure fields beyond, were apparently regarded not as brick residences, pollards, meadows; but as human dwellings in the abstract, vegetation and the wide dark world.[7]

The arrival of this singular child transforms the lives of Jude and Sue and eventually brings about the melodramatic catastrophe that destroys their existence.

Initially, the unexpected apparition of the child did not disturb their domestic arrangements as much as they had feared. In fact, at the beginning they were hardly conscious of his presence. Occasionally, however, when they had practically forgotten him, his small, slow voice would emerge from what seemed to be subterranean depths, and a brief remark would make them realize that he had absorbed everything they had said, and worse yet seen through everything. Such instances naturally provoked in them an acute sense of discomfort. Then, too, because of the child's yearning look, Jude and Sue sensed in him a desperate need to be loved, and yet they were frustrated in all of their attempts at demonstrating their affection toward him. All of their efforts at making him happy failed, and they found themselves faced with the unsatisfying task of dealing with a boy "singularly deficient in all the usual hopes of childhood."[8] Thus he acts as a constraining factor that to some degree spoils their happy-go-lucky relationship. When they go on an outing, they take Little Father Time with them, but the shadow of his company casts somewhat of a pall over their excursion: "Not regardful of themselves alone, they had taken care to bring Father Time, to try every means of making him kindle and laugh like other boys, though he was to some extent a hindrance to the delightfully unreserved intercourse in their pilgrimages which they so much enjoyed."[9] This inability to disturb the child's fundamental indifference troubles and saddens them. But most frightening of all is the child's own awareness of the effect he has on others and his frank admission of this awareness:

"I feel we have returned to Greek joyousness, and have blinded ourselves to sickness and sorrow. ... There is one immediate shadow, however — only one." And she [Sue] looked at the aged child, whom, though they had taken him to everything likely to attract a young intelligence, they had utterly failed to interest. He knew what they were saying and thinking. "I am very, very sorry, Father and Mother," he said. "But please don't mind! — I can't help it."[10]

The two adults are thwarted in their attempts to deal with the preternaturally aged child, they are disturbed by his lucid intuition, and this combined frustration and fear sours their own relationship.

The effect of the child's presence on the inner life of the couple is insidious and subtle, but the impact on their social existence is dramatic and obvious. For it is the child who forces upon them a crucial decision regarding their own intercourse. Until his arrival, the two cousins had lived together, but maintained a Platonic relationship. They are still too intellectually honest to overcome their aversion to the institution of marriage and thus refuse to go through the ceremony of legalizing their union. However, they do decide to live together as man and wife. Furthermore, in order to give their union a semblance of respectability that might make the child's life easier, they pretend that they had indeed been joined in wedlock. Thus the presence of Little Father Time has forced them to the very hypocrisy typical of the Victorian Age in which they lived and that they had always condemned. The results of their cohabitation are two children, a boy and a girl, whose existence places additional constraints on the already rather restricted life-style of the family. Compelled to move to another city they seek lodgings, but find only a temporary refuge for the mother and children. Before putting the young ones to sleep in a narrow closet, Sue speaks openly to Little Father Time of the problems they face, problems exacerbated by the fact that she is again pregnant. His responses, as usual, are laconic. The next morning, when Sue opens the closet door to awaken the children, she finds all three with cords tied around their necks, hanging strangled from the garment hooks. Near the elder boy the mother sees an overturned chair. The following evening the child Sue was carrying was stillborn.

This climactic catastrophe results in a denouement in which the relationship between Jude and Sue disintegrates completely and in which both Jude and Sue as individuals are destroyed. The lurid disaster seems in one sense unprecipitated. Critics are tempted to reproach the author for presenting an unprepared climax. Yet it is the very unexpected nature of the double murder and suicide that gives the reader an insight into the nature of the enigmatic child. Because he is unable to communicate and because his inner existence remains impenetrable to the adult intelligence, his actions are totally unpredictable. And yet his unexpected crime is no *acte gratuit*, even though the motivation of the child remains unclear. Nor does it suffice to say that the carnage he wreaks simply represents a revolt either against the human condition or against a society that is based on injustice. The doctor who had attempted unsuccessfully to resuscitate the three little corpses does make an effort at a reasonable explanation:

The doctor says there are such boys springing up amongst us — boys of a sort unknown in the last generation — the outcome of new views of life. They seem to see all its terrors before they are old enough to have staying power to resist them. He says it is the beginning of the coming universal wish not to live.[11]

Hardy himself offers yet another interpretation:

On that little shape had converged all the inauspiciousness and shadow which had darkened the first union of Jude, and all the accidents, mistakes, fears, errors of the last. He was their nodal point, their focus, their expression in a single term. For the rashness of those parents he had groaned, for their ill assortment he had quaked, and for the misfortunes of these he had died.[12]

Perhaps Little Father Time's own poignant justification is, in its simplicity, closest to the truth. Before hanging himself, he had scrawled in pencil on a piece of paper: "*Done because we are too menny.*"[13] In other words, Little Father Time, like Sartre's Roquentin, is conscious of the superfluity of his presence on earth. His own existence, as well as that of others, is *de trop*, and thus the simplest expedient is to rid the world of some of these excessive and suffering beings. Despite these attempts at rationalization, the fact of the matter remains that the enigmatic child represents a menace to human life and to human institutions, because his is a perception of them that is radically different, and thus his reactions to them appear unmotivated, but only because they are for adults unpredictable. The enigmatic child has the uncanny knack of seeing life as it is, stripped of all human pretenses and conventions. Since man cannot stand too much reality, the unwavering stare of the child which reveals it in a stark and unadorned fashion is a serious threat. It is Sue who sensed all of this when she saw Little Father Time first and compared his tragic face to the mask of Melpomene.

The fictional universe of Kafka is not one usually associated with children, and the very atmosphere of sordid inns and wretched tenements that serve as thresholds to labyrinthian bureaucracies seems hostile to the innocence normally associated with childhood. And yet there are children everywhere in the fragmentary novels and short stories of Kafka. The apartments of the buildings in which the offices of the tribunal are lodged teem with them, and they are omnipresent in the hovels of the village that lies at the foot of the Castle. Because of its complexity, the symbolic import of these children is extremely difficult to assess. At the very least, they represent an annoying intrusion upon personal life: their presence makes it impossible, in *The Castle*, for the land surveyor and for Frieda to settle down to a domestic life in the schoolhouse. As for K's two assistants (who are repeatedly portrayed as children), their destructive and irresponsible pranks are a constant hindrance. At the same time, children represent something far more difficult to define. When for the first time K attempts to communicate with the Castle, the receiver of the telephone emits a buzzing of a type K had never heard before. The eerie hum is compounded of the echoes of children singing at an infinite distance, and K is literally entranced by their incomprehensible voices, and is enraged when the sound is interrupted by the comprehensible words of adults. The children's murmuring is here equated with the song of sirens, a song that is perhaps responsible for luring him into a continuation of his hopeless quest. Even more difficult to interpret is the moribund little boy in the short story 'A Country Doctor.' His gaping wound is alive with writhing worms: it is an unanswerable challenge to the doctor's healing powers and its youthful victim is a threat to the physician's very existence. All of these children, whether seductive or inimical, are, thanks to their hermetic nature, seemingly sinister forces.

In *The Trial* we find the same basic hostility between the child and the adult world of accepted social conventions as in *Jude the Obscure*, only in this instance the resultant conflict is stripped of all melodrama and presented with a stark realism that accentuates the nightmare that Kafka's protagonist attempts unsuccessfully to elucidate. The initial encounter between Joseph K and a group of children takes place when for the first time he is attempting to obtain a hearing before the tribunal. He has located the building in which the courtroom is presumably located and where the interrogation is to take place, but he is at a loss as to which of the many entries of the complex structure might lead him to the proper room. He picks a door at random and begins to go up the staircase. His way is temporarily blocked by a horde of children who say nothing to him:

In going up, he disturbed a lot of children who where playing on the stairs and who looked up at him with anger as he strode through their ranks. He said to himself, "The next time I come here again, I must bring with me either candy to win them over or a cane to beat them with." Just before reaching the second floor he even had to wait a moment, until a playing marble had completed its rolling course; in the meanwhile two little boys with the queer faces of grown up vagabonds held him fast by the legs of his trousers; had he wanted to shake them off, he would have had to hurt them, and he was afraid of their cries.[14]

Here again we have taciturn boys, older than their years, and they inspire fear in Joseph K as they hinder his progress. They seem to be sinister agents of obstruction, the sullen guardians of the portals to the antechambers of the Law.

Joseph K's other encounter wth children again takes place on a staircase. This time he is looking for the studio of the painter Titorelli who, so he hopes, might give him some useful information concerning the minor judges whose portraits he paints. There are obvious similarities between the group of girls he meets this time and the boys who had previously hindered his passage. The girls, too, are an unprepossessing lot. Their leader is slightly hunchbacked, and they are all described as being corrupted despite their youth: "All their faces represented a blend of childishness and depravity."[15] They, too, are initially taciturn, and, when K asks them whether they know where the painter Titorelli lives, they stare at him "with a piercing, provocative gaze."[16] But the subsequent attitude of the girls toward K is very different from that of the boys. Instead of making it difficult for him to proceed, it is really only thanks to them that he finds his way. As he is about to take a wrong turn, their leader actually points out the right corridor to him. In fact, as he approaches, they make a concerted effort to speed him on his way: "The hunchbacked girl clapped her hands, and the rest of them crowded behind K in order to push him forward more quickly."[17] It is as if these strange girls were intent on pushing him more rapidly toward his doom. Titorelli, during the course of his interview with K, had to lock them out of his atelier. Nonetheless, their presence is constantly felt, and their ceaseless chatter behind the door accompanies the conversation of the two men. Furthermore, they seek to attract the attention of the interlocutors by all sorts of distracting means — one of them, for example, pushes a straw through the crevisse between the door and the sill and moves it back and forth with maddening regularity. K, who is having enough trouble as it is concentrating on the intricate meanderings of Titorelli, becomes even more bewildered because of their antics and finds it increasingly difficult to follow the discourse of the painter. But when K inquires about the girls, Titorelli leaves no doubt in K's mind concerning their allegiance. They belong to the tribunal, he informs him, and this revelation serves to make K even more uneasy. In fact, he is so afraid of encountering them again that he finally asks his host to be allowed to leave by a different door in the back of the room. As he passes through the back entrance, he finds himself in the corridors of the offices of the tribunal — and at once runs into the same bevy of girls whom he had hoped to avoid. The conclusion is clear. The enigmatic children in *The Trial* represent a menace; they belong to the tribunal. If they seek to hinder the progress of the protagonist, it is in order to make his existence more difficult. If they seek to further his progress. It is to help him on the way to his execution. Whichever of these two courses they pursue, their mere presence, whether in its sinister taciturnity or in its irritating loquacity, is a constant mockery and torment.

The enigmatic child appears as a beneficent element just as frequently as in the guise of a menacing presence. The concept of the child as a symbol of resurrection predates Christianity and is already contained in the famous prophecy of the birth of Immanuel, the child nourished on butter and honey and who will reject evil in order to choose the good (Isaiah, 7:15). Virgil sees in the figure of a baby boy the herald of a Golden Age. One of the most striking portraits of the enigmatic child as a redeemer is that of Joas (or Eliacin as he is called until his identity is revealed) in Racine's biblical tragedy, *Athalie* (1691). Upon the death of her son, Athalie, in order to assure herself of the throne, follows the blood-thirsty tradition of her family and stabs all of her grandchildren. Unknown to her, one of the infants had survived this carnage and was raised by the high priest and his wife. The first encounter — a silent one — between the murderous queen, now a worshipper of Baal, and Joas takes place when he is nine years old. She sees him by chance in the temple that she had penetrated in order to confront the high priest. The aging queen recognizes him, but not as her offspring. For her he is the incarnation of the apparition of the gentle and modest boy who, in a dream, had come to assassinate her. The sight of the real child seems to paralyze her. Although her mouth is open to speak, she can utter not a sound; her wild and terrified eyes cannot turn away from the child. From this mute scene there emanates a sense of horror that can be attributed to the chilling presence of the incomprehensible child.

Although words are exchanged during the entire course of their second meeting, it is equally disorienting for Athalie. Hoping to elucidate the mystery of the child, she has summoned him in order to interrogate him. She is convinced that the child cannot help but reveal the truth. Indeed, Joas's curt and simple responses provide answers to everything; at the same time they reveal nothing. Just a few examples of these questions and answers suffice to demonstrate how candor only deepens the mystery:

ATHALIE: What are you called?
JOAS: My name is Eliacin.
ATHALIE: Your Father?
JOAS: I am, so they say, an orphan.
ATHALIE: You have no parents?
JOAS: They abandoned me.
ATHALIE: How and since when?
JOAS: Since I was born.

(II, vii, 633–38)

As Athalie probes deeper into his existence, his replies, while losing nothing in simplicity, become more ambiguous and seem designed to wound her. For example, when she wishes to know what he is learning he paraphrases the law — but precisely that part of it that deals with the punishment of homicides. Whether such parries are the result of the calculated cruelty of a child seeking to draw the blood of an aging woman or of the spontaneity of an innocent boy who does not know what he is saying is a question that remains forever unanswered. Joas is as much an enigma for us as for Athalie.

From the very onset, Joas represents a menace for Athalie. His mere presence suffices to sap the willpower of the queen and to transform her from a calculating and decisive tyrant into a wavering and sentimental old woman. He is her nemesis, and when, at the end of the play, his pathetic and broken grandmother is taken away to be executed, he stands by silent. Whatever he might represent for Athalie, this child is the symbol of salvation for the Hebrews. When the play opens, they have practically deserted the Temple and lost all hope. They are a "broken people" who have lost their faith and no longer have the courage to oppose the military might of Athalie. In the last act, when the true identity of the child Joas is revealed to them and when he is crowned their king, the children of Israel are immediately reanimated, they find their former zeal and put the armies of Baal to route.

The theme of the enigmatic child is a complex one, and if there is any writer who has done full justice to its richness it is Thomas Mann. His early novella *Tristan* already contains all of the later elements of his masterpieces: the interrelationship of art and disease, the dangers of music, and the confrontation between the practical world and the aesthetic one. Even the cadre of the story, the sanatorium, is one that the author was to develop later in *The Magic Mountain*. The realm of art in *Tristan* is represented by the novelist Spinell who persuades his fellow inmate Gabriele Klöterjahn, the wife of a businessman, to play the piano for him. She yields to this temptation, and her desperate plunge into the world of romantic music is the equivalent of an act of treason against the world of commerce personified by her husband. It is more than that. It represents an act of adultery and finally suicide, a sort of Wagnerian love death. Her betrayal has profound consequences not only on the symbolic but on the realistic plane: the emotional upheaval caused by the intensity of the music is too much for her frail constitution. It leads to a serious deterioration of her already fragile health and eventually to her death. And just prior to her passing away, it results in a violent clash between two antithetical ideologies. This confrontation is in the form of a bitter debate between her husband and Spinell, an argument between an ineffectual visionary and a materialistic activist that is interrupted by the announcement of the demise of Gabriele Klöterjahn and thus left temporarily in suspension. The resolution comes only at the end of the story, in the form of a dramatic encounter between Spinell and the now motherless Klöterjahn baby who is being wheeled in his pram by his nurse. There is something insolent in the appearance of this extraordinarily healthy infant who is described as "plump ... with chubby cheeks, superb and well-proportioned."[18] And the eyes of the child that meet those of the novelist are "cheerful and infallible." As Spinell is about to avert his gaze, something "horrendous" occurs. The child begins to laugh triumphantly, and this inexplicable, jubilant screeching is sinister. Spinell becomes even more uneasy when young Anton Klöterjahn shakes his rattle and his pacifier at him as if he wanted to mock him and to drive him off. It is then that Spinell turns on his heels and walks off " ... with the powerfully hesitant steps of someone who wants to hide the fact that inwardly he is running away."[19] It is with these words, describing the controlled panic of flight, that *Tristan* ends. Spinell, whose aestheticism is allied with sickness and death, has encountered the future in the form of a grotesquely healthy and incomprehensible child.

Gustav Aschenbach, the author-protagonist of *Death in Venice* (1912), is the very antithesis of Spinell, just as the child who tempts him is the very opposite of young Klöterjahn. Aschenbach is a successful writer, and his prose is "lucid and powerful."[20] His "sober conscienciousness" had brought into check the "darker, fiery impulses." His is an art that has by force of will overcome any decadent tendencies. He is not, like Spinell, a tempter; he is the one who is tempted and succumbs. The story of the last months of his life is the account of a progressive self-abandonment, the degenerative movement from a Spartan existence to the disintegration caused by self-indulgence. His nemesis is Tadzio, the Polish boy whom he encounters but never meets in Venice. The youth is the incarnation of physical perfection, of godlike beauty. There is something supernatural in his "expression of fair and divine earnestness."[21] He is clearly a being apart. From the very first, when Tadzio turns to gaze at him in the lobby of their hotel, Aschenbach is fascinated. Already in the early stages of his infatuation, the presence of the child makes Aschenbach conscious of his own decline. As he looks in the mirror, only moments after his initial encounter with Tadzio, he for the first time becomes aware of his gray hair, of his flabby and tired face. The more his mind is occupied by the vision of Tadzio, the more a mortal lethargy seems to overcome the once vigorous writer. Simultaneously, his thoughts turn to illness, and he seems to detect in Tadzio a sickly element, an unnatural, ivorylike pallor. It is at this time, too, that he realizes how deleterious the stagnant air of Venice is for his own health. However, his own halfhearted attempt at flight is aborted, more because of his own lack of volition than because of the misdispatched trunk that he uses as a convenient excuse to prolong his stay. Several weeks later, Aschenbach is irremediably overcome, when the child secretively looks at him with a mysterious smile that the aging writer accepts as a "fearful gift." It is immediately thereafter that Aschenbach obtains the first intimations of the mortal sickness that is infecting all of Venice. But the potential cholera epidemic that frightens off most of the tourists exerts a dreadful fascination on him, a fascination that is closely linked with the seductive influence of Tadzio's existence. Aschenbach's obsession is a compound of "fear and ecstasy and a terrified curiosity."[22] This unhealthy preoccupation with sickness and love leads to the desperate attempt on the part of Aschenbach to make himself more attractive through the pathetic recourse to cosmetics. On the day of Tadzio's planned departure, Aschenbach, seated in his chair on the beach, watches Tadzio for the last time as he wades into the sea. He thinks that Tadzio is beckoning to him and he wants to follow him toward the "awful region full of promise"[23] that Tadzio's graceful gesture indicates. But it is during this vision, as he gives himself over completely to the enchantment that Tadzio exercizes over him, that death overtakes him. The encounter with the enigmatic Tadzio had been a fleeting one, with not a word ever exchanged. It is the encounter with Eros that had come too late, or perhaps the revenge of Eros on the writer who had tried to deny him.

Anton Klöterjahn is the frightful harbinger of a new world order inimical to aesthetic values; Tadzio is the ultimate incarnation of the type of child who in *The Hymns of the Night* had promised salvation through death. Nepomuk Schneidewein, in *Dr. Faustus* (1947), is as enigmatic as any of his predecessors. He is also quite obviously a symbol of possible redemption

THE ENIGMATIC CHILD IN LITERATURE 259

through innocence, and far less obviously the personification of a future that is more uncertain and that lies beyond the one announced by young Klöterjahn. The symbolic import of this subject is further complicated by the fact that in his case innocence proves impotent when confronted with the powers of the demonic. The young Nepomuk is doomed from the outset to be destroyed by the forces of evil with whom his uncle, Adrian Leverkühn, had made such a fateful pact. Nepomuk arrives at the country home of the composer after a series of catastrophes had culminated in the disintegration of Leverkühn's marriage plans and in the melodramatic death of his closest friend, Rudolph Schwerdtfeger. The life of the musician, plagued by ever-worsening migraine headaches, had reached a low point just before his five-year-old nephew came to live with him. There are, in the explanation of the child's arrival, unmistakable reminiscences of *Tristan*. Nepomuk's mother, like Gabriele Klöterjahn, had been seriously weakened by her pregnancy. She, too, suffers from a recurrent lung disease, and it is her confinement to a sanatorium that makes it necessary to find a temporary home for her youngest child. Nepomuk's appearance at Pfeiffering has an immediate impact, and everyone, including the menials and the farmers greet him with delight and with cries of astonishment. There is something of the miracle in his advent. There is also something "magical in his restrained smile that gave promise of a precious lesson and message."[24] The Klöterjahn baby had not yet attained the powers of speech, and Tadzio conversed only in Polish which Aschenbach could not comprehend. The enigma of Nepomuk is heightened by the strange speech he employs. It is accented and contains many colloquialisms and unusual dialectical expressions, as well as archaic words from a long-forgotten language. This unique composite lingo makes him difficult to comprehend. As if aware of his very special nature, he refers to himself always in the third person, and usually as "Echo." This usage adds a further element of artificiality to his language. But the overall effect is an enchanting one, and no one is immune to the charm of this ethereal child. Only the pedagogically inclined narrator, Zeitblom, momentarily attempts to deal with him as he would with one of his ordinary pupils. But when he speaks rather gruffly and condescendingly to Nepomuk, the child attempts to suppress his laughter. There is a confrontation between candor and pomposity that makes the adult feel extremely foolish. So Zeitblom, sensitive to the absurdity of his attitude, gives up his halfhearted attempts at sternness and recognizes Nepomuk for what he is, "a little emissary from the realm of children and elves."[25]

While Zeitblom was fully conscious of the extraordinary quality of this child, he was also aware of the transient nature of the enchantment embodied by him. As an educator, Zeitblom was concerned because he knew what the effects of time would be on Nepomuk. He was, like all mortals, destined to mature and to fall prey to the earthly. The innocence of his blue eyes would become clouded by impurity and it would be all too soon before the poetry of his being would be transmogrified into prose, before his elfine essence would take on the form of a more or less ordinary boy. Even the down to earth Zeitblom revolts against the tyranny of the temporal and tries to justify his refusal to accept the brutal power of time over "this blessed apparition" by an explanation of what Nepomuk represents:

> ... his rare quality of being self-enclosed, his validity as a representative of *the child* on earth, the feeling he inspired of having descended from on high (and I repeat myself) in order to deliver a precious message.... His character could not deny the inevitability of growth, but took refuge in an imaginary sphere of the mythic-atemporal, of simultaneity and existence in a continuum in which the adult form of man is not in contradiction with the child in the arms of the mother.[26]

The presence of Nepomuk has transformed the practical and sober Zeitblom into an enthusiast.

It is on Leverkühn that Nepomuk exerts the greatest influence. The in many respects inhuman composer becomes an adoring uncle who tells the child stories, buys him toys, lets him look at his work, and with touching admiration repeats to his acquaintances everyone of his infantile sayings. Nepomuk's sojourn represents for Leverkühn an interlude, a charming domestic respite within his own tormented existence. However, there is far more to Leverkühn's feelings than the pure enjoyment of the familial idyll so dear to Hugo. The man who knows he is cursed is especially fond of those moments when Nepomuk says his evening prayers. After one such session, he and Zeitblom analyze the child's orison:

> "You are right." I replied. "He keeps his prayers in the realm of the selfless because he does not plead merely for himself, but rather for all of us."
> "Yes, for all of us," said Adrian quietly.[27]

Leverkühn sees in Nepomuk a figure who might save him from damnation, who might absolve him of his pact with the devil. But Leverkühn's hopes are illusory, and it is he who will be the cause of the destruction of Nepomuk.

Nepomuk, after having captured the hearts of all those who surround him, falls victim to cerebral meningitis. The delicacy of his physical system is no match for the infectious venereal disease that is ravaging his uncle's body, just as the fragility of his spiritual system is incapable of resisting the demonic powers that Leverkühn had purchased from the devil at the price of his soul. So Nepomuk must die, and Leverkühn is perfectly aware of his responsibility for this death. To Zeitblom he explains: "'... what guilt, what a sin, what a crime ... to have let him come here, to have left him in my vicinity, to have let me feast my eyes on him! You must know that children are delicate creatures, all too susceptible to poisonous influences.'"[28] Shortly after Nepomuk's death, Leverkühn plunges into the composition of his last work, the monstrous and despairing "Lamentation of Dr. Faustus," a composition whose fundamental theme is constructed out of the letters forming the name of the woman who had infected him. The first audition of this apocalyptic oratorium is a disaster, and soon thereafter Leverkühn lapses into the imbecility in which he lives out the remainder of his days on earth. In an ironic counterpoint typical of the entire novel, Zeitblom's depiction of Nepomuk's death is contemporaneous with the Wagnerian disintegration of the Third Reich.

Music is the key to everything in *Dr. Faustus*, and it is Zeitblom's description of the chorus in Leverkühn's "Apocalypsis cum Figuris," the last work the composer wrote before hsi encounter with Nepomuk, that helps to explain the role of the enigmatic child in the novel:

The infernal laughter with which the first part ends has its counterpart in the wondrously supernatural children's chorus ... that opens the second part, a piece of cosmic music of the spheres, icy, clear, transparent as glass, astringently dissonant and yet imbued with a tonal sweetness that, so to speak, is unattainable and supernatural, foreign, filling the heart with a longing without hope.[29]

When Zeitblom finds truly astonishing in this children's chorus is that it is formed out of exactly the same notes as the laughter that had arisen from hell. The angelic tones are those of Satan, only they have been orchestrated differently and given a new rhythm. The enigmatic child in Mann's masterpiece is the ultimate in complexity and ambiguity. The angelic being is of the same material as his infernal counterpart. Although unable to stand up to the demonic, he is, even in his death, the harbinger of the end of the epoch of bourgeois humanism and the herald of the advent of a new stage in life, a stage "when a mutation is taking place during which the world will find itself under a new, still nameless constellation."[30] After the era of art promised by Tadzio's gesture and after the age of commercialism announced by the Klöterjahn baby will come the unknown and unnamed epoch heralded by Nepomuk.

The enigmatic child, as he appears in literature from antiquity to modern times, assumes a multiplicity of forms. His presence may be threatening or beneficent, or neither of these. He may, like Joas in *Athalie*, play the role of intercessor or, like Little Father Time in *Jude the Obscure*, of destroyer, or, as in Gide's *Voyage of Urien*, he may play no real role at all. Despite all of the seemingly disparate incarnations of this protean figure, the various personages have so many traits in common that it is possible to speak of the enigmatic child in generic terms. First of all, this child's universe represents a self-enclosed, nonreferential system. Both Chaucer's child and Mann's Nepomuk are strangers to the world and are independent of it. Secondly, communication between the child and the adult is virtually impossible. This is symbolized by the linguistic deficiency of so many of these children. The boy in "The Tale of the Prioress" does not even comprehend the words he sings, the boy in Goethe's *Novella* must have recourse to the flute, and Tadzio, in *Death in Venice* speaks a language incomprehensible to Aschenbach. Thirdly, despite his taciturnity, the enigmatic child has a message to deliver; in this respect the boy in Beckett's *Waiting for Godot* is typical. He is an emissary, even if we do not know from where or from whom. It is here that we see the fundamental paradox of the enigmatic child: although a message-bearer, he is inarticulate, or at least incapable of making himself understood. Fourthly, these children, of whom James's Maisie is a prototype, are all wise beyond their years, and thus present a combination of youth and age that adds to their mystery. Finally, these children are all doomed, for while their being is atemporal, they are subject to the vicissitudes of time. In summary, the enigmatic child is a stranger to this world, sufficient unto himself, incapable of communication and yet the bearer of important tidings. Thus it is his presence that is significant, even if we can never decipher its meaning.

The constant reemergence of the enigmatic child in literatures of various periods, nations, and genres seems to indicate that it is an archetypal theme deeply embedded in the psyche of the human race. Seeming to slumber in the human consciousness, the child awakens periodically to make his appearance in works of art and literature. Like a somnambulist, he traverses our horizon briefly to disappear again in the void from which he had emerged. His transient appearance is a disturbing one, because the enigmatic child is uncomfortably similar to all children, even the most ordinary. He is a composite of the elements which define the essence of all those who have not yet reached maturity. For to a certain degree all children are autonomous universes; to some extent all of them speak their own language and seem to have a message to convey that they forget just as soon as they are old enough to transmit it.

Without overemphasizing the Jungian implications of the preceding interpretation, one could say that the theme of the enigmatic child seems to respond to a deeply felt universal need that is indefinable in anything but symbolic or poetic terms. It is possible to see in this theme the expression of a combination of nostalgia for the past and a longing for a new and transcendental future. Out of this amalgam there emerges the verbalization of a far more complex aspiration, of the desire to suspend time, that implacable enemy of childhood. And so there is inherent in the existence of the enigmatic child a profound tragedy, for on the one hand he is eternal and on the other hand constantly menaced by extinction. Like Nepomuk, he must die, for if he did not he would grow up to be an ordinary mortal. Like Saint-Exupery's Little Prince, he cannot prolong for too long his sojourn in a world that is not his natural habitat. He must return to the celestial regions from whence he came, his message unspoken or misunderstood and leaving behind the adults to their hopeless yearning after innocence.

NOTES

[1] André Gide, *Romans* (Paris: Gallimard, 1958), p. 31.
[2] Henry James, *What Maisie Knew* (Garden City, N.Y.: Doubleday, 1954), p. 280.
[3] J. W. von Goethe, *Werke* (Stuttgart: Union Deutsche Verlagsgesellschaft, 1904), XIV, 188.
[4] Ibid., p. 195.
[5] Thomas Hardy, *Jude the Obscure* (New York: Airmont, 1966), p. 217.
[6] Ibid., pp. 217f.
[7] Ibid., pp. 218f.
[8] Ibid., p. 227.
[9] Ibid., p. 229.
[10] Ibid., p. 234.
[11] Ibid., p. 264.
[12] Ibid.
[13] Ibid.
[14] Franz Kafka, *Der Prozess* (Berlin: Schoken, 1955), p. 49.
[15] Ibid., p. 171.
[16] Ibid., p. 170.
[17] Ibid., p. 172.
[18] Thomas Mann, *Gesammelte Werke* (Frankfurt: S. Fischer, 1974), VIII, 262.
[19] Ibid.
[20] Ibid., p. 450.

[21] Ibid., p. 469.
[22] Ibid., p. 516.
[23] Ibid., p. 525.
[24] Ibid., VI, 611–12.
[25] Ibid., p. 618.
[26] Ibid., p. 619.
[27] Ibid., p. 626.
[28] Ibid., p. 633.
[29] Ibid., pp. 502–3.
[30] Ibid., p. 469.

PART III

THE GIFT OF NATURE: MAN AND THE LITERARY WORK OF ART

JEFFNER ALLEN

HOMECOMING IN HEIDEGGER AND HEBEL

The movement of homecoming, which is motivated by a friendship with the "friend of the house,"[1] appears as one of the fundamental movements of human existence in both the writings of Johann Peter Hebel (1760–1826) and in the essay 'Hebel — der Hausfreund' by Martin Heidegger. To better understand the nature of this primordial movement, we will offer a Heideggerian elucidation of those fundamental images of the human being which come to light within the context of such a friendship.

Our elucidation may indirectly serve to indicate a new direction for studies of both Heidegger and Hebel, for it will point to Heidegger's understanding of the role of the inaugural language of dialect. It will also point to a generally neglected unity that pervades all of Hebel's writings — his Alemannic poems, his almanacs, and even his Bible stories — all of which may be viewed as the work of the "friend of the house," or poet. However, what may perhaps be more significant is that our elucidation may elicit on our part a desire for friendship with the *Hausfreund*, and may thereby motivate us to reflect upon our own movement of homecoming.

Accordingly, with the intent of making more transparent the nature of homecoming, we will examine the fundamental images of the human being (1) which are disclosed by Hebel the *Hausfreund*, (2) which are endangered and hidden by the essence of the modern age, in which the *Hausfreund* is absent, and (3) which may be revealed and reappropriated by a new relatedness to the *Hausfreund*.

I. THE FRIEND OF THE HOUSE

Hebel, the "friend of the house," names the deeper relations of human existence, and does so especially in the following: "We are plants which — whether we like to admit it to ourselves or not — must with our roots rise out of the earth in order to bloom in the ether and to bear fruit."[2] It is through such a naming that the *Hausfreund* becomes who he is, and it is through an appropriation of what is named here that we can become what we are. Yet, what is being named?

If we take Hebel's statement as our guide, we may turn first to the image of the earth. In so doing, we may view our own becoming as intertwined with that of the earth in which we are rooted and from which we arise, for the earth houses us. The earth bestows on us that space in which our historical being is founded and unfolds — in which we may come to be "at home."[3]

For Hebel, the earth is the sensuous domain of nature, of all that we can see, touch, and, as embodied beings, experience. It is also the marketplace of the world, within which everyday adventures and misadventures, noble deeds and feats of craft and cunning, transpire.

In that the earth needs the *Hausfreund* and the *Hausfreund* needs the earth, their becoming is co-originary and inseparable. That is, Hebel can be the *"earthly* friend of the house"[4] only by caring for the earth; the earth can maintain its earthliness only by sustaining, and being sustained by, the *Hausfreund*. The earth is entrusted to the *Hausfreund* who, although often hidden and unseen, illuminates what is essential on the earth.[5] Such illumination, which arises through Hebel's language and writings, is not like that of a glaring searchlight that forcibly intrudes into every hidden crook and cranny. Rather, it is a soft light[6] that reveals our earthly roots while, at the same time, allowing them the darkness and concealment needed so that they can thrive. Nor is such a light of the earthly *Hausfreund's* own self-willed making. It is, on the contrary, but a reflection of the light of the moon.

As we turn from the earth, the sensuous realm that is illuminated by Hebel, the "earthly friend of the house," to the image of the sky, we can see that our being is also inseparable from that of the sky. The sky, which is, at least in part, the abode of the moon and the moon's light, is that realm in which we flourish.

For Hebel, the sky is the nonsensuous, free and open, realm of everything that pertains to the mind and spirit. The sky of nature also forms the holy heavens of the gods, angels, and other spirits.

The sky and the "earthly *Hausfreund*" are together, for the light that Hebel receives and passes on first radiates from the sky. The light flows, in particular, from the moon, which is the *"true* friend of the house."[7] Hebel receives such an influx of light as the moon marks off the seasons, thereby serving as the "first" maker of almanacs[8] and educating Hebel as the "earthly" maker of almanacs. Like those poets

> Whom no master alone, whom she, wonderfully
> All-present, educates in a light embrace
> The powerful, divinely beautiful nature.[9]

Hebel is educated by the moon of the evening skies. The moon, which governs the seasons and the tides, the times for sowing and the times for harvesting, is present in everything that Hebel writes about. The "light embrace" with which its quiet light surrounds Hebel is not impotent, but rather has and bestows a far-reaching power of its own. Moreover, Hebel also stands under the soft, though scarcely felt light when it emanates from the moon which, as the "highest head night watchman,"[10] guards the sleep of mortals. Like Hebel's light, the moon's light is not its own. It is, instead, a reflection of the light of the sun.[11]

Thus, we are beings who, somewhat Neoplatonically, share in the light of the "earthly *Hausfreund*," who receives and reflects the overflow of light from the moon, which indirectly mirrors and thereby participates in the light of the sun. Yet, how may we think the image of the open space that ranges from the depths of our roots in the earth to the highest heights of heavenly ether?

The realm of this "between" may be said to be measured by language. It is language that keeps open the *Spielraum*,[12] the region of interplay, between the earth and the sky. Moreover, if we wish to determine what we mean here by "language," we may ask: What is special about Hebel's naming of the fundamental relations of human existence?

We may first note that the language of Hebel's naming is not just any everyday language, or idle chatter; rather, Hebel's language is frequently the inaugural language of dialect, and even his high German is the "echo" of his dialect.[13] Hebel's writings clearly show that dialect is not an inferior form of discourse that speaks merely of its own limited world. Quite to the contrary, it is rooted in the homesoil[14] and, in virtue of its rootedness in what is most originary, is universal in its import. Dialect is not only Hebel's mother tongue; it is, even more fundamentally, the mother of language.[15] Since Hebel's language is rooted in the all-encompassing dialect of the homeland, it is near to what may be said to be concealed by the "spirit of language" (*Sprachgeist*), namely, ". . . the inconspicuous but sustaining relations to God, to the world, to men and their works."[16] In fact, it is because Hebel's language shelters and conceals within itself the fundamental relations of human existence that Hebel's naming is able to keep open the play-space of everything that is essential, though unsaid.

Hebel's inaugural language is also that of the poet. His language distinguishes him from those poets of somewhat narrower vision who use language only once it has become highly structured and formalized. Such poets are like those who "study established forms" and, in consequence, "can only repeat the once real spirit, for they place themselves like birds on a branch of the tree of language and rock themselves gently on it, according to the primordial rhythm which lies in its roots."[17] In contrast, Hebel's kinship with the more fluid and originary language of dialect makes him a poet whose poetizing has a more fundamental character. His inaugural language enables him to encompass all that is present, from the deepest roots of the "tree of language" up to its highest reaches or, in other words, to "soar like a spirited eagle hatched by the living spirit of language."[18] Furthermore, Hebel the poet gathers together the earth and sky by both revealing them and preserving them in their concealedness.[19] As a poet, he "preaches" — announces, makes known, and praises[20] — the open space in which earth and sky, gods and mortals, belong together. By virtue of his poethood, the "friend of the house" stands in the "between," in the center that mediates the fourfold of earth, sky, gods, and men. Needed and used by the round dance of this fourfold, Hebel's poetic language and images illuminate the very essence of human existence and thereby offer to the mortals the possibility of understanding their place in the cosmos.

Nevertheless, although the realm of the "between" is measured by Hebel's poetic dialect, language is not the invention of the "earthly friend of the house." Just as the *Hausfreund's* light radiates from the moon, so too, language speaks to the *Hausfreund*.[21] Hebel, the "friend of the house," conveys this saying by speaking to us, and it is for us, in turn, to listen to the *Hausfreund* and to co-respond to his saying.

As we listen to the "friend of the house," we find that he is naming a certain autochthony (*Bodenständigkeit*), or rootedness in the earth. Or, in other words, Hebel's saying names what we may call a "poetic dwelling," or the primordial "way in which men carry out their journey on the earth, under the sky, from birth to death."[22] It names the nature of our abode in the house of which Hebel is the friend, namely, the "house of the world."[23] By thinking such a poetic dwelling, we may enter into friendship with the poet, or *Hausfreund*, and thereby begin to participate in the movement of homecoming.

II. THE ABSENCE OF THE FRIEND OF THE HOUSE

In what sense does Hebel's view of human existence hold true today? Perhaps only in a negative sense, as that which we are not. Today, the "friend of the house" is missing.[24] But perhaps we have not even noticed the *Hausfreund*'s absence?

Symptoms of the *Hausfreund*'s absence may be found everywhere. Earth and sky are, at present, held under the domination of the essence of modern technology. Language, which we have attempted to control by calculative thinking, has become a mere instrument for the communication of factual information. Language is no longer poetic, for it is severed from its source in dialect and the homeland. Insofar as it is measured by uniform standardized language, the open play-space of the fourfold has become closed. The house, whose meaning has been leveled down by our mechanical existence and "everydayness," has become a receptacle, devoid of meaning, in which we subsist. The houses in which we live are empty, and thus, we are homeless. With Hebel, nature as experienced in poetic dwelling amid the rising and setting of the sun and moon, and nature as understood by the Enlightenment's laws of modern natural science, were two aspects of the same. In contrast, whenever we briefly catch sight of the "naturalness of nature" (*die Natürlichkeit der Natur*),[25] we find it to be distant from, and at war with, the nature of natural science. The naturalness of nature has been virtually eclipsed by a scientifically calculable and technologically masterable "nature."

When we look further we find that all of these symptoms are lodged in the figure of nihilism. It is nihilism, not the *Hausfreund*, who presently stands at the door: "Nihilism stands at the door: whence comes this uncanniest [*unheimlichsten*] of all guests?"[26] Nihilism is the strangest, or uncanniest of all guests, for it wants to stand at the door of our homelessness, it wants homelessness as such. Should we wish to rid ourselves of this unwelcome guest we would immediately find ourselves unable to do so, for nihilism is rooted in our modern world view, that is, in our ever-present Christian-moral metaphysics. In place of the *Hausfreund* and his almanac of astronomical, scientific, and historical information, of poems, short stories, and jokes, which was frequently read nightly in the company of family and friends, we find the television. In fact, we may find ourselves glued to the television set and caught up in the antics of the nightly weatherman not because the television is truly our friend, but rather because we have no friends.

If we penetrate to the heart of the symptoms that crowd around the figure of nihilism, we find that they arise from a common source: the essence of technology, *Gestell*. Here, everything is dominated by *Gestell* and is converted into a fund (*Bestand*) that we seemingly create, control, and use until it is exhausted. Within the context of such a state of affairs there also arises the complete domination of calculative thinking over meditative thinking and, accordingly, the domination of the poet by the spirit of technique.[27]

It is as a result of such a domination that, in the present age, we seek our *Boden-ständigkeit*, our rootedness in the soil or ground, by means of a calculative thinking that seeks the rational ground (*Grund*), the reason "why," for each of our actions.[28] In that our most fundamental relations of human existence are now reduced to those which are most accessible to calculative thinking, we have come to view ourselves as a fund of energies and skills — human resources — which must be channeled so that they can be used in the most efficient way possible. Now, when we search for our roots, or proper image, we usually first fabricate an abstract idea of ourselves and then look around to determine whether we concretely embody the idea. Amazingly, the idea frequently just happens to fit and thus, to be "correct." As we constantly become more adept at fabricating rational grounds for why we are as we are, the poetic ground and fundamental poetic images of the human being become endangered and hidden; our poetic dwelling in the world is threatened at its very core.

III. HOMECOMING: THE RETURN OF THE FRIEND OF THE HOUSE

Must the *Hausfreund*, who names the fundamental relations of human existence, continue to be absent?

Certainly we cannot return to the country life that is portrayed in Hebel's stories and poems. Nevertheless, perhaps by an enactment of something like Hegel's dialectic, which Hegel himself likens to the fluid and organic growth of the plant,[29] the present ground of calculative thinking, on which we now stand, and whose complete domination endangers our very being, may be sublated (*aufgehoben*).

Such an enactment centers around a reappropriation of our *Bodenständigkeit*, that is, around the movement of homecoming which ensues from our experience of the absence of the *Hausfreund* and our loss of the homeland. In homecoming we preserve and transform, overcome and reunite, the soil of poetic dwelling, whose truth is disclosed by the *Hausfreund*, and our present rational ground, whose truth is posited by calculative thinking. Moreover, in homecoming there is a "change of key,"[30] a leap of thought which "does not abandon its point of departure, but appropriates it in a more original way."[31] The leap is not a mere progression. It can never be bridged over. It is, rather, a genuine leap into a new ground in which we may strike new roots, and from which we may enter into a new relatedness to the *Hausfreund* and to ourselves. More specifically, the leap is a releasement (*Gelassenheit*) toward a ground that is without why. Whereas in the period in which the *Hausfreund* is absent, calculative thinking reigns supreme and nothing is without why,[32] in the leap of homecoming, earth and sky, gods and mortals, are thought meditatively and are without why. Here one may say of nature what Angelus Silesius (1624–1677) says of the rose: "The rose is without why; it blooms because it blooms."[33] Here one may also say that "... man, in the most concealed ground of his being, truly is for the first time only when he is, in his own way, like the rose — without why."[34]

By a Hegelian *Aufhebung*, and a leap into a new form of ground, thinking, and dwelling, we may, perhaps, enter into a homecoming in which we renew in a different manner, and in another age, the truth of Hebel's poetic saying. In fact, there is a saying that already addresses us and that may help to prepare the way for such a homecoming. It speaks to us as the meditative thought of a thinker, Heidegger, who already stands in a friendly relatedness to the poet, the friend of the house of the world:

> ... growing means: to open oneself to the expanse of the sky and, at the same time, to strike root in the darkness of the earth; that everything that is genuine thrives only if man is both in right measure: ready for the appeal of highest heaven and *aufgehoben* into the protection of the sustaining earth.[35]

Heidegger's saying does not address us in order to posit a Platonic and Western differentiation between two separate realms: suprasensuous heaven and sensuous earth. It does not offer a prospectus for, or claim to, a cure for the ills of the modern age. Nor is it the expression of a deep pessimism or a search for a reassuring optimism. Rather, Heidegger's saying suggests a possibility which may lead us into greater danger and/or to a saving. The thinker and friend of the poet cannot help us effect any immediate alteration in the present world condition. In fact, no purely human activity can accomplish this, for "the world is what it is and how it is neither through man nor without man."[36] We may prepare for a possible return of the *Hausfreund*, but we must also remember that even if the *Hausfreund* comes in some way to stand again at our door we may still not have completed our homecoming.

NOTES

[1] Although the word *Hausfreund* is frequently translated "friend of the family," we have chosen to translate it by "friend of the house." Our choice has been based on the fact that, in many cases, the word "family" is not appropriate. For instance, Hebel is the "earthly friend of the house," the moon is the "true friend of the house," and the house itself is the "house of the world." The term *Hausfreund*, which Hebel chose as the name of the almanac he was editing (*Der Rheinländische Hausfreund*), also names the essence of the poet. In all of these instances, the stress is not on the family as a social unit, but rather on the house as the place of a more originary poetic dwelling.

[2] Martin Heidegger, *Discourse on Thinking*, trans. John M. Anderson and E. Hans Freund (New York: Harper & Row, 1966), p. 47. See also, J. P. Hebel, *Werke*, ed. Wilhelm Altwegg (Zürich: Atlantis-Verlag, 1940), III, p. 314.
[3] Heidegger, *Erläuterungen zu Hölderlins Dichtung*, 4th ed. (Frankfurt: Klostermann, 1971), p. 16 (*Elucidations of Hölderlin's Poetry*, trans. Keith Hoeller [University: University of Alabama Press, 1977], forthcoming).
[4] Heidegger, *Hebel-der Hausfreund* (Pfullingen: Neske, 1957), p. 21; hereafter cited as *HHf*.
[5] Ibid., pp. 21, 23.
[6] Ibid., p. 21.
[7] Ibid., p. 19.
[8] Ibid., p. 20. See also, Hebel, *Werke*, I, pp. 326ff.
[9] Heidegger, *Erläuterungen zu Hölderlins Dichtung*, p. 52.
[10] Heidegger, *HHf*, p. 20. See also, Hebel, *Werke*, I, pp. 326ff.
[11] Heidegger, *HHf*, p. 21.
[12] Ibid., p. 38.
[13] Ibid., p. 34.
[14] Heidegger, Unsere Mundart: Die Sprache Johann Peter Hebels,' *Der Lichtgang* 5 (1955), 4.
[15] Heidegger, 'Sprache und Heimat,' *Hebel-Jahrbuch* (1960), p. 28.
[16] Heidegger, *HHf*, p. 10.
[17] Bettina von Arnim, *Sämtliche Werke*, ed. W. Oehlke, II, p. 345, in Heidegger, *Erläuterungen zu Hölderlins Dichtung*, p. 154.
[18] Ibid.
[19] Heidegger, *HHf*, pp. 16, 25.
[20] Ibid., pp. 25, 26.
[21] Ibid., p. 34.
[22] Ibid., p. 17.
[23] Ibid., p. 18.
[24] Ibid., p. 31.
[25] Ibid., p. 28.
[26] Nietzsche, *The Will to Power*, trans. Walter Kaufmann and R. J. Hollingdale (New York: Random House, 1967), p. 7; Heidegger, *The Question of Being*, trans. William Kluback and Jean T. Wilde (New York: Twayne, 1958), p. 37.
[27] Heidegger, 'Der Frage nach der Technik,' in *Vorträge und Aufsätze* (Pfullingen: Neske, 1954).
[28] Heidegger, *Der Satz vom Grund* (Pfullingen: Neske, 1957); hereafter cited as *SG*. See also 'The Principle of Ground,' trans. Keith Hoeller, *Man and World* 7 (1974), 207–22.
[29] G. W. F. Hegel, 'Preface' to the *Phenomenology of Mind*, in *Hegel: Texts and Commentary*, trans. Walter Kaufmann (New York, Doubleday, 1965), p. 8. "The bud disappears as the blossom bursts forth, and one could say that the former is refuted by the latter. In the same way, the fruit declares the blossom to be a false existence of the plant, and the fruit supplants the blossom as the truth of the plant. These forms do not only differ, they also displace each other because they are incompatible. Their fluid nature, however, makes them at the same time, elements of an organic unity in which

they not only do not conflict, but in which one is as necessary as the other; and it is only this equal necessity that constitutes the life of the whole."

[30] Heidegger, *SG*, pp. 94, 95.
[31] Ibid., p. 107.
[32] Ibid., p. 67.
[33] Ibid., p. 68.
[34] Ibid., pp. 72, 73,
[35] Heidegger, 'Der Feldweg,' in *Martin Heidegger: Zum 80. Geburtstag* (Frankfurt: Klostermann, 1969), p. 12.
[36] Heidegger, 'Nur noch ein Gott kann uns retten,' *Der Spiegel*, 31 May 1976, pp. 208, 209, 212.

MARGARET COLLINS WEITZ

PASTORAL PARADOXES

What is one to make of a literary genre that, according to some critics, encompasses works as varied in style and in period as the *Song of Solomon* and *Alice in Wonderland* — a genre, moreover, that has attracted such towering figures in Western literature as Cervantes, Shakespeare, Milton, and Goethe. The variety of works included under the rubric "pastoral" seems limitless: and when one examines works designated pastoral in order to ascertain what one common property these works have, by virtue of which they are termed pastoral, the single, unequivocal quality proves protean indeed — and elusive. Perhaps we have in the pastoral yet another candidate for Wittgenstein's notion of an open concept; that is, one that is perennially open and able to accommodate new works of art which we accept as "pastoral," even if we are unable to spell out definitively the nature of the "pastoral." In order to analyze the pastoral, one must concentrate on a particular literary form: for the pastoral includes poetry, prose, and drama. Accordingly, this study will focus on pastoral prose.

The development of the modern novel was in part made possible by the tentative efforts at psychological exploration of that most pervasive of human emotions — love. And in particular, the pastoral prose tradition of the French Renaissance will be studied. Pastoral fiction flourished in France somewhat after it did so in Italy and Spain: in general, it was an adaptation of these two pastoral prose traditions tempered by the native Gallic contribution. Not only are the same problems of *how* the pastoral developed and what the pastoral *is* found here; literary history and literary theory — at times in concord, at times in conflict — but this corpus of Renaissance French prose has the additional advantage for our study of being the least known and studied of the literatures which contributed to the flowering of the pastoral.

One of the paradoxes to which the title alludes has already been indicated in the suggestion of the difficulties inherent in attempting to ascertain just what constitutes the pastoral. Before pursuing that problem further, let us first examine the critical paradox; that is, the ambivalence of the recent critics who have written about the pastoral, presumably because of some attraction to it, yet whose comments reveal a puzzling ambiguity, at times confusion and even antipathy. Gustave Reynier, in one of the few studies yet devoted to Renaissance French prose fiction, speaks of Montemayor's shepherds who "chantent étérnellement leurs peines sans rien faire pour les guérir."[1] Another French critic, Jules Marsan, gives not only unsympathetic, but at times inaccurate judgments and finds the *Diana* monotonous and empty.[2] C. Wedgwood classes this pastoral novel with the Amadis and Palmerin tales, and refers to *L'Astrée* (1607) as that "asphyxiating pastoral romance from France."[3] In his authoritative study *Pastoral Poetry and Pastoral Drama*, W. W. Greg concludes with this summary: "It remained throughout nerveless and diffuse, and, *in spite of* much incidental beauty, was habitually wanting in interest, except in so far as it *renounced* its pastoral nature."[4]

In other words, the pastoral was only "successful" when it was not pastoral. The refined, polished style which today's critics regret must be seen in part as an attempt to reflect the refined, purified, and idealized passion which the pastoral depicted. But paradoxical attitudes are not restricted to the more recent works. What is generally considered to be one of the finest pastoral poems, Milton's "Lycidas," is condemned in strong terms by Johnson: "In this poem there is no nature, for there is no truth: there is no art, for there is nothing new.... Its form is that of a pastoral, easy, vulgar, and therefore disgusting; whatever images it can supply, are long ago exhausted; and its inherent improbability always forces dissatisfaction on the mind."[5]

It is in these writers of the past several centuries that we find yet another aspect of the pastoral paradox. For in the majority of instances, the very critics of the pastoral were themselves writers of pastorals. Fontenelle perhaps is a good example of this love/hate attitude in the French tradition: he rails against, yet succumbs to, the charms of the pastoral.[6] The best known of those authors whose works reveal this paradoxical attitude toward the pastoral is Cervantes. His life evidences a continuing love of the pastoral even as he wrote some of its most stringent criticism. The *Galatea* (1585) is generally dismissed as a youthful extravagance. But this is belied by the fact that on at least five occasions Cervantes promised a conclusion to the *Galatea*. In 1615, the year before his death, he states (in the preface to the second part of *Don Quixote*) that he is finishing a second part for *Galatea*. In the Lucian-like *Colloquy of the Dogs* one of the dogs comments at length on the lack of truth in "certain books" read by his master's lady:

All of them treated of shepherds and shepherdesses, saying that they spend their whole lives in singing and playing on bagpipes, fifes, rebecs and other wind instruments. Well, all the reflections I have mentioned, and many more, opened my eyes to the glaring differences between the lives and habits of my shepherds and those shepherds in the countryside that I had heard read about in books.[7]

It will be remembered that *Don Quixote* itself contains many pastoral interludes. The priest and the barber save the *Diana* from the inquisitional bonfire of the Don's library. The late Renato Poggioli has suggested that the ambivalent attitude toward pastoral literature of Cervantes parallels the ambiguity of his outlook toward the chivalric romances manifested in his masterpiece.[8] The Don's niece begs that even the few pastoral works to be found on his library's shelves not be spared from the *auto da fé* for she fears: "Once my uncle is cured of his disease of chivalry, he might very likely read those books and take it into his head to turn shepherd and roam about the woods and fields, singing and piping and, even worse, turn poet, for the disease is incurable and catching so they say."[9] This tension between the attraction to the world of action and to the realm of letters – on the part of the Don, like his creator, Cervantes – is most revealing. It exemplifies what the Renaissance understood to be the profounder implications of the shepherd's life: the poetic calling.

Just what did the Renaissance consider under the rubric pastoral? The first vernacular theorist in the French tradition to consider the pastoral in any detail was Vauquelin de la Fresnaye. Since he was the author of the *Foresteries* (1555) and the *Idyllies* (published in 1605, although written considerably earlier) he was speaking as a practitioneer as well. Vauquelin's poetic is essentially a rhymed version and extension of Horace's *Epistula ad Pisones* (generally referred to as the *Ars poetica*). He provided no rules relative to the pastoral. Book 3 traces the origins of the genre and includes the claim that he, Vauquelin, not Belleau, introduced the eclogue in France.[10] Pierre de Laudun d'Aigeliers discusses the bucolic genre in his *Art poétique français,* published in 1597, as well as in the prefaces to several collections of his poems. In his poetic, all the various manifestations of the genre known to the sixteenth century are enumerated, for Laudun was principally a literary historian and was well acquainted with the works of his contemporaries. His principal merit lies in his lack of legislation. Laudun enumerates the different forms the pastoral or *bergerie* may take. He claims that because of its "sens moral," the eclogue is better designated a *moralité*: "Les personnes de l'Eglogue sont bergers, et gens rustiques qui discourent le plus souvent de leurs calamitez, de leurs ayses, et bien souvent il y a un grand sens moral à L'Eglogue."[11] This poetic attests on the one hand to the popularity of the pastoral genre at the end of the sixteenth century, and on the other, to the confusion in denomination and the open nature of the terms employed.

To all the Renaissance theorists, the absence of any mention of the pastoral genre in Aristotle's *Poetics*, and, subsequently in Horace and Vida, may have caused some concern. However, this very lack of poetic theory was to prove beneficial to the humanist authors of the period, for it permitted both freedom in adaptation and the assimilation of other genres. The pastoral mode became a suitable vehicle for whatever the author's intention dictated.

Throughout the century the moral and educative potential of the pastoral was noted. This emphasis is particularly discernible in the neo-Latin tradition. Badius Ascanius proclaims the pastoral genre as being distinguished by this educational potential in the preface to the *Bucolia* in his important *Opera Virgiliana* (1500–1501). Pastoral works serve to purify the soul and character of youth, hence are necessary to any *programme de collège*. Further, they are capable of accommodating allusions to contemporary events, and indirectly, praise of patrons. Julius-Caesar Scaliger's *Poetices libri septem* (1561) reveals the dilemmas faced by those trying to adhere strictly to the ancients and to rationalize subsequent developments. For Scaliger, the shepherd's life was one of leisure. This observation leads him to expand upon the favorite commonplace of the humanists: leisure is the parent of luxury and wantonness. Scantily clothed and carefree youth and maidens (*pace* Cervantes) could occupy their leisure only with love — presumably therefore the earliest pastoral theme. A century later, Colletet in his *Poëme Bucolique* (1657), the first French vernacular study devoted exclusively to the pastoral, repeats this idea: "comme l'oisiveté est la mère de la volupté, les Bergers traitent le plus souvent de l'amour, qui est la passion la plus conforme à la nature."[12] The majority of the vernacular poetics, however, focused on what the pastoral had come to be, not what it had been to the ancients. This freed them of the difficulties encountered by Scaliger in attempting to accommodate all the known traditions and terms of the pastoral. At times it led him into ingenious explanations. An example would be his gloss of the term "eclogue." Since the word in Greek means "selection" ($\dot{\epsilon}\kappa\lambda o\gamma\dot{\eta}$), he decided that its origins are to be found in certain "superior" poets (presumably Theocritus, Bion, and Moschus) who decided to destroy their hurried productions and "select" an anthology of their best works.

Where the pastoral could in fact select, was in the style and moral lessons to be conveyed, given the lack of legislation in the ancients. Nevertheless, a tacit agreement on the subject matter of the French Renaissance pastoral had been formed, deriving ultimately from Theocritus and then from Virgil. In an outdoor setting, various topics — *merae rusticae* — were discussed by shepherds and shepherdesses: artistic, political, and religious matters might be discussed, but the predominant subject was love. The expression of this sentiment was not overtly sensual and often tended to sadness and melancholy.

Given the theorists' view of pastoral prose fiction, let us turn to some of the pastoral prose of the Renaissance, noting that it has been suggested that some of the strange vitality that the pastoral exercised over European letters may in part be ascribed to the frequent contradictions and inconsistencies — the paradoxes — that ensued from this lack of legislation.

The admixture of poetry and prose found in the Renaissance prose pastoral had antecedents in the works of such figures as Boëthius and Dante. Yet the commonly acknowledged model for the pastoral romance was Boccaccio's *Ameto* (written ca. 1341, hence prior to the *Decameron*). Here Boccaccio links a series of tales with *terza rima* verse, thereafter the standard verse form for the Italian pastoral. Like the *Divine Comedy* and the *Faery Queene*, the *Ameto* is a complex work which functions on several levels. Its main contribution to the pastoral, in addition to the form, was the emphasis on love. The form of Sannazaro's *Arcadia* (1502–1504) is based upon that of the *Ameto*. Noteworthy in the *Arcadia* is a genuine fondness for nature. It was Sannazaro's emphasis and appreciation of the classical writers' sentiment toward nature which were to have the strongest impact upon the Renaissance pastoral tradition.[13] Occasional critics have tended to denigrate the work as a mosaic of literary passages artfully arranged. Such a view neglects the unity of tone and effect, and above all, attempts to apply aesthetic and critical canons of imitation not pertinent to the Renaissance. Here then is yet another paradox: the literary works of the Renaissance are judged by twentieth-century standards. Yet imitation was a literary principle of the period: Virgil's example had set the precedent for the pastoral. Sannazaro organized and elaborated a disparate tradition. Since Petrarch did not know Greek and Boccaccio turned mainly to Virgil, it was Sannazaro the humanist poet-scholar who brought together the classical pastoral motifs. Whether acknowledged or not, Scève, Belleau, Ronsard, and others found in his work both inspiration and sources.[14]

Spain was the first country to respond to the Italian Renaissance, at least so far as literature was concerned. The comparative similarity of the two languages facilitated the transfer and assimilation of the Italian works. However, the main reason for this geographical direction was historical. From the late thirteenth century, the claims of Aragon had brought Spanish occupation to Sicily. According to Rennert in his *The Spanish Pastoral Novel* (1892), the influence of Italy upon Spanish poets was immense: it is found on almost every great figure without exception from the Marquis de Santillona to Lope de Vega.[15] Among the new forms imported from Italy was the pastoral in which poetry and prose were mingled.

It is not the purpose of this paper to outline the development of pastoral prose in Spain, nor to study its greatest exponent, Jorge de Montemayor. Suffice it to say, that here once again, extant studies reveal some surprising critical judgments. Rennert, for example concludes: "It must be admitted, however, that though some of the Spanish pastoral romances attained a very high degree of excellence, they are generally wanting in that idyllic simplicity and truth to nature which we find in the *Arcadia* of Sannazaro. They often indulge in the utmost extravagances and inconsistencies, introducing courtiers in the guise of shepherds."[16] Some critics contend that the pastoral was not intended for the *profanum vulgus*. Its popularity would seem to contradict this. And for Rennert to render a Taine-like judgment, that the pastoral could only flourish in a warm climate among impressionable people, is to ignore Sydney's *Arcadia*, among other Elizabethan masterpieces.

The success of Montemayor's *Diana* (1559) was immediate and international. But in spite of numerous editions and a host of imitations, some features of the *Diana* were criticized long before Rennert. Apart from the disparity between the real and the poetic view of the lives of the shepherds, concern was expressed for the "alarming" knowledge of subjects ranging from literature and philosophy to theology exhibited by Montemayor's nymphs and shepherds.

As we have seen, the claim of introducing pastoral prose to France was disputed. Vauquelin says that he was the first to "marry verse to prose." However, the prose of his *Foresteries* does little more than serve as a short introduction to his poems, which themselves are juxtaposed without artistic integration. Belleau stated that his *Bergerie* (1565) introduced a "nouvelle façon d'escrire qui n'a encores esté pratiquée ny congnue en nostre France."[17] In point of fact, it was François de Belleforest who wrote the first pastoral novel in French, *La Pyrénée* (1571). The Gascon author modeled his work after the *Diana*: he elaborated the *topos* of the pastor/leader, and, rather than centering his work on the celebration of love and the praise of great women, as did Montemayor, Belleforest expanded the patriotic theme and made it as important as love. In a France torn by civil wars, the patriotic theme had an obvious and pressing appeal.

The exhortation to write a pastoral had earlier been voiced by the Pléiade:

"Chante moy d'une musette bien resonnante & d'une fluste bien jointe ces plaisantes ecclogues rustiques, à l'exemple de Théocrit & de Virgile: marines à l'exemple de Sennazar" proclaimed DuBellay in the *Deffense et Illustration de la langue francoyse* (II, iv). A further impetus to the growing popularity of the pastoral mode in the period of the 1560s was that Catherine de Médicis, "la bonne Cybèle," had taken off her mourning.

In the pastoral, the metaphor of the shepherd, spiritual leader of his flock, finds its natural place. This metaphor was used by Christ: "I am the good shepherd: the good shepherd giveth his life for the sheep" (John 10:11). A ninth-century *Egloga* by Radbert in honor of a deceased abbot extended the use of the conceit. With Petrarch and his followers it became an effective weapon in church satire. Mantuan (Spagnuoli) had satirized the Roman Curia in *Falco, sive de moribus Romanae*, an eclogue which became the model for Protestant attacks on the pope. Both Spenser and Milton were indebted to it. In this first French pastoral novel, *La Pyrénée*, it was perhaps inevitable that royalty, such as Charles IX, and his mother Catherine de Médicis would be spoken of as the principal shepherds and shepherdesses of France, and that this comparison would be extended to the other great heroes as they had been in the central part of the *Diana*. The culmination of the French pastoral prose is found in d'Urfé's masterpiece, *L'Astrée*, which is dedicated to Henri IV, the "pasteur souverain."

It has not been sufficiently noted that Sannazaro, Montemayor, and d'Urfé, as well as Puritans such as Marvel and Milton, wrote devotional as well as pastoral works. And once more we discern a paradox in the discussion of the *roman pastoral* by a recent French critic, Antoine Adam. He finds it strange that d'Urfé, who showed such deep religious convictions, would be attracted to such a "frivolous" genre as the pastoral.[18] Such a judgment is more a reflection on us than it is on those writers. The French of the sixteenth century saw no clear distinction between the genres. The pastoral, like devotional works, could treat of moral edification even as it delighted: the rhetorical style and moral concerns remained in various humanist literary presentations. It was the form which differed. The pastoral novels of the Renaissance are a witness to the concerns of French society during a complex period of its development. The important role played by women in that society and a growing concern with social behavior are seen in their preoccupations. These pastorals might be salons in an outdoor setting. The numerous *trattati d'amore* and manuals of conversation such as Guazzo's *Civil Conversazione* (which Belleforest translated) attest to this growing interest in refining manners and improving conversation.

There is no need to call attention here to the success during this same period of the pastoral — poetry, prose, and theater — in England. Shakespeare's romantic comedies were variously titled "pastoral plays." His 'Green World' has been the subject of various studies. Of the many pastoral prose works, Sir Philip Sidney's *Arcadia* (1509) and Sir John Fletcher's *The Faithful Shepherdesses* (1608–1609) were the two most successful and had considerable influence.

This brief survey of Renaissance pastoral fiction reveals yet another paradox, namely, that pastoral literature was popular during a period when one would least expect such success. On a literal level, the pastoral world affords no place for the erudition and love of learning that are generally considered to be the hallmarks of the Renaissance. Closer examination, however, has revealed the tapestry of literary reminiscences to be found in these pastorals, starting with the *Arcadia*. The contrast between the simplicity implicit in the pastoral life in a period concerned with rhetoric and poetics accounted not only for the dilemma of the theoreticians, but probably contributed to the failure as well of some of the less successful writers who concerned themselves with the "ornaments," that is to say, the superficial aspects of the pastoral tradition. The pastoral involves itself with an occupation common to most cultures, and centers upon that most prevalent and permanent of human passions — love.

In presenting his love discussions, the Renaissance author had diverse material from which to draw, in addition to the outstanding pastoral models. In France, for example, to the native tradition of courtly love with the *tenson*, a love debate, had been added the rich Italian contributions. The recent love manuals, particularly Ebreo's *Dialoghi d'Amore* (which Montemayor utilized for the love discussions in the *Diana*) and works on manners such as *Gli Azolani* and *Il Cortegiano*, extended the Neoplatonic vogue. Petrarch's presence was pervasive, as was that of Boccaccio. The *questioni d'amore* in the latter's *Filocopo* became a popular feature of sentimental rhetoric. Native works such as the *Tractatus amoris* (ca. 1180) of Andreas Capellanus and the even more popular *Aresta amorum* (ca. 1460) of Martial d'Auvergne further influenced the thorough and extensive dissertations on the casuistry and metaphysics of love in this period. No wonder at the concern over the knowledge of pastoral folk such as Montemayor's. Rhetorical expertise apart, many of these discussions are still but tentative perceptions at exploring human motives and conduct and the hidden recesses of the human psyche. Nevertheless, today's psychological preoccupations have very real roots in this period. The strong emphasis on chastity found in both the French and Spanish pastorals befits writers concerned with morality and religion. It is undoubtedly more than just a coincidence that this concern for sexual morality takes place in the period of the Counter-Reformation.

Commenting upon the Renaissance pastoral, Empson perceives: "The essential trick of the old pastoral, which was felt to imply a beautiful relation between rich and poor, was to make simple people express strong feelings (felt as the most universal subject) in learned and fashionable language (so that you wrote about the best subject in the best way."[19] The verbal trick, as it has been called, was the essence of the pastoral process; putting the complex into the simple. And, although after the Renaissance there was a decline in "formal" pastoral works, the spirit persisted. Today's concerns with ecology and "back to nature," communes and "flower children," all reveal some of its various manifestations paralleled in literary forms too numerous to pursue further here. One might cite Goëthe as yet another example of the continuing, complex, paradoxical attitude toward the pastoral: mocking and parodic at times, he, like Cervantes, nevertheless put an important episode centering upon an attempt to find pastoral peace at the center of his masterpiece, *Faust* (act 2).

After analyzing the formal aspects of the pastoral, let us conclude with a few observations on the setting, which, perhaps not surprisingly, present yet other pastoral paradoxes. Curtius has examined and detailed the *locus amoenus* in *European Literature and the Latin Middle Ages*. His research revealed the etymological confusion which had connected the word *amor* with *amoenus* ("pleasant," "lovely"). While descriptions of nature tended to become rhetorical exercises, it must be remembered that part of the earlier pastoral tradition presumed the existence of unspoiled, uncultivated land. Rousseau — certainly a "pastoral" writer — decried the introduction of the notion of property and the enclosure of these lands.[20] Harry Levin's masterful study, *The Myth of the Golden Age in the Renaissance*, analyzes the ramifications and transmutations of the myth of the Golden Age as it became associated and transposed into the Arcadian setting. The conflation of Virgil's *Fourth Eclogue* and the biblical accounts of Christ's nativity helped bring the Christian pastoral setting, the Garden of Eden, into the pastoral tradition. Regret for lost innocence reinforced the emphasis on youth and youthful love found in the pastoral. The satyr was generally the symbol of age, the older man: youth must be reminded, as Panofsky has shown, of the presence of death, even in Arcadia: *Et in Arcadia Ego*. The location of this ideal landscape was further complicated by the introduction of Utopia or never-never land. Was this an

Eden to be regretted or a brave new world to be sought among the Renaissance discoveries such as the Americas or the South Seas. Several centuries later the socialists would attempt to find this ideal setting in a new social order.

Further attempts to spatially locate this Arcadian world would entail extensive excursions, for many routes have been attempted since the Renaissance. And, more pertinent, such explorations would not really "lead" anywhere. The pastoral paintings of artists such as Titian and Giorgione suggest an inner vision of these realms. *Don Quixote* expressed the Renaissance view of the pastoral life as one of contemplation. This was a prevalent view. Ficino, for example, saw the *Phaedrus* (which takes place outside the "fatiguing" city in a bucolic setting by a brook) as an allegory of the academic life, an idea which Milton repeats in *Paradise Regained*. The pastoral paradoxes may be better understood when we turn from geographic locations — remembered or future paradises — and consider what may account for many of the contradictions we have noted: How can an ideal nature be reconciled with a less than ideal occupant? It was Thoreau who wrote: "The lament for a golden age is only a lament for golden man." Granted the rhetorical excesses of some of the many pastorals of the Renaissance, the strong and perennial appeal this genre has exerted indicates that there is something more than literary bravura and challenge involved in the part of the poet who is drawn to the pastoral; something that defies those attempts to legislate it that have produced some of the paradoxical views we have examined.

The two literary genres which exercised the strongest influence on the Renaissance were the epic and the pastoral; and the epic was, without exception, acclaimed the greatest. It was attempted by every literary figure heedful to the call of antiquity. Yet the numerous French neoclassical epics, from Ronsard's *Franciade* to Voltaire's *Henriade*, were artistic failures. Admittedly the ancients were not bothered by the problem of the Christian *merveilleux*. Nor did they have the daunting genealogy such as that of the *rois fainéants* which faced Ronsard. Louis XIV's crossing of the Rhine did not measure up to the dimensions of the Homeric deeds. When the heroic trumpet gave way to the slender reed, the poets of the Renaissance found a subject which brought out their best qualities.

The pastoral would appear to address itself to the basic and pervasive human concerns; the complexities and contradictions of man. The epic concerns itself with the larger than life, while the pastoral reflects the inner tensions of man. Some of the paradoxes herein discussed arise from the attempt to reconcile tensions such as those between the world of action and the world of contemplation. In the pastoral, the celebration of the poetic calling, man steps back to survey the landscape of the mind. There is more in the pastoral than, as some would have it, a town/country opposition. Sannazaro's pastoral which gave the very name *Arcadia* to that ideal land terminated with his return to the city, Naples. (The work is all the more poignant in that it was written while he was in exile in France.) From time to time man has need of a pastoral retreat but, like Candide who eventually departed from El Dorado, cannot remain there. Arcadia should be viewed like Utopia, as a never-never land; that is, in no specific geographic realm. Rather, it would seem to be an interior world, a permanent part of the human psyche. Nor should the pastoral world be viewed as an attempt at escapism such as Baudelaire's "Anywhere out of this world." It is a temporary withdrawal, a turning inward in order to attain better self-knowledge and solace for the soul.

Analysis of the French Renaissance pastorals revealed the conflict between the theoreticians — both among themselves and with the poets. Great art should not be limited by legislation: it must remain perennially open so that is may evolve with the times. The pastoral, unlike the epic, benefited from the lack of legislation on the part of the ancients. It could accommodate an ever-widening sphere which paralleled the ever-expanding discoveries of man. The application of the term pastoral today to works of such variety may be viewed as indicative of its abiding concern to express the profound and complex nature of man. The pastoral poet pays tribute to the irreducible mystery and complexities of man. It is not merely a quest for lost innocence: here is a mythographic representation of man, stripped of the veneer of civilization. The paradoxes we have discerned arise from rational attempts to analyze the pastoral phenomenon. The emphasis on youth may, in a larger sense, be perceived as a symbolic expression of the eternal potential of the human spirit. The humanists sensed the educative value but its portent was far greater than they realized. The pastoral further provides an escape value against the increasing pressures of the collective life. Man himself, like the pastoral lands, does not wish to be completely enclosed. The pastoral flute celebrates both the poet's art and his subject. If, as a literary genre, it reveals many paradoxes, these paradoxes I suggest are those of the human condition.

NOTES

[1] Gustave Reynier, *Le Roman sentimental avant "l'Astrée"* (Paris: Colin, 1908), p. 312.
[2] Jules Marsan, *La Pastorale dramatique en France à la fin du XVIe et au commencement*

du XVIIe siècle (Paris: Hachette, 1905). At times one wonders just what version of the *Diana* Marsan read. He says not only that "le roman demeure monotone et vide," but that the characters "ignorent la jalousie" (p. 119). This is contradicted by that of Diana, her husband, and the story of Amarillis and Filemon which illustrates the theme of jealousy.

[3] C. Wedgwood, *Seventeenth Century English Literature* (London: Home University Library, 1950), pp. 24–25.

[4] Walter W. Greg, *Pastoral Poetry and Pastoral Drama* (1905; reprint ed., New York: Russell and Russell, 1959), p. 154; my italics.

[5] Samuel Johnson, *Lives of the English Poets* (London: Oxford University Press, 1955), p. 112.

[6] Charles Fontenelle, *Poésies pastorales avec un Traité sur la nature de l'Eglogue & une Digression sur les Anciens & les Modernes* (London: Paul & Isaac Vaillant, 1707). Like Cervantes, he comments mockingly on the problem of *vraisemblance*. He notes that the true life of shepherds is one of dirt and toil ("viles occupations"), while in the pastoral works they are always idealized; "spirituels, délicats et galans," thus actually as "fabuleux" as the characters of the *Amadis* tales. "L'illusion & en même temps l'agrément des Bergeries consiste donc à n'offrir aux yeux que la tranquilité de la vie Pastorale, dont on dissimule la bassesse; on en laisse voir la simplicité." The vogue of the pastoral novel was parodied, most amusingly, in Sorel's *Berger Extravagant* (1628). Among sixteenth-century Italian writers, Folengo had ridiculed the politeness of the pastoral by showing the crudities of actual rusticity in *Zanitonella*: Berni had given a realistic picture of village love which made a travesty of courtly traditions. The comic effect was deliberately sought by both.

[7] Miguel de Cervantes, *The Deceitful Marriage and Other Exemplary Novels*, trans. Walter Starkie (New York: New American Library, Signet Classic, 1963).

[8] Renato Poggioli, *The Oaten Flute* (Cambridge: Harvard University Press, 1975).

[9] Miguel de Cervantes, *Don Quixote*, trans. J. M. Cohen (Baltimore: Penguin Books, 1950), p. 61.

[10] Vauquelin de la Fresnaye, *L'Art poétique*, ed. G. Pellissier (Paris: Garnier, 1885).

[11] Pierre de Laudun d'Aigeliers, *L'Art Poétique français*, ed. J. Dedieu (Toulouse: Facultés libres, 1909), pp. 124–25.

[12] Guillaume Colletet, *Discours de Poème Bucolique où il est traité de l'Eglogue, de l'Idylle et de la Bergerie* (1657; reprint ed., Geneva: Slatkine Reprints, 1970), p. 12.

[13] Jacopo Sannazaro, *L'Arcadia*, ed. M. Scherillo, (Turin, 1888). At Paris, in 1544, appeared *L'Arcadie de Messire Iaques Sannazar, gentil homme Napolitain, excellent Poete entre les modernes, mise d'italien en Francoys par Iehan Martin*. . . .

[14] Francesco Torraca, *Gl'imitatori stranieri di J. Sannazaro* (Rome, 1882).

[15] Hugo Rennert, *The Spanish Pastoral Romances* (1892; reprint ed., New York: Biblo and Tannen, 1968), p. 13.

[16] Rennert, p. 16.

[17] Remy Belleau, *La Bergerie*, ed. D. Delacourcelle (Geneva: Droz, 1954), p. 26.

[18] Antoine Adam, *Histoire de la littérature française au XVIIe siècle* (1949; reprint ed., Paris: Editions mondiales, 1962).

[19] William Empson, *Some Versions of Pastoral* (1935; reprint ed., New York: New Directions, 1950), p. 11.

[20] One is not surprised to learn of Rousseau's youthful admiration for *L'Astrée* which he reveals in the opening pages of his *Confessions*. With conscious rivalry Voltaire celebrates material progress in *Le Mondain*.

JEAN BRUNEAU

REALITY AND TRUTH IN *LA COMÉDIE HUMAINE*

Since the nineteenth century criticism has been divided concerning the nature of Balzac's work. For some critics, like Gautier or Baudelaire, Balzac was a "voyant,"[1] a "visionary."[2] Others, like Taine, admired *La Comédie humaine* as "le plus grand magasin de documents" in literature, ranking with Shakespeare and Saint-Simon.[3] These contradictions in the critical literature which are, moreover, encountered as well in this century, are, I believe, only apparent.

It was Balzac himself who was the origin of this diversity of opinion. In the "Avant-propos" of *La Comédie humaine* he wrote: "La Société française allait être l'historien, je ne devais être que le secrétaire."[4] And, again, in the prospectus to the Furne edition of the work, he wrote: "Le plan de l'auteur consiste à tracer dans ses détails infinis, la fidèle histoire, le tableau *exact* des moeurs de notre société moderne."[5] Balzac's work would thus have been written "sous la dictée" of the nineteenth century[6] — the author's genius playing no role in it. Frequently, however, Balzac envisages his creative activity in a very different way. According to him, the artist must transform reality for religious and moral reasons, for philosophical and aesthetic reasons, and, to employ Goethe's terminology, for "demonic" reasons.

To Balzac's eyes, realistic novels, no less than his own time, lacked moral value. It is up to the artist to punish vice and to reward virtue: "Les actions blâmables, les fautes, les crimes, depuis les plus légers jusqu'aux plus graves, y trouvent toujours [dans *La Comédie humaine*] leur punition humaine ou divine, éclatante ou secrète. J'ai mieux fait que l'historien, je suis plus libre."[7] We must, moreover, qualify this judgment which only applies to *Ursule Mirouet*, to *Médecin de Campagne*, to *Curé de Village*, and to *Études philosophiques*. It is obvious that in the *Études de Moeurs*, the society which Balzac denounces favors the wicked to the detriment of the good. We find so many virtuous victims when we begin to read *La Comédie humaine*! And we find so many triumphant executioners as well! The celebrated phrase of Eugène de Rastignac at the end of *Père Goriot*, "A nous deux, Paris," expresses his decision to abandon the camp of the vanquished to join that of the victors — Vautrin's lesson. The society about which Balzac dreamed, founded on the Roman Church and on monarchy — not royalty,

291

but rather, the rule of a *single person* — would not tolerate such injustices. This is precisely why, in the *Études de Moeurs*, Balzac refuses to idealize nature. The "secretary" must become a judge.

In the second place, following Aristotle, Balzac clearly distinguishes history and poetry; in the *Poetics*, Aristotle wrote that "poetry is something more philosophic and of graver import than history, since its statements are of the nature rather of universals, whereas those of history are singulars."[8] Balzac himself echoes this by very clearly dissociating reality from what he calls "la vraie réalité," which we might call "truth." Thus, in his dialogue with Vidocq, as recounted by Léon Gozlan:

> Vous disiez donc, monsieur Vidocq?
> Je disais que vous vous donnez beaucoup de mal, monsieur de Balzac. pour créer des histoires de l'autre monde, quand la réalité est là, devant vos yeux, près de votre oreille, sous votre main.
> Ah! vous croyez à la réalité!... Allons donc! C'est nous qui la faisons, la réalité.
> Non, monsieur de Balzac.
> Si, monsieur Vidocq. Voyez-vous, la vraie réalité, c'est cette belle pêche de Montreuil. Celle que vous appelez réelle, vous, celle-là pousse naturellement dans la forêt, sur le sauvageon. Eh bien, celle-là ne vaut rien, elle est petite, aigre, amère, impossible à manger. ... Cette pêche exquise, c'est nous qui l'avons faite: elle est la seule réelle. Même procédé chez moi. J'obtiens la réalité dans mes romans, comme Montreuil obtient la réalité dans ses pêches. Je suis jardinier en livres...."[9]

The artist must thus reach truth by starting from reality: Balzac very clearly exposes his method in the preface of *Cabinet des Antiques*:

> ... le *fait vrai* qui a servi à l'auteur dans la composition du *Cabinet des Antiques* a eu quelque chose d'horrible. Le jeune homme a paru en cour d'assises, a été condamné, a été marqué; mais il s'est présenté dans une autre circonstance, ... des détails, moins dramatiques peut-être, mais qui peignaient mieux la vie de province. Ainsi, le commencement d'un fait et la fin d'un autre ont composé ce tout. Cette manière de procéder doit être celle d'un historien des moeurs: sa tâche consiste à fondre les faits analogues dans un seul tableau; n'est-il pas tenu de donner plutôt l'esprit que la lettre des événements? il les synthétise. Souvent il est nécessaire de prendre plusieurs caractères semblables pour arriver à en composer un seul, de même qu'il se rencontre des originaux où le ridicule abonde si bien, qu'en les dédoublant, ils fournissent deux personnages. Souvent la tête d'un drame est très éloignée de sa queue. La nature, qui avait très bien commencé son oeuvre à Paris, et l'y avait finie d'une manière vulgaire, l'a supérieurement achevée ailleurs. ... La littérature se sert du procédé qu'emploie la peinture, qui, pour faire une belle figure, prend les mains de tel modèle, le pied de tel autre, la poitrine de celui-ci, les épaules de celui-là. L'affaire du peintre est de donner la vie à ces membres choisis et de la rendre probable.[10] S'il vous copiait une femme vraie, vous détourneriez la tête.[11]

Balzac's philosophical vocabulary is quite imprecise; he employs somewhat haphazardly the terms "nature," "reality," and "truth." However, his conception of art is perfectly clear: the novelist must be not content to "copier la vie", he should also discover and stress its laws, in short give its meaning to life.

Finally, Balzac draws heavily on the "daemonic" element in his artistic creation. In the preface to *La Peau de Chagrin*, he writes: " . . . il se passe chez les poètes ou chez les écrivains réellement philosophes, un phénomène moral, inexplicable, inouï, dont la science peut difficilement rendre compte. C'est une sorte de seconde vue qui leur permet de deviner la vérité dans toutes les situations possibles; ou, mieux encore, je ne sais quelle puissance qui les transporte là où ils doivent être, où ils veulent être. Ils inventent le vrai, par analogie. . . . "[12] The same theme may be found at the beginning of *Facino Cane*: "En entendant ces gens [a worker's family returning from l'Ambigu-Comique], je pouvais épouser leur vie, je me sentais leurs guenilles sur le dos, je marchais les pieds dans leurs souliers percés; leurs désirs, leurs besoins, tout passait dans mon âme, ou mon âme passait dans la leur. C'était le rêve d'un homme éveillé. . . . À quoi dois-je ce don? Est-ce une seconde vue? Est-ce une de ces qualités, dont l'abus mènerait à la folie? Je n'ai jamais recherché les causes de cette puissance; je la possède et m'en sers, voilà tout."[13] In the hierarchy of beings, the "écrivains réellement philosophes" are situated between the sphere of the "abstractifs" and that of the "spécialistes," to employ the terms used by Louis Lambert-Balzac.[14]

The role of "secretary" of society is thus far from being passive. The historian of morals must discover and bring to light the hidden sense of obscure reality. Only the artists who are graced with intuition [15] are able to accomplish this task. But, if all great geniuses attain this, they do not do so in the same way, nor with the same procedures. I wish now to study those procedures which are particular to Balzac.

I propose the following working hypothesis: *Balzac never invents*. He always begins from an experienced, observed, narrated, or read fact. He affirms this, and I believe him: "Ah! sachez-le: ce drame [le Père Goriot] n'est ni une fiction, ni un roman. *All is true*. . . . "[16] He repeats this five years later in the preface of *Cabinet des Antiques*: "Quelques lecteurs ont traité *le Père Goriot* comme une calomnie envers les enfants; mais l'événement qui a servi de modèle offrait des circonstances affreuses, et comme il ne s'en présente pas chez les Cannibales; le pauvre père a crié pendant vingt heures d'agonie pour avoir à boire, sans que personne arrivât à son secours, et ses deux filles étaient, l'une au bal, l'autre au spectacle, quoiqu'elles n'ignorassent pas l'état de leur père. Ce vrai-là n'eût pas été croyable."[17] Sometimes the critical literature has discovered these "models", and sometimes it has not. But I do not believe that, in the *Études de Moeurs*, Balzac created anything from pure imagination.

And yet the Balzacian world evidently differs from the real world and this is due to the will of its author. I give one proof among the many possible: the absence of any historical characters in *La Comédie humaine*. In this Balzac parts company with his master Sir Walter Scott who, in his novels, mixed genuine persons with invented ones; for example, Richard the Lion-Hearted and Wilfrid of Ivanhoe. The world of Balzac is parallel to the real world; he recounts in it the history of France from the Great Revolution to that of 1848, utilizing characters of his own invention, from peasant to minister. Similarly, he utilizes certain procedures to pass from nature to the work of art, from the real to the Balzacian universe.

The first of these procedures is hyperbole. Balzac's characters are all *extra*ordinary in vice or virtue, in intelligence or dullness, in beauty or ugliness. This magnification is not at all an idealization, since such beings exist. But monsters are rare in life, while they abound in the Balzacian universe. This explains Balzac's frequent employment of superlatives. Consider this example from among so many others: "A cette époque florissait une société de jeunes gens riches ou pauvres, tous désoeuvrés, appelés *viveurs*, et qui vivaient en effet avec une incroyable insouciance, intrépides mangeurs, buveurs plus intrépides encore. Tous bourreaux d'argent et mêlant les plus rudes plaisanteries à cette existence, non pas folle, mais enragée, ils ne reculaient devant aucune impossibilité, se faisaient gloire de leurs méfaits, contenus néanmoins dans de certaines bornes: l'esprit le plus original couvrait leurs escapades, il était impossible de ne pas les leur pardonner."[18] To provide another example: at the beginning of *Un Grand Homme de province à Paris*, Coralie is presented as "une des plus charmantes et des plus délicieuses actrices de Paris."[19] Several pages later, she becomes "la plus jolie, la plus belle actrice de Paris."[20] Sometimes the hyperbole is carried to the point of absurdity, as in this phrase of Rastignac in a letter to his mother: "Ma chère mère, vois si tu n'as pas une troisième mammelle à t'ouvrir pour moi."[21]

Balzacian hyperbole is not limited to characters; it also affects the situations in which the novelist places them. In *La Comédie humaine* events are almost always more dramatic than the vast majority of actual events. The entire work illustrates the famous Roman proverb: "Tarquin's rock is close to the Capitol." I will provide but a single example since they are innumerable: the death of Coralie in *Illusions perdues*. This death is already dramatic in itself, since Coralie is young, beautiful, and loving. However, in order to make the funeral vigil more pathetic, Lucien has to compose drinking songs to pay for his mistress' burial. The two circumstances, the death of a beloved young woman and the obligation to write drinking songs in conditions of need or misfortune, are perfectly natural. But not their collocation, which alone makes possible this hyperbolic epitaph: "Heureux ceux qui trouvent l'Enfer ici-bas, dit gravement le prêtre."[22] Through this frequent use of hyperbole, Balzacian "realism" does not at all herald such works as *Germinie Lacerteux, l'Education sentimentale*, or that masterpiece of the nineteenth-century antinovel, *Une Belle journée* by Henri Céard.

The Balzacian universe is as filled with antitheses as it is with exaggerations. Here again examples abound. Consider the titles of some of his works such as *Grandeur et décadence de César Birotteau* or *Splendeurs et misères des courtisanes*. Consider the mutually opposed characters such as David Séchard and Lucien de Rubempré: "Le contraste produit par l'opposition de ces deux caractères et de ces deux figures fut alors si vigoureusement accusé, qu'il aurait séduit la brosse d'un grand peintre."[23] Or consider Madame de Bargeton and Coralie, the great lady with a cold heart and the warm-hearted little *actrice*. There are also characters who are opposed to a group, to a body, like Rastignac at the end of *Père Goriot*: "À nous deux, Paris." There are opposed parties, like the Crussotins and the Grassinistes in *Eugénie Grandet*. What is more, since Balzac believed in the interaction of milieus and personalities, the lodgings of his characters are themselves frequently antithetical, like the rooms of d'Arthez and Lousteau in *Illusions Perdues*,[24] or the contradictions in the world of litterature: "La vie littéraire, depuis deux mois si pauvre, si dénuée à ses yeux [Lucien], si horrible dans la chambre de Lousteau, si humble et si insolente à la fois aux Galeries-de-Bois, se déroulait avec d'étranges magnificences et sous des aspects singuliers. Ce mélange de hauts et de bas, de compromis avec la conscience, de suprématies et de lâchetés, de trahisons et de plaisirs, de grandeurs et de servitudes, le rendait hébété comme un homme attentif à un spectacle inoui."[25] The Balzacian universe could be reduced to two opposing camps, executioners and victims, wolves and sheep. As in the fable, the sheep always lose, but the wolves often destroy each other, if de Marsay becomes first minister, Maxime de Trailles fails; and how would the great duel between Corentin and Vautrin have ended, if Balzac had lived to write it.

Antithesis is encountered in situations where it is often mixed with hyperbole, as in the Roman proverb cited above: the final ball of Madame de Beauséant in *Père Goriot*, the death of Coralie in *Illusions Perdues*, the death of Esther in *Splendeurs et misères des courtisanes*. The Balzacian universe teems with antitheses which, again, involve no invention, idealization, or deformation on Balzac's part. Antitheses exist in nature and in history; Napoléon against Europe, the Resistance and the Movement, the peasants and the large landowners. However, in their *pure state*, these antitheses are rare in life. By multiplying them, Balzac makes them stand out. Marx and Engels were not mistaken in the esteem which they had for *La Comédie humaine*, in which the struggles and the contradictions of the society of the time are so powerfully described.

Balzac's third great procedure — and I can see no others — is the acceleration of time. In Balzac's novels events follow upon one another with astonishing rapidity. Here again examples are legion. The destiny of the heroes seems to be ruled by touches of a magic wand, as in *Le Père Goriot*: "Ce coup de baguette, dù à la puissante intervention d'un nom [Madame de Beauséant], ouvrit trente cases dans le cerveau du méridional [Rastignac], et lui rendit l'esprit qu'il avait préparé. Une soudaine lumière lui fit voir clair dans l'atmosphère de la haute société parisienne. . . . "[26] There is another wave of the wand in an inverse sense a page later: "En prononçant le nom du père Goriot, Eugène avait donné un coup de baguette magique. . . . "[27] One more example may be cited, from *Illusions Perdues*, namely, David Séchard's coming to trial:

Imitons ... le style des bulletins de la Grande-Armée; car, pour l'intelligence du récit, plus rapide sera l'énoncé des faits et des gestes de Petit-Claud, meilleure sera cette page exclusivement judiciaire.

Assigné, le 3 Juillet, au tribunal de commerce d'Angoulême, David fit défaut; le jugement lui fut signifié le 8.

Le 10, Doublon lança un commandement et tenta, le 12, une saisie à laquelle s'opposa Petit-Claud en réassignant Métivier à quinze jours.

De son côté, Métivier trouva ce temps trop long, réassigna le lendemain à bref délai, et obtint, le 19, un jugement qui débouta Séchard de son opposition.

Ce jugement, signifié roide le 21, autorisa un commandement le 22, une signification de contrainte par corps le 23, et un procès-verbal de saisie le 24.[28]

If we are familiar with the customs of French justice at that time — and in ours — we could believe ourselves to be dreaming!

The destinies of Balzacian heroes are thus more rapid than they would be in nature or, rather, they are always rapid, while life offers but a few examples of such prompt successes or reversals. Let us again recall the Roman proverb, "Tarquin's rock is close to the Capitol," which could serve as an epitaph for *La Comédie humaine*, since it is precisely composed of the three elements which we have considered: hyperbole, antithesis, and the acceleration of time. There is no better illustration of this proverb than this sentence concerning Lucien in *Illusions Perdues*: "En quelques mois sa vie avait *si brusquement* changé d'aspect, il était si *promptement* passé de l'*extrême misère* à l'*extrême opulence*, que par moments il lui prenait des inquiétudes comme aux gens qui, tout en rêvant, se savent endormis."[29]

The Balzacian universe is thus real, since all its elements arise from nature. However, it differs from the *total* real to the extent that Balzac selects, for his heroes, destinies composed of significant events. For better or for worse, Balzac's creatures are "hors du commun"; the oppositions between them are more clear-cut than in life; time runs more quickly for them than it does for us. Hyperbole, antithesis, and acceleration of time − it is with the aid of these three procedures that Balzac transforms the society of his time into his own universe.

There is, therefore, no contradiction between the judgments of Gautier, Baudelaire, and Taine. *La Comédie humaine* represents a vast endeavor to describe and to explicate the French society of the nineteenth century. In order to make it intelligible to his readers, Balzac simplifies and dramatizes it, but he does not change it one iota. He goes directly to significant, exemplary situations. This is not to say that, like Ronsard, Boileau, or Racine, he derives the "vraisemblable" or the "probable" from the true. He employs only the real, but it is the typical real, which is to say, the true. It is thus that he is able to give an object lesson to his readers who, without his "intuition" and his experience, would have difficulty in comprehending their own destiny. Balzac saw only a jungle in the society of his time; he reveals it, showing its principles and its structures. Meanwhile, he dreams of a different social order which would be its antithesis, founded on the double religious and temporal authority of the pope and the monarch. Without this authority − which would not allow any delegation − man can only live in this animalistic jungle, where the wolves devour the sheep and themselves as well. This is no one's fault, moreover, except that of this lawless society. The highest creation of Balzac, the character most gifted with genius, force, and cunning − Vautrin − is not condemned by Balzac; quite to the contrary! And it was through these very qualities for survival − whether virtues or vices − that Honoré de Balzac became, as Baudelaire said so well, "le plus héroïque, le plus singulier, le plus romantique et le plus poétique parmi les personnages [in *La Comédie humaine*]."[30]

A final word: these three principal characteristics of *La Comédie humaine* − hyperbole, antithesis, and acceleration of time − are precisely those of epic poetry. In it there are the most valiant of heroes, the most beautiful of women, the opposed camps, the multitude of glorious actions; it is Homer's

Iliad. It is not by chance that, for Rastignac, "Madame de Beauséant avait les proportions des déesses de l'*Iliade*,"[31] or that l'abbé Carlos Herrera called out his wish for a modern epic poem — which is no other than *La Comédie humaine*: "Il y a là tout un poème à faire, qui serait l'avant-scène du *Paradis perdu*, qui n'est que l'apologie de la Révolte. — Celui-là serait l'*Iliade* de la corruption, dit Lucien."[32] It is not like the *Odyssey*, the *Aeneid*, *Jerusalem Delivered*, or the *Lusiads* where the good triumph over the evil with the aid of the gods or of God; but rather, it is similar to the *Iliad* where gods and men are divided into two enemy camps for futile reasons, a world without moral law which dissolves with the death of the generous Hector and the enslavement of the Trojans. *La Comédie humaine* is the *Iliad*[33] of our time.[34]

NOTES

[1] Théophile Gautier, article on Balzac in *Portraits contemporains*, 3d ed., (Paris: Charpentier, 1874), p. 63: "Quoique cela semble singulier à dire en plein XIXe siècle, Balzac fut un 'voyant.'"

[2] Charles Baudelaire, article on Gautier (1859), in *Oeuvres Complètes* (Paris: Éditions du Seuil, 1968), p. 465: "J'ai maintes fois été étonné que la grande gloire de Balzac fût de passer pour un observateur; il m'avait toujours semblé que son principal mérite était d'être visionnaire, et visionnaire passionné."

[3] Hippolyte Taine, *Nouveaux Essais de critique et d'histoire*, 5th ed. (Paris: Hachette, 1892), p. 140.

[4] Honoré de Balzac, *La Comédie humaine* (Paris: Éditions du Seuil, 1965), I, p. 52. Except when indicated to the contrary, all references are to this edition.

[5] This passage is cited by André Billy, in *Vie de Balzac* (Paris: Flammarion, 1944), II, p. 72.

[6] Honoré de Balzac, "Théorie de la démarche," in *Oeuvres diverses*, ed. Marcel Bouteron and Henri Longnon (Paris: Conard, 1938), II, pp. 625–26: "Il y a dans tous les temps un homme de génie qui se fait le secrétaire de son époque. Homère, Aristote, Tacite, Shakespeare, l'Arétin, Machiavel, Rabelais, Bacon, Molière, Voltaire, ont tenu la plume sous la dictée de leurs siècles." Should we not include Aristophanes?

[7] Balzac, "Avant-propos" of *La Comédie humaine*, I, p. 54.

[8] Aristotle, *Poetics*, in *Introduction to Aristotle*, ed. Richard McKeon (New York: Random House, Modern Library, 1947), p. 636. Professor Bruneau cites the French edition of J. Hardy, *Poétique* (Paris: Les Belles lettres, 1961), p. 42.

[9] Cited by André Billy, *Vie de Balzac*, II, pp. 162–63.

[10] "Probable": a translation of the Aristotelian τὸ εἶχος. Note that Balzac takes all the elements of his narratives from reality, but that he selects them.

[11] Preface to the first edition of *Cabinet des Antiques* (1839), in *La Comédie humaine*, III, p. 626.

LA COMÉDIE HUMAINE   299

[12] Preface to the first edition of *La Peau de Chagrin* (1831), in *La Comédie humaine*, VI, p. 708.
[13] *Facino Cane*, in *La Comédie humaine*, IV, pp. 257–58.
[14] Balzac, *Louis Lambert*, in *La Comédie Humaine*, VII, p. 322: "Entre la Sphère du Spécialisme et celle de l'Abstractivité se trouvent . . . des êtres chez lesquels les divers attributs des deux règnes se confondent et produisent les mixtes: les hommes de génie."
[15] *Facino Cane*, in *La Comédie Humaine*, IV, 257: "Chez moi l'observation était devenue intuitive, elle pénétrait l'âme sans négliger le corps; ou plutôt elle saisissait si bien les détails extérieurs, qu'elle allait sur-le-champ au-delà; elle me donnait la faculté de vivre la vie de l'individu sur laquelle elle s'exerçait, en me permettant de me substituer à lui. . . ."
[16] *Le Père Goriot*, in *La Comédie Humaine*, II, p. 217.
[17] Preface to the first edition of *Cabinet des Antiques* (1839), in *La Comédie Humaine*, III, p. 626.
[18] *Illusions Perdues*, in *La Comédie Humaine*, III, p. 524.
[19] Ibid., p. 485.
[20] Ibid., p. 493.
[21] *Le Père Goriot*, in *La Comédie Humaine*, II, p. 243.
[22] *Illusions Perdues*, in *La Comédie Humaine*, III, p. 546.
[23] Ibid., p. 395.
[24] Ibid., pp. 457, 471.
[25] Ibid., p. 482.
[26] *Le Père Goriot*, in *La Comédie Humaine*, II, p. 236.
[27] Ibid., p. 237.
[28] *Illusions Perdues*, in *La Comédie Humaine*, III, p. 569.
[29] Ibid., p. 517; italics are mine.
[30] Baudelaire, *Salon de 1846*, in *Oeuvres Complètes*, p. 261.
[31] *Le Père Goriot*, in *La Comédie Humaine*, II, p. 298.
[32] *Illusions Perdues*, in *La Comédie Humaine*, III, p. 607.
[33] *Cf.* Baudelaire, *Salon de 1846*, in *Oeuvres complètes*, p. 261: "Car les héros de l'*Iliade* ne vont qu'à votre cheville, ô Vautrin, ô Rastignac, ô Birotteau. . . ."
[34] This essay was written in French. The author declines all responsibility for the English translation.

MARIA DA PENHA PETIT VILLELA DE CARVALHO

MAN AND NATURE: DOES THE HUSSERLIAN ANALYSIS OF PRE-PREDICATIVE EXPERIENCE SHED LIGHT ON THE EMERGENCE OF NATURE IN THE WORK OF ART?

Understanding notions such as man and nature, and especially understanding the "and," that is, the tie which binds them together, is strictly dependent on the ontological horizon situating this statement.

For Western man this "and" has been an "and" of disjunction confirming an opposition, a relation of *being confronted with*; this has been the case throughout the development and the interweaving of some of the dominant traditions down to what has been termed the modern age. Nature, then, is considered an "outside" which is presented to a determination of the mind, to intellection.

The reply to this separation, which is the corollary of a split in being itself — with being as mind on the one side and being as nature considered *res extensa* on the other — could only be sought in terms of method, of the ways in which man as mind managed to grasp in an objective (that is, mathematizable) manner natural being as it *is* beyond its sensuous appearance.

This split was abolished, it seems to me, neither by the idealist unification in the dialectic nor by the positivist reduction, which, despite its claims, abolished the split only at the expense of a reduction which sacrificed the essential dimensions of man while at the same time it failed to recognize nature itself in its original givenness.

All this, as we know, is the topic Husserl addresses in the *Crisis*, when he writes: " ... we must realize that the conception of the new idea of 'nature' as an encapsuled, really, and theoretically self-enclosed world of bodies soon brings about a complete transformation of the idea of the world in general. The world splits, so to speak, into two worlds: nature and the psychic world."[1] And this split, we might add, affects man himself through and through.

From this point of view, scientific practices, in the course of their remarkable development, only consolidate the "dualist split"; yet when science is considered from the perspective of invention, as is more and more the case in contemporary epistemology, indications of an unsuspected overcoming of this split quickly become apparent.

Now, if we allude to these questions — which, of course, cannot be considered here — this is because it should be seen — if only glimpsed — that even

A.-T. Tymieniecka (ed.), The Philosophical Reflection of Man in Literature, 301–311.
Copyright © *1982 by D. Reidel Publishing Company.*

where the split was most firmly entrenched due to the dominance of the objectivist attitude, it nevertheless shows signs of giving way. In addition, on the level of scientific objectification itself, we cannot fail to notice that the biological sciences are undermining the overly influential position of a "purely physical conception of nature," to use Husserl's own terms.

Whatever the case may be, the giving way at issue here, when considered at the radical level this notion demands, is in large part the work of phenomenology and its attendant schools of thought, which, we should remember, preceded the emergence in the collective opinion of an ecological consciousness.

However, to adhere to Husserl's position, we might ask ourselves whether the Husserlian notion of intentionality, just as his egology, did not place new obstacles in the way of what was elsewhere coming to light and leading to the transformation of this "and" in the expression "man and nature."

Although we must content ourselves with only a partial outline of what motivates this question, we are nevertheless tempted to say in advance that there is no single exclusive answer; every answer, it seems to us, can only repeat the oscillation written into the Husserlian text itself, and this is perhaps a sign of its refusal to allow itself to be fixed into a system.

Let us be more specific. The question is whether the place and the orientation given by Husserl to the notion of intentionality, by his analysis of intentional acts, do not make us miss the original character of the opening out onto the world which every intentional aim presupposes in one way or another.

We must therefore naturally return to the Husserlian analyses dealing with prepredicative experience, and I am particularly reminded of three chapters which make up the first section of *Experience and Judgment* (*Erfahrung und Urteil*).

There is no doubt that the guiding thread in the phenomenological description here is the object, the *Gegenstand*, that is, the substrate, the individual, sensuous object taken as the correlative of an active objective orientation and presentation of sense by which the ego breaks a path through the original passivity of sensuous data.

The point I should like to draw attention to today does not directly concern the question of activity and passivity. On several occasions, Husserl points to the passivity of this ultimate founding level, which is nothing other than the temporal flow of consciousness itself, on which all receptivity, in the final analysis, rests.

The point in question concerns the centering of the prepredicative phenomenological analysis around the grasp of the object, the individual, sensuous thing aimed at in perception. Not that Husserl's analyses at this level are deficient. They are instead astonishingly rich and enlightening in many respects. As proof of this, should there be any need for proof, we shall take only one example chosen for the interest currently shown in the referential power of language, and of poetic language in particular.

Listen to this descriptive résumé provided by Husserl: "Thus *with regard to content the most general synthesis of sensuous data raised to prominence within a field*, data which at any given moment are united in the living present of a consciousness, are those in conformity with *affinity (homogeneity)* and *strangeness (heterogeneity)*."[2] And there follows a section on the analogy of singular givens which provides the basis for a meditation on metaphor. Let us limit ourselves to the synthesis of sensuous contents in conformity with strangeness, heterogeneity. Do we not find here a referential pole of experience for this recourse to poetic language consisting in bringing together various "images" which belong to heterogeneous sensible fields? Such as, for example, Ezra Pound's "inhale its colours," when he writes in his translation of "Hagoromo":

> All these are no common things,
> nor is this cloak that hangs upon the pine-tree.
> As I approach to *inhale its colours*,
> I am aware of mystery
> its *colour-smell* is mysterious.

In the same way, the deviation of poetic syntax in relation to the syntax of ordinary language could be compared to what Husserl says of the unity of connection (*verbundene Einheit*), that is, of what is not a categorial unity produced in a creative spontaneity. This is not to deny poetic creativity nor the linking of sound and sense, by which language attests in its own way to the indissoluable connection of man and nature. We want instead to indicate the level of experience which is precisely revealed by this very creativity, when it tears language away from the attraction exerted by logical predication.

It must, however, be noted that despite their many pertinent features, the Husserlian analyses are directed more toward seizing or grasping the object and making its inner horizons more explicit, that is, toward unfolding its constitutive moments considered either as pieces or as dependent moments than toward the surprise of being there between heaven and earth at the emergence of the world.

Of course, Husserl often explicitely refers to the world as the universal ground and fundamental presupposition, just as he stresses the external horizons, the frame presupposed by every object which affects us and becomes capable of attracting our attention.

But consciousness of a horizon is itself understood as consciousness of the external horizon of the object, that is, as consciousness of the plurality of mutually affecting objects which are part of the givenness of the object, as that which at every instance surpasses its intention. So it is always the object which is the center around which the whole is then articulated.

And consciousness of this mutually affecting plurality does not yet allow the world to be glimpsed as a ground, or rather as advent, thus allowing the advent of a certain space which would not be merely that involving relations of coexistence between objects but the space where something surges forth and where my own being takes on its sense and I am compelled to orient myself.

In making these remarks we are not forgetting the essential recognition of the unity of the field of givenness belonging to the mutually affecting plurality which appears in § 35; we should like to quote the following passage in this regard: "This unity of the field, on the basis of which any orientation of apprehension toward individual objectivities affecting us, as well as their explication and reciprocal putting-in-relation, is first possible, has been, up to this point, simply presupposed; and it has only been mentioned that these are achievements of the passive synthesis of time consciousness, by means of which such unity becomes fundamentally possible."[3]

This recognition of the fact that beings exist together simultaneously or successively, therefore, noetically has a unity of connection which is constituted, in the final analysis, by the unity of the passive field in its original temporal structure.

But this deepening of the Husserlian meditation in the direction of an increased recognition of the unity of the field of original givenness does not, however, lead him to a realization of all the consequences due to the noematic emphasis of his analyses of prepredicative experience. Husserl always allows himself to be guided by the appearing of objectivities, objects as individual things, even when they are given originally in a unity of connection. This, it seems to us, is not without influence on the manner in which the unity of the field is itself understood, and it leads Husserl to avoid analyses which would deal with *how* the world appears, with the meaning of its affective tonality and the opening of its various directions of sense.

And yet, it is indeed in a kind of feeling, in a *Befindlichkeit*, which in a certain way is prior to all perception, to all *Wahrnehmung*, that the world originally, that is, from all time and in every instance, gives itself.

We are not concerned here with developing the ontological implications of what we can now only briefly outline in the margins of our reading of *Erfahrung und Urteil*, although these implications are already shown to lead in the direction of the experience of a de-substantification of being. We should simply like to suggest that this obliteration perhaps cannot help but make itself felt in the way in which, in *Erfahrung und Urteil*, we are led to understand this pair: man and nature, which, by all it implies, haunts our colloquium.

For this reason, should we not be surprised that in this section dealing with prepredicative experience, not once is there any mention of the countryside, its atmosphere, or the dynamism of its rhythms? But what place could there be for an experience the ego might have in an encounter with the sky, the sea, the light playing on the leaves of a tree, or even the nonsubstantial flight of a bird written across the sky, if, as Husserl says, "all that relates to the external world is given to us as a body within spatio-temporal nature in a sensuous perception?"

To guide himself in these analyses of what is thus to be understood as simple nature, is Husserl not placing too great a confidence in a certain type of object — an object which can be described as a body and apprehended as a substrate — so that he then neglects all that appears, as a manifestation of *physis*, first as "milieu," as spark or as rhythm, and which unfolds, at what we might call an infra-objective level, a meaning essential for being.

In order to move beyond the generality of these remarks, we now propose to point out certain aspects in which the primacy of the object-substrate, as it is imposed on the understanding of the pre-predicative sphere, may prove to be misleading with regard to our experience of the world's appearing, as it is presented to us, and in what way this sphere is hinted at in the metamorphosis to which it is submitted in the work of art. The choice of art seems to flow naturally from the question of man and nature as such. For it has always been in art that the tie linking man and nature, assuming the most varied forms imaginable — where the very fate of form is itself at stake — never ceases to manifest itself. It is therefore starting from what is witnessed in art that I will reinterpret some passages from *Experience and Judgment* which deal with color, edges, and surfaces in order to then prolong this reflection with a consideration of the manifestation of nature in the work of Paul Klee.

Having said this, let me note in passing that we are conscious of the fact that it is possible to find pictorial equivalents for the analyses of perception undertaken by Husserl in certain modes of representation which have marked modern art. I am thinking in particular of the analogy which has been suggested between phenomenology and cubism — an analogy which has been developed recently in a stimulating article by Hintikka,[4] who uses the Husserlian notion of noema to understand the problem of representation in Cubist painting.[5] He suggests here that "Just as a Husserlian noema may contain at one and the same time expectations as to what an object or a person would look like from many different perspectives, in the same way cubists often depicted the same subject from several different angles at one and the same time."[6]

Be that as it may, Hintikka's entire reflection in this essay — which, moreover, concludes with another analogy, this time between cubism and the interpretation of the representative relations of language provided by a theory of models — revolves around the concept of representation. The problem, however, is precisely that of knowing whether the relations of representation exhaust the question of the presence of nature in the work of art, and in particular in pictorial work.

Finally, when Hintikka attempts to clarify what is meant by noema, the central notion of his comparison between phenomenology and cubism, he refers to Dagfinn Føllesdal's chapter "Phenomenology," in *The Handbook of Perception*, where we read: "To take an example from perception, let us consider the act of seeing a tree. When we see a tree, we do not see a collection of colored spots ... distributed in a certain way; we see a tree, a material object with a back, with sides, and so forth. ..."[7]

But are Føllesdal and Hintikka aware that what this description of perception omits is precisely the celebrated definition of a painting given by Maurice Denis?[8] How, then, can this serve to shed light on painting, unless one holds that painting is nothing more than the representation of objects?

Of course, our purpose here is not to provide a detailed analysis of Hintikka's essay. If we borrow this quote from him, which, moreover, does not seem to me to echo Husserl's own statements, it is because what he says directly involves the phenomenological situation of the relations between dependent moments and perceptual synthesis, a situation I shall now attempt to examine through the pages of *Erfahrung und Urteil*.

I have already alluded to the explication of the various moments of the object — moments in which the apprehension literally unfolds as the perceptual attention sustains our glance as it moves toward a penetration of the internal horizons of the object.

This is the case of color which as quality is given as a part, or rather as a moment of the object as a whole, the object-substrate of which the color is, consequently, only one determination. Color, whether the red of the ashtray or the white of the paper, is therefore constituted as a moment of the object's appearing and, although it belongs to the noema, the color is directly transcended by the intention as it apprehends the object.

This sort of analysis is probably valid on the level of everyday perceptions, which are almost completely bound up with the expectations imposed by the necessities of daily life. But it is not valid absolutely. And this is so because the analysis neglects that moment in the encounter with a splash of color or with a colored atmosphere when the color focuses the world for us (as does the white of the paper for someone who is looking for "inspiration" in order to write), for we who live in these colors, open to the meaning of their pulsations, their radiance, their play. The preceding analysis circumvents the significance of the sensuous event in order to go straight to the sense of the object. Is this not a way for the analyses to refuse the testimony offerred by the painter's vision, which by means of the action of painting never dissociates the sense from the sensible means by which the trajectory of sense is constituted.

Henri Maldiney, who has long meditated on esthesis, likes to evoke these words of Cézanne, attentive to the lessons of Courbet: "Remember Courbet and his story of the bundle of sticks, Cézanne repeated. He applied his hue, without knowing it was a bundle of sticks. He asked what he was representing there. Someone went to see. The objects were sticks. And so it is in the world, in this vast world. To paint it in its essence, you must have the eyes of a painter who in the color alone sees the object." [9]

Again, there is Cézanne's exclamation: "Look. Those blues! Those blues over there under the pines" – blues which Maldiney calls "the organs of his communication with the world, the pathic moment of his total presence to their appearing." [10] The list of such expressions from painters is long, from the "high yellow note of this summer," of which Van Gogh spoke to his brother Theo, to Matisse's green which is not yet grass.

Might not Husserl have caught a glimmer of this dimension of color when he wrote: " ... if I pass from the color, which caught my eye and which I have first made my object, to the paper? ... " [11] But, precisely, he moves too quickly over the whiteness which has struck him, interpreting it as immediateness, to the object in which he then encloses it. Whence the rest of his sentence: "the latter [the paper] is still a 'whole' relative to the white. In this way I include something 'more' in my glance, just as when I pass from the base of an ashtray, taken as a part, to the whole ashtray. In both cases, it is a transition from explicate to substrate." [12]

Indeed, we might add, this is a return movement which shifts from the event in the world, taking place in accordance with the climatic and directional dimensions of color or form — the trace of which is recorded by the painter's work on the canvas — to its resolution in the reassuring solidity of a world of permanent objects, the subject-substrates of predication.

Now, nature, to be present in art, does not necessarily have to be represented by means of the appearance of these object-substrates, correlatives of perception. Moreover, even in representative art, nature is really present only when its evocation, its presentation, creates a unique space, the space of the work, which transcends its representation.

To give substance to these remarks, I will be guided by some brief glimpses into the artistic enterprise of Paul Klee. Why Paul Klee? I can give only two reasons here. The first has to do with the importance of the work of this artist, who, involved in the abstract movement, throughout his life unceasingly interrogated both nature and art.

The second reason is that Paul Klee had a special tie to Sicily, where he resided on several occasions; and what he has said of the area which welcomes us today gives us an inkling of a possible answer to the question of the presence of nature in the work of art beyond representative efforts, which are inherently equivocal, since out of ignorance or misunderstanding of the specificity of artistic form, one can mistake what is represented for the essential part of the work itself.

Let us listen to what Klee remarked in 1929 in his notebook: "I have never lived a precise experience nor do I wish to do so: I carry inside myself the mountains and the sun of Sicily. I simply think of the countryside from a purely abstract point of view and something begins to take shape; for the past two days I have begun to paint again...."

What does this signify, if not the rejection of the anecdote in simple representation, which postcards are enough to capture; on the contrary, this signifies the profound feeling of the country through its light, lines of force, rhythms, that is, something quite different from individual things perceived as substrates.[13] There is also the evidence of Jürg Spiller, who collected and annotated Klee's writings on art to form the volume entitled *Das bildnerische Denken*, that Klee "possessed numerous collections which enabled him to study the form, the composition and the very essence of living things including the most varied types of organisms.[14]

Let us leave these details and move on; for what I am concerned with here is the fact that for Klee the study of all these natural forms was indispensable to his work, to his research – highly aware of the properly pictorial means of expression open to the painter. In his *Creative Confession*, Klee reflects on the formal elements of drawing, namely, the point, the energy of the line, the plane, space and plastic elements – lines, light and dark contrast, color. Color made him feel the same exaltation as the artists quoted above, and of it he wrote: "Color possesses me. I have no need to catch it in flight. It possesses me for always and it knows this. This is the meaning of the minute of bliss: color and I are one and the same thing. I am a painter."

Are we to smile at this pathos, mock this painterly wave of pantheism? But what if through this pathos, something more decisive were thereby asserted, namely, that the visibility of the world surpassses and cuts through the mere vision of things? From this arises the possibility of understanding the liberation, which art achieves through the graphic and plastic means it possesses, from that which Husserl understood only as dependent moments of the object: colors, edges, surfaces, variations in light.[15]

Klee belongs to those who, like Heidegger and Binswanger after him, know the significance of these "moments" in the world's appearing. And his work itself bears witness to the fact that with sensible dimensions commences the adventure of meaning, for through them arise those meanings which men can decipher beyond and before the senses constituted in the noematic sphere. Was it not Klee who said, for example, of "the feeling of verticalness [that] it corresponds to a living reality: it prevents man from falling?"[16]

Now we can ask what he was trying to discover through his untiring dialogue with nature, which he insisted was "a sine qua non condition" for the painter.

Was it not to move from forms which had reached their completion, their definition, back to the very movement which engendered them, as if he were looking beyond appearances to the forces which give rise to them or, to borrow one of Klee's own expressions, "the forces of the created." Not what is visible, he liked to say, but what makes things visible. In artistic creation the effort is of the same order: "As movement of form," he wrote, "the genesis is the essence of the work."

This is no different in the case of the feeling of space which appears in his painting; this is not the feeling of the space of co-existing objects, in which their dimensions would be represented by means of a more or less static arrangement of planes, but instead that of a dynamic space, that of a field made of the forces and stresses traversing it, initiating directions as well as places in which configurations are drawn.

Let us content ourselves with these few glimpses of Klee's undertaking; my purpose here, in any event, cannot be an analysis of these works which would take into account the singularity of what takes place and come to be in each of them. But perhaps, through these glimpses and through the remarkable endeavor to which they refer, I have been able to indicate with sufficient clarity that all understanding of the emergence of nature in art requires a reflection which does not reduce pre-predicative experience — or rather the contact with *physis* — to the experience of objects of perception and their relations. Only in this way can we glimpses the necessary belonging and the necessary tension linking man and nature in the work of art.

Translated by Kathleen McLaughlin

NOTES

[1] Cf. Edmund Husserl, *The Crisis of European Sciences and Transcendental Phenomenology*, trans. David Carr (Evanston: Northwestern University Press, 1970), p. 60.
[2] Husserl, *Experience and Judgment*, trans. James S. Churchill and Karl Ameriks (Evanston: Northwestern University Press, 1973), p. 74.
[3] Cf. Ibid., p. 156; § 35, "The question of the essence of the unity establishing the relation" ("Frage nach dem Wesen der Beziehung begründenden Einheit").
[4] Cf. Jaakko Hintikka, "Concept as Vision: On the problem of representation in modern art and in modern philosophy," in *The Intentions of Intentionality and other new Models for Modalities* (Dordrecht: Reidel, 1975).
[5] This comparison is not new. As Hintikka writes, "Probably the first to have called attention to it is Ortega y Gasset. The most careful statement on record of the analogy is by Guy Habasque. ... " However, he adds, "in earlier discussions of the relation between cubism and phenomenology the emphasis has typically been on the mental operations involved in the two activities, cubist painting and phenomenology, respectively, rather than on the semantics of the representational or intentional situation" (pp. 230–31). It is, therefore, in the latter direction that Hintikka develops the analogy in an effort to show that "the cubists were dealing with the noemata of objects, not with objects as such" (p. 231).
[6] Ibid., p. 232.
[7] Quoted by Hintikka, p. 231.
[8] Cf. Maurice Denis, 'Définition du Néo-traditionnisme,' in *Art et Critique*, 1 (23–30 August 1890): "Remember that a painting — before it is a warhorse, a nude woman, or any sort of anecdote — is basically a flat surface covered with colors applied in a certain order."

[9] Cf. Henri Maldiney, *L'Equivoque de l'Image dans la Peinture*, in *Regard, Parole, Espace* (Geneva: L'Age d'Homme, 1973).

[10] Cf. Maldiney, p. 138.

[11] Cf. Husserl, *Experience and Judgment*, p. 141.

[12] Ibid.

[13] This dynamic and nonobjective feeling of the Sicilian countryside is transcribed using the names of cities, mountain ranges, and the indication of direction in the astonishing drawing, Klee's inner landscape of Sicily, entitled: 'Mountain chains near Taormina' (1924) (in the Kunstmuseum, Bern).

[14] During his sojourns along the Baltic coast, Spiller continues, "Klee collected algae which, after they were dried, were glued between two plates of glass with the inscription 'Forest of the Baltic Sea.' From Sicily and the Mediterranean he brought back sea urchins, sea horses, coral, and molluscs. In addition to butterflies, he also had a rock collection: crystals, fossils, yellow amber, variegated sandstone, quartz, and mice. He was interested in stratification, the transparent effect of the layers and the color composition. In tree bark he studied the movements of the bostryches." Cf. Paul Klee, *La Pensée Créatrice*, trans. Sylvie Gerard (Paris: Dessain and Tolra, 1973), p. 12.

[15] Moreover, understanding this autonomy, this significance of the "part," of the "dependent moment" in relation to the sense of the "whole" (of the thing perceived) can also clear the way for understanding not only certain aspects of creative fantasy but also "fantasizing" as the term is taken in psychoanalysis.

[16] Cf. Klee, p. 147.

ALPHONSO LINGIS

THE LANGUAGE OF *THE GAY SCIENCE*

To Nietzsche, the logical axioms of identity, non-contradiction and excluded middle and the metaphysical categories of entity, being, enduring entity have but anthropological and not ontological value. They are the simplifications, the falsifications that have long ago been "incorporated," embodied in the very way our sense organs pattern the flux about us. They have enabled the species to survive; perhaps countless nascent species that did not conclude that similar things were identical, that situations recurred the same, that properties or dispositions were enduring perished.

The Nietzschean ontology of Becoming excludes, then, a conception of language such that its truth would lie in the correspondence between its signs and identifiable objects of perception. Words of language can no longer be taken to be ideally self-identical terms which signify cores of identity that endure and recur in the perceived field.

When Nietzsche boasts that he is the first philosopher to have put into question the value of truth, one of the things he means is that he has put into question the validity of the "true world," that is, the order of identities, essences, beings, taken to endure over and beyond the flux of the phenomenal field, and taken to be designated by the fixed terms of language. But he also means to subordinate the value of truth to the value of life. This is not a simple pragmatist position, as Arthur Danto somewhat misleadingly calls it. For the logico-ontological "fundamental errors" which make possible the representation of the "true world" has precisely anthropological value; it has enabled this species' life to survive.

The life to which truth is to be subordinated, then, is not the species' life, and it is not the exigency for survival of that life. It is the highest kind of life, that which is sovereign of itself, sovereignly individual. It belongs to the meaning of the sovereignty of life that it not act out of needs and not need to endure, as though endurance would bring it a perfection that its present existence does not have. Such a life is not under the Darwinian compulsion to adjust to its environment; it is the source of a production of surplus force, which has to be discharged. It is active and not reactive. Such production of an excess, such sovereign individuality is natural, solar, like the sun, hub of nature, pouring its gold on the seas as it sets, as it sinks to its death, and happiest when the poorest fisherman — that can repay nothing — rows with golden oars.[1] This life is also not under the compulsion to adjust its mental processes to reality, it rather enhances reality with the surplus of its own force. In this sense beauty is for it a higher value than truth. This also is not to be given the pragmatist sense Heidegger has given it, when he understands Nietzsche's phrase that beauty is a stimulant to life. Life does not gild the world with the gold of its values in order to excite itself to accumulate ever more power; its solar economy is governed by the law of squandering.

What kind of conception of the origin and operations of language does such an ontological and axiological position entail? The principle Nietzschean theses are assembled in a key text in *Zarathustra*:

"O my animals," replied Zarathustra, "chatter on like this and let me listen. It is so refreshing for me to hear you chattering: where there is chattering, there the world lies before me like a garden. How lovely it is that there are words and sounds! Are not words and sounds rainbows and illusive bridges between things which are eternally apart?

"To every soul there belongs another world; for every soul, every other soul is an afterworld. Precisely between what is most similar, illusion lies most beautifully; for the smallest cleft is the hardest to bridge.

"For me — how should there be any outside-myself? There is no outside. But all sounds make us forget this; how lovely it is that we forget. Have not names and sounds been given to things that men might find things refreshing? Speaking is beautiful folly; with that man dances over all things. How lovely is all talking, and all the deception of sounds! With sounds our love dances on many-hued rainbows." . . .

"O Zarathustra," the animals said, "to those who think as we do, all things themselves are dancing; they come and offer their hands and laugh and flee — and come back. Everything goes, everything comes back; eternally rolls the wheel of being. Everything dies, everything blossoms again; eternally runs the year of being. Everything breaks, everything is joined anew; eternally the same house of being is built. Everything parts, everything greets every other thing again; eternally the ring of being remains faithful to itself. In every Now, being begins; round every Here rolls the sphere There. The center is everywhere. Bent is the path of eternity."[2]

A first position denies language its truth value — words are illusive bridges — so as to recognize its Apollonian function: duplicating reality, words function to enhance it, to invest it with the clarity and distinction of the beautiful.

Nietzsche has traced speech back to the organs for producing sounds in insects and birds, where vocalization first occurs in sexual excitement. With Nietzschean malice the highest — the speech that conveys thought — is explained by the lowest. (The insect law of life, metamorphosis, set forth as higher than rational maturity is another example.) Vocalization in its primitive and fundamental form is not utilitarian, but seductive. The valorizing usage of sound is prior to the cognitive or representational.

This seductive state must also not be understood pragmatically, as the set of means devised by penury and need, a "need for a partner," an inadequacy of solitude. On the contrary, the state of lustful frenzy, "the brain charged with sexual energy," is a state of excess power; in this state "one gives to beings, one *forces* them to accept from us, this is called *idealizing*."[3] The excess energy puts the whole psychomotor apparatus in a vibrant state of expressivity and susceptibility. The abbreviated exclusively linguistic signs owe their extension and subtleness to this fullest state of life.

> The aesthetic state has a superabundance of *means of communication*, as well as an extreme *receptivity* for solicitations and signs. It is the summit of communicativity and transmissibility among living beings – it is the source of languages.
>
> It is there that languages have been born: the languages of sounds as well as the languages of gestures and looks. It is the fullest phenomenon which always constitutes the beginning: our powers as civilized men are reductions of richer powers. But today still one understands with the muscles, one even still reads with the muscles.
>
> At the base of every mature art, there is a profusion of convention: in the measure that it is a language. Convention is the condition for great art, *not* its prevention . . .
>
> Every elevation of life intensifies the force of communication, and at the same time the force of comprehension of man. *To identify oneself* with other souls is originally nothing moral, but is due to a physiological excitability of suggestion: "sympathy" or what one calls "altruism" are but deformations of this psychomotor relationship which one credits to intellectuality (*psychomotor induction*, affirms Ch. Fère). One never communicates thoughts, one communicates movements, mimic signs, which are reinterpreted by us as thoughts . . .[4]

Nietzsche founds the apophantic judgment on the proposition that posits "this as that," but this "as" is not the Heideggerian hermeneutical "as." It is the evaluative "as." And the first evaluation was not comparative, work of a reckoning, a rational calculation. All the noble words of language, Nietzsche teaches in the first essay of *On the Genealogy of Morals*, were invented by the strong, the conquering, the joyous. They were formed in exultant utterances: "This is good! healthy! strong! beautiful!" They issued out of a life that so knew itself, and so consecrated itself: "How happy we are, how strong we are, how blessed we are!" It is in these exclamatory acts that the words acquired their use and their meaning. They refer then not to objects compared, but to inner surges of force, which are blissful and exultant. Such speech acts function not to report facts, but to consecrate and enhance the states they cover, intensifying them. It is thus strong and beneficent states of will that have invented the noble words of language. "How lovely is all talking, and all the deception of sounds! With sounds our love dances over many-hued rainbows."

But it is at this point precisely that language corresponds to the world. For language that moves exultantly over the things reproduces in its own inner tempos the incessant nonteleological recurrencies of the world. " 'O Zarathustra,' the animals said, 'to those who think as we do, all things themselves are dancing; they come and offer their hands and laugh and flee — and come back.' " The movements of speech capture and reflect the patterns of flux of the world, the Dionysian format of the world.

The world in which "all things themselves are dancing" was depicted in the dances of Dionysian ritual, but it is also depicted in the modern scientific representation of the universe, in the science become science of forces and no longer science of forms.

The aesthetic state, state of heightened expressivity and communicability, conventionalizes itself, Nietzsche has said. The great participationist rituals of the Orphic cults, which captured the rhythms and recurrencies and metamorphoses of nature in their dances, masks and transvestism, yielded a representation of the cosmic Becoming. The Dionsyian dance was the very movement of a thought. The original form of the thought that represents the ontological format of the world was this dance driven by the inwardly felt propulsion of the universal becoming of things, the eternal recurrence of all things. Nietzsche conceives all veritable thought, and the language in which it is articulated, not as intuition but as dance. "Thinking wants to be learned as dancing, *like* a kind of dancing."[5] Thought is not the immediacy and immobility of intuition, but movement that captures and follows the movements of phenomena. It is not teleological, aimed at the end, at the end-product, the finished state, the final state, the telos. Thought penetrates; its insight is not in terms and formulas but in movement that can continue. Veritable thought represents the world articulated in universal becoming and not in ends, terms, final states.

This would require a conception of language such that the meaning does not lie deposited in terms, but lies in the movement of speech. The analysis of linguistic means is only fragmentary in Nietzsche's work. Yet the emphasis throughout on style, the conviction that it is the turns of speech and the patterns of composition and not the vocabulary or the system that are what is telling in an account, embodies this conception. The style then is not arbitrary; there is a fit style, and inauthentic, sham, deluded and deluding styles. "O you buffons and barrel organs! . . . Have you already made a hurdy-gurdy song of this?"

Yet in this response of Zarathustra there is perhaps more than a criticism of the style; there is perhaps a reserve with regard to all speech.

When the animals had spoken these words they were silent and waited for Zarathustra to say something to them; but Zarathustra did not hear that they were silent. Rather he lay still with his eyes closed, like one sleeping, although he was not asleep; for he was conversing with his soul.[6]

Indeed there is a vice that lies in the very passage to speech, essentially common speech. The vision of eternal recurrence is the most singular doctrine, doctrine of the most singular soul; its sense is not something seen in the coherence of the terms of language. The Dionysian law of eternal recurrence is not first known by the contemplation of external nature, it is not a truth of the animals. Although it is indeed a cosmological doctrine, doctrine of the structure of the Dionysian universe, and is an ontological doctrine, doctrine of the essence of Being as Becoming, it is first a singular, singularizing truth, truth in the first person singular, law of the structure and movement of the Zarathustran soul. It is in the nature of his own exultant will − "Joy wills eternity, deep, deep eternity!" − that Zarathustra discerns the law of the cosmic will that is particularized in himself. Its evidence is already betrayed in being expressed, exteriorized, in being found in the words.

The transcription into language undoes the very meaning it institutes. It turns into a hurdy-gurdy song, a round, a circulation which empties of its significance in the measure that it becomes commonplace. For it is not the forms of language that circulate that are the essential, but the inward vision and the force that issues it.

We no longer esteem ourselves sufficiently when we communicate ourselves. Our true experiences are not at all garrulous. They could not communicate themselves even if they tried. That is because they lack the right word. Whatever we have words for, that we have already got beyond. In all talk there is a grain of contempt. Language, it seems, was invented only for what is average, medium, communicable. With language the speaker immediately vulgarizes himself. Out of a morality for deaf-mutes and other philosophers.[7]

Nietzsche also says it is only the shallowest and worst part of ourselves that can be put into words.[8] For words are words in common − herd signals. That which got formulated in ourselves, in "consciousness," was also that which needed to get formulated − our needs and wants, for which we appealed to others. We put them then in terms they could understand, expressing them as common wants and needs. But our lives in their positive reality are force and not emptiness, self-intensifying power and not wants. What is deep in us is the positive force of life; the wants and needs are surface, exposed to the outside. And they are the worst part of ourselves – the negative part. And formulating our lives as needs and wants, addressing ourselves to others, makes us ever more shallow and negative. The will to set ourselves forth as dependent, the servile will, arises here. The others respond to our needs and wants with their own power and will to power; in seeing us as wants and needs they see us as servile, they are motivated to enslave us. And, in Nietzsche, those who have been made slaves are those who have first made themselves slaves.

The core of ourselves, that which makes one a positive thrust of life and a singular life, is never expressed with the common terms of language, the herd signals. It remains ineffable, inarticulated in signs, unconscious.

This is the conclusion inasmuch as Nietzsche centers on the common structure of the terms of conventionalized language. Yet the terms of language did not originally become significant as conventionalized terms of a code. Words are not significant because they are conventionalized; they can be conventionalized because, resonating the singular movements of a life, they are significant. They are first articulated in the song of self-consecration with which the joyous vibrancy of life returns back over itself, heightening itself. And in a few moments, Zarathustra comes out of his silence, to utter the great song of self-consecration that closes the third book of *Zarathustra*.

The Nietzschean thoughts about language, about the practice of the language he considered significant, his own and that of science, are incohative only. They anticipate 20th century language philosophies which are no longer bound to substantive metaphysical conceptions, but the essential Nietzschean theses have not been elaborated in those philosophical movements. Nietzsche's own aphoristic composition, exploiting effects of suspended movement, surprise, irony, sarcasm, inversion, syncopation, abbreviation, he considered essential to the meaning of what it was to convey. But in addition he recognized the algorithmic notation held intrinsic to modern mathematized science to enunciate a nonsubstantive relational representation of the universe, representing once more the Dionysian ontology. That these two modes of representation should convey the same image is no doubt a paradox, which Nietzsche did not expressly deal with.

NOTES

[1] Friedrich Nietzsche, *The Gay Science*, trans. Walter Kaufmann (New York, 1974), § 337.

[2] Friedrich Nietzsche, *Thus Spoke Zarathustra*, in *The Portable Nietzsche*, trans. Walter Kaufmann (New York, 1977), III, 13.
[3] Friedrich Nietzsche, *Twilight of the Idols*, in *The Portable Nietzsche*, trans. Walter Kaufmann (New York, 1977), p. 518.
[4] Friedrich Nietzsche, *The Will to Power*, trans. Walter Kaufmann (New York, 1967), 14 [119].
[5] Friedrich Nietzsche, *Twilight of the Idols, op. cit.*, p. 512.
[6] Friedrich Nietzsche, *Thus Spoke Zarathustra, op. cit.*, III, 13.
[7] Friedrich Nietzsche, *Twilight of the Idols, op. cit.*, pp. 530–31.
[8] Friedrich Nietzsche, *The Gay Science, op. cit.*, § 354.

HENNY WENKART

SANTAYANA ON BEAUTY

Santayana is often ignored by analytic philosophers as a "merely literary" figure, and until recently relatively little serious work has been done to explicate his views. It was his misfortune that he wrote so supremely well that his work was often classified as literature "rather than" as philosophy both by literary critics and (pejoratively) by philosophers. Yet he did hold important and original views on many of the topics which occupy philosophers to this day; he was in touch with all the philosophical concerns current through his long lifetime; and his observations upon the writings of other philosophers bear witness to an analytical capacity which cuts through rhetoric to the crux of what a writer is saying, with his own precisely articulated view as the vantage point for his analysis. This is not to imply that there are no shifts, particularly between his earlier and his later work. But these shifts are in the nature of adjustments rather than complete turnabouts or contradictions.

When philosophers say that Santayana's work is "merely" literary, they mean that a thoroughgoing analysis of his views and of the terms he employs to express them would be unfruitful, yielding neither a consistent philosophical system nor even any identifiably uniform approach. It is commonplace among those philosophers who speak of Santayana at all to say that while his critique of other thinkers is always acute, suggestive, and finely articulated, his systematic philosophical work is fuzzy and "impressionistic." Such remarks ought to arouse suspicion. If a philosopher produces precise analysis of the ideas of others, is he really likely to lose this precision when engaged on work of his own? And unless he wrote that work hastily under the pressure of time (which we know as a biographical fact that Santayana did not) is it likely that he would become *inadvertently* obscure in systematic works which he prepared over the course of many years?

It is my position that Santayana's systematic philosophical work is indeed literary, and that it is so in a deliberate and very serious way. He consciously employs especially selected literary devices as philosophical tools. He will sometimes use the *form* of a device as the symbol for the *content* of a philosophical assertion, as when he uses a parable, what I call his parable of the realms of discourse, as an icon for the relationship between mind and matter.

Of Santayana's major works his book on aesthetics, *The Sense of Beauty*, is very early. None of his later systematic books deals specifically with beauty, and yet in the profoundest sense all of his philosophy, in particular his moral philosophy, is concerned with individual choice, individual taste: a kind of all-inclusive aesthetics. He is often criticized by moralists on account of his detached, aesthetic stance toward matters of ethics. They make the mistake of thinking that such a stance implies that he does not take ethical matters seriously, when in fact, on the contrary, it is these very matters of basically aesthetic import, matters concerning morality *and* beauty, that he takes with the deepest seriousness.

Given his biography it was almost inevitable that he should take an observer's stance and an observer's delight in life. His father was Spanish but his mother lived in America; he spent his early boyhood with his father in Spain, and was then educated in his adolescence and young manhood in Boston. At the end of his autobiography he says that he was, by chance, a foreigner where he was educated; that the world was his host in its busy, animated establishment, where he was a stranger. It was natural that diversity itself should become and remain for him an intrinsic value in its own right.

In 1896, when *The Sense of Beauty* appeared, Santayana was thirty-three years old. He classified himself at that time as a naturalist in metaphysics. Here is how he defined beauty then: "It is an affection of the soul, a consciousness of joy and security, a pang, a dream, a pure pleasure. It suffuses an object without telling why. . . . It is an experience."[1]

In order to be beautiful an object must satisfy the principles of both utility and purity. He ascribes the origin of beauty to a harmony between nature at large and the nature of our own sensibility. Perception is pleasure when our sense and imagination find what they crave — when the world either shapes itself or molds the mind in such a way that world and mind are in perfect harmony. To feel such pleasure at all, the body must be in well-functioning health: the consciousness of its own perfect function is, then, an occasion for finding the world which it encounters in harmony with itself — and hence beautiful. Beauty is, by definition, the transformation of an element of sensation into the quality of a thing. Moreover, in addition to being good in itself, it bears witness to the possibility of good, yea, of perfection, in the world as a whole.

In his mature philosophy, a thoroughgoing skepticism led Santayana out of naturalism and into epiphenomenalism in metaphysics. He insisted that everything that happens in the material world has strictly material causes. There is no mutuality in the interaction between mind and the physical world.

The material world is what causes everything that is in the mind. A felt purpose is merely the feeling that accompanies the actual purpose, which itself is physical in every respect.

Such a view may seem fatal to any importance that might be assigned to the beautiful. But the exact opposite is the outcome. This is because *importance itself* is not a physical thing, but is a matter of what is felt to be important by someone in particular. Importance is in the realm of consciousness — what Santayana calls the "realm of spirit" — which is a realm lacking all power, force, causation, and yet the realm wherein reside all significance and all value. The spirit can not bring about the existence of anything, beautiful or otherwise. But what is beautiful is so only insofar as it delights the spirit.

What follows is that the beautiful occupies the highest position in Santayana's value system. It is supreme in ethics, in political theory, and in religion, as well as in art.

The material world of Santayana's mature philosophy is Heraclitean: its basic natural state is motion and change. It is rest, rather than motion, which requires explanation. Matter consists of a "flux" whose atomic units are internally controlled "natural moments" of varying duration. Each natural moment, containing no internal temporal diversity, has internal tension and external material continuity with other moments, which generate it and which it generates. The essence of a natural moment is like that of a valve. It contains a reference to the direction in which matter may flow through it. It exists only in act, yielding to unilateral pressure from one side and opening out into the other.

Matter is never merely inert. Where there is no motion and change, there is at least tension in certain directions, a tension capable of bringing about certain changes and not others. This tension is called "potentiality." A blind impulse and need to shift is basic to existence. Matter falls into repetitive patterns which in their turn fall into superpatterns, until these are sufficiently involuted and complex to form organisms which spin off part of themselves as seeds, seeds which will repeat their own specific pattern in fresh organisms of the same species. Within the seed which an organism spins off there lies great organising power. The pattern of the organism's behavior, inherent and potential in the seed, active in the successive events in the life of the resulting organism, Santayana calls the "psyche."

The psyche's interaction with surrounding events is largely a reaching out into them: in answer to a felt need, the psyche causes new organs to develop in the organism, which are at the same time an iconic expression of the felt need in fleshly form and a means of working toward the fulfillment of the need. Among the impulses which compose the psyche, intent (which Santayana sometimes identifies with intelligence) is that assurance and expectancy which causes sense organs to develop in the body, to fulfill the need for information about the environment. Intent is the expression of the organism's "animal faith" that the environment is there to be explored. The psyche reaches out into the environment by generating sense organs; the specific result of this reaching out is knowledge.

However, the mental aspect of this knowledge, what Santayana calls its "spiritual" component, is not itself useful in the gathering of information — or in any other way. It is utterly superfluous, although it develops spontaneously into a "spiritual life" of great elaboration and beauty. At first it is a simple ground tone, a conscious *continuo* associated with that life rhythm which is the psyche. In consciousness it is perceived not as a rhythm, but rather as a relative steadiness and permanence, a blank feeling of duration which corresponds on the physical side with the basic ground rhythm of life, just as a simple tone we hear corresponds on the physical level with a vibration in the air.

As the psyche's rhythms become involuted and elaborate themselves into complex organic functions, the ground tone becomes "clearer" and more precisely articulated, and worthy of the name of "intuition." This is that function of the mind which is the most certain and the least subject to doubt. It is the faculty of *presence*. It is indubitable, since it makes no statements whatever. It occurs whenever the psyche makes a physical synthesis; the more physical elements there are which have been successfully synthesized, the "clearer" the accompanying intuition will be.

The initial role of the environment is the clarification of the vague inherited preferences of the psyche. Later, depending upon the environment in which it has happened to find itself, the psyche develops fixed and precise needs. The tensions within the psyche evoke various kinds of emotional aura — the greater the tension, the stronger the aura. Consciousness is the way it feels to be a vortex in a material flux. The relative stability of the vortex, or its instability, results in a desire for self-preservation and an "interest" in anything which promises to further such self-preservation. The whole of life is a prolonged predicament, says Santayana, and the whole of mind is the variously modulated cry which that predicament wrings from the psyche.

Physically, the mind is an act of attention on the part of the physical psyche. Phenomenologically it is focused upon essence, isolating in thought the form and substance from its flux. The "exclamations" take the form of essences which the psyche evokes; they express the set and the needs of the psyche. Conscious intent expresses the fact that the psyche is not a mere accidental nexus of rhythms which *happens* to take a certain shape, but rather a powerful complex of real potentialities already present in the seed from which the organism has developed. The psyche is "wound up" to go on in a specific form of movements, growth, reaction patterns. Suspense outward, toward objects not part of her organism, is habitual to the psyche.

All the practical work goes on in the body. From the point of view of the real work of the world, conscious mind is superfluous. Then why need the psyche become conscious at all? Ah – it need not! *Therefore*, the fact that it does so anyway is particularly poignant and beautiful. Mind is the means to nothing else. It is *the* end in itself. It has no prudential usefulness, but is part of that poetic superfluity, that generosity with which nature is filled. Useless in the production of anything further, intuitions are in fact *the only valuable things*, precisely because value has to be *value for some specific consciousness*.

The chief thing that distinguishes the work of the psyche from that of a machine is the fact that the psyche is a living interest, developing organs which are this interest made manifest. Now, these organs in their turn are not machines, and they do not go mechanically about fulfilling their intended functions; they are alive, and once constituted become subcenters of interest in their own right. All life is filled with this element of generosity, of surplus, of poetry. The psyche, in outrunning mere usefulness, becomes a poet. It makes that gratuitous comment upon its situation which is the entire mental world – a world which serves no utilitarian purpose, which does nothing to help preserve the organism, but exists purely as an intrinsic good in itself.

The organs of mental activity themselves outrun their intended function, which was merely to report truly about the environment. In the process of developing sensitivity to light, sound, and so forth, they develop an independent interest in the continuation and further elaboration of conscious data, for the pure sake of the enjoyment of these data – of music, of sensuous visual forms, and so forth. Thus arises the joy in hearing music and seeing colors and shapes. This joy is projected outward and is perceived as a quality of its objects: the joy is perceived as beauty residing in the objects of the organs of perception.

Thus animal roots are a precondition for all feeling; it is feeling which is the supreme aim of living. The development of our animal roots goes forward by an inner necessity which is perceived by consciousness as absolute freedom. A pleasurable emotion accompanies it. Whatever happens well is seen by primitive consciousness as the result of its power. An aesthetic glow accompanies some of this experience, but we pay little attention to it unless the aesthetic ingredient becomes predominant. Then we exclaim, "How beautiful!"

The equilibrium which is life is maintained by accepting modification sometimes, and at other times by imposing changes. When man changes natural objects in such a way as to make them congenial to his mind, he has humanized them and made them *art*. No idea governs the production of a work of art: invention is not the child of necessity, but of abundance. What achieves art is a kind of genius, a vital premonition and groping on the part of the psyche, which at the same time as it achieves the physical object also achieves the ideas which express it (without affecting its physical production).

Images and satisfactions come from a blind craving, of themselves. This blind craving then recognizes its object, and turns into pleasure. Instantly as the object may be welcomed, it could not have been summoned, because it was not known beforehand.

This, then, is Santayana's view of beauty, early and late: whatever is supremely harmonious with the motions of the living self, whatever touches it intensely and sublimely, is beautiful. Nothing can be more important than the beautiful, for it alone gives value to life. All of mind is really a generous surplus of life in nature: unnecessary for the continuation of life, and glorious precisely because of its superfluity. Beauty, or the good, for on this peak the two are interchangeable, is the bonus *par excellence*. If the entire conscious life itself is a cry on the part of the living creature, then that experience which constitutes beauty is the cry of joy.

NOTE

[1] George Santayana, *The Sense of Beauty* (New York: Scribner's, 1896), p. 203.

PART IV

GENESIS OF THE AESTHETIC REALITY: WAYS AND MEANS

BEVERLY ANN SCHLACK

HEROISM AND CREATIVITY IN LITERATURE: SOME ETHICAL AND AESTHETIC ASPECTS

> What is a hero? The exceptional individual. How is he recognized, whether in life or in books? By the degree of interest he arouses in the spectator or the reader. A comparative study, therefore, of the kinds of individuals which writers in various periods have chosen for their heroes often provides a useful clue to the attitudes and preoccupations of each age, ... the hero and his story are simultaneously a stating and a solving of the problem ...

As W. H. Auden suggests in the quotation above,[1] it is in the hero figure, a dramatic extreme of the human condition, that important issues, such as existence and essence, thought and action, creativity and morality, are presented most vividly. This paper is an attempt to examine some of the more salient features of the nature of the hero.

There have been many attempts to define the various types of hero, one of the most interesting by Northrop Frye, who relates the hero's power of action to ordinary men and to nature. The mythic hero, a divine being or god, is superior in kind both to other men and to the environment. If the hero is superior only in *degree* to others and to his environment, he is one of those demi-god heroes of romance; i.e., although he is not a god, his acts are those miraculous deeds of folklore and legend in which the normal laws of nature are contravened by his behavior. The hero of epic or tragedy is superior in degree to other men, but not to his environment. He is the leader figure, greater than most men but subject to the censure of others and to the order of nature. This point of achieved heroism is midway between godlike heroism and human limitation, where superior stature still does not negate the fact of mortality. The hero of comedy and realistic fiction, superior to neither his environment nor other men, is measured by the same canons of probability and humanity as the rest of mankind. If he is inferior to other men, the hero belongs to an ironic mode of presentation in which we may look down upon his frustrations and ineptness. Frye sees the modes as both cyclic, with irony tending back toward myth again, and as historical, with literary presentations of the hero moving downward into the nonhero or antihero (type five) who has populated the last hundred years of modern literature.[2]

The quintessential hero is characterized by courage, nobility of purpose, or special achievement in the face of extreme difficulty or death itself. Every hero, male or female, is a seeker with an imperative quest or mission. The aim of the Homeric hero was glory and honor, fame, renown, reputation — those factors which entered into the blind but heroic wrath with which Achilles protested his lack of proper recognition. Achilles saw that anonymity is antiheroic: suppose the world will little note nor long remember what he did here? Glory is to the warrior-hero what the work of art is to the artist-hero: immortality in the memory of his survivors, an ultimate victory over blank oblivion. In the splendid isolation of his superiority, the hero's only reward is glory while alive and/or praise and honor after death. These needs are related to his highest sense of self and to his struggle to be deemed worthy; they are not mere vanity or arrogance (although the tragic hero's classic mistake *is* hubris — proper pride *exaggerated* out of proper proportion). Negation of the recognition the hero seeks cannot be endured, because it would deny the very point of his life quest — the consummate *beingness* which results from his combination of self-discovery and self-creation. In short, the problem of personal identity may be a classic philosophic puzzle, but it is also a primary dilemma for the literary hero.

From Achilles on, every hero has tested the limits of the heroic stance. To become what he will be, the hero has to risk what he has, what already *is*, including, possibly, his very life. A conflict avoided makes an expedient, even clever, but ordinary person. The fortitude to confront the crisis makes the hero. The heroic battle of and for self-realization involves a crisis of awareness and a difficult choice among several alternatives, always including the unheroic possibility of doing nothing and holding one's peace. In addition to a clear comprehension of what is needed, the hero must alter his life in accordance with the bidding of his conscience. His destiny is not felt as an imposed duty, but as personal conviction, as the outward expression of inner values. At the crucial point, the hero cannot permit himself to compromise his principles or be deflected. Antigone *must* bury Polynices. Thus, the hero *becomes* his relentless moral imperative; his will to act and the act itself merge. Yeats asked how one could "know the dancer from the dance," and Wallace Stevens in a poem on the hero has observed: "He is the heroic/Actor and act but not divided."[3]

Comparing the hero to the common man, the better to understand the essence of heroism, is not an original approach; one can trace it back to the

Poetics (2, II), in which Aristotle points out that the hero must stand in some comparative relation to other men, regarding degrees of virtue and vice; i.e., the hero is the same, better, or worse than others. Many writers, however, have presented the hero as different from the average man only in degree; he is Frye's type four, or common-man hero, an average person who rises to a special occasion and enacts an extraordinary deed. For example, there is Dickens' Sydney Carton uttering those famous words which conclude *A Tale of Two Cities*: "It is a far, far better thing that I do, than I have ever done." But entrance into the most select domain of heroism must be held to more exacting admission requirements. Carton, however noble or briefly tragic, is not so much heroic as a man whom circumstance has put into the right place at the right time to do something right. While the effects of his action may be extraordinary, his capacities are not. Random chance, or accident of position, do not a genuine hero make, although they do make a man whose imitation of heroic ideals has fleetingly inspired his ordinary life.

Because he is set apart in some unique, qualitative way, the hero's reaction to the common man is one of opposition to and dissociation from ordinary goals. In the words of Wallace Stevens, "the classic hero/And the bourgeois, are different, much." The Hero "walks with a defter/And lither stride. His arms are heavy/And his breast is greatness." His success cannot be achieved in an indiscriminate way. Lesser men may act out of a spontaneous instinct that briefly overrules prudence, or that confuses sudden, stupid bravery with deliberate and informed action; the genuine hero is required to act out of deliberate commitment, out of an *a priori* belief in certain values before they are put to a particular test. His thoughts, according to Stevens, are "begotten at clear sources," and his acts express his essential character and his focused will, not chance or the random luck of being in the right place at the right time.

The hero is not the common man intensified. Genius is not compounded talent, nor heroism the common man cubed. Five ordinary soldiers do not make an Achilles, nor do four minor poets add up to Shakespeare. The quality of existence is radically different. It is the difference between exceptional and mundane reality, between a transfiguring potential and brute actuality. Stevens ends a poem called 'The Common Life' with the observation that "The men have no shadows/And the women have only one side." In *Heroes and Hero Worship* Carlyle spoke of "the struggle of men intent on the real essence of things, against men intent on the semblance and forms of things."

The hero must dare; his or her opposite lives an existence which lacks depth of experience, precisely because it ventures no risks. Henry James' character, John Marcher, realizes just this heroic lack at the conclusion of 'The Beast in the Jungle,' when he confronts what James calls "the sounded void of his life." The Beast lurking in the jungle of Marcher's life finally springs at him, revealing "that all the while [he had been waiting for something significant to happen] the wait was itself his portion." James has captured a telling aspect of the average man's life: that his freedom is often negatively expressed as freedom to do nothing, and that his use of conscious will is in the service of acquiescence, not mastery or defiance. In contrast, the hero's appetite for experience and intensity of purpose touch him with passionate courage. Heroic rebellion is justified in threatening equilibrium and stability — those nonheroic values, those subtle excuses for oppression by which Creon, for example, demands unthinking obedience from Antigone — because heroic rebellion brings not chaos and destruction, but improvement, development and renewal. Not the stubborn *status quo* of the average man, expressed in face-saving generalizations that one cannot fight city hall (for heroes always do), but the recognition that some inadequate or unjust reality must be replaced — that is the basis of the heroic vision.

Eschewing common sense values, the hero scorns self-protective action, living and dying in an ambiance of agony and glory seldom reached or desired by the common man. Hazlitt saw the heroic as "the fanaticism of common life: it is the contempt of danger, of pain, of death, in pursuit of a favorite idea."[4] Hazlitt knows that passion and indiscretion are components of heroic action, but he is careful to distinguish these traits from their average cousins, recklessness and desperation. He notes that many who defy danger and dire consequences are false heroes whose motives are not sufficiently impersonal: "The abstracted, the *ideal*, is necessary to the true heroic." "To have an object always in view dearer to one than one's self," Hazlitt declares, "to cling to a principle in contempt of danger, of interest, of the opinion of the world, — this is the true *ideal*, the high and heroic state of man."

The heroic life seems to invite a dramatic and painful destiny. As Northrop Frye observes, "Great trees are more likely to be struck by lightning than a clump of grass" (*Anatomy of Criticism*). The common man is drawn to no such life-style or fate. From his perspective, defiance costs too much in struggle, suffering, unpopularity, and other sundry interferences with his personal comfort. Content with mediocrity, secure in his sense of finite boundaries — for limitation and restriction can be comforting — the average man is part of Shelley's "trembling throng whose sails were never to the tempest given" (*Adonais*). He does not want to sail "far from the shore" or be "borne darkly, fearfully afar." His preference for snug shelter exempts him from heroic existential risk. Security is the common goal; e.g., the guard who delivers Antigone to Creon says of her heroic disobedience: "All such things are of less account to me than mine own safety."[5] Where the common man lives out his constrained destiny, the hero tries to launch out into freedom of thought and action. An Edna St. Vincent Millay sonnet, 'Euclid Alone Has Looked on Beauty Bare,' observes that "heroes seek release/From dusty bondage into luminous air." There are not bound by what Carlyle called society's "Smooth-shaven Respectabilities" (*H&HW*).

The Homeric hero, Achilles in particular, formed the standard against which subsequent heroes were measured. Like Achilles, the heroes of old romances embodied a dream of adventure and action. These knights were fair-minded as well as strong; a certain equation of physical prowess with moral stature was suggested. The triumph of these heroes did have a moral dimension — even a socio-political one — if there values were shared by the larger community. This sort of hero was more than an athlete or warrior; he could be the hero as patriot, defending or establishing a just social order. The heroes of romances sought such opportunities for heroism as slaying dragons and conquering armies; victory as palpable result seemed more "real" than abstractions like glory or fame or recognition.

If none but the brave deserves the fair, another goal of the heroic quest can be love. The Renaissance remaking of the ancient heroic tradition added the subject of love, which became a disorder fit for heroes, and Spenser and Sidney, among others, devoted much attention to its nobility.[6] Yet the mythical Hercules was also a great hero-figure for the sixteenth century, and that same century saw a certain Machiavelli answer those who wished to end warfare by demanding where, then, men would be able to assert their honor.

In an increasingly complex social and political world, the metaphysics of heroism underwent complex development. The battlefield of the ancient epic, the proving ground of heroes like Achilles, evolved into some more subtle, less physical milieu. The ethical hero who refuses to betray his or her principles can face death with as much courage and nobility as a soldier. *Thought* may be a proving ground for the exceptional person, as Hamlet's world of ethical distinction and semantic rigor suggests. Shakespeare's heroes are still quite active, but they are far more reflective, meditative and analytic than their ancestors. A battle, after all, can be an inner struggle of reason and passion, right and wrong; glory can be self-education and awareness; defeat can be failure to achieve selfhood; triumphs can be of an intellectual, spiritual or aesthetic nature. When the Prince of Denmark says "Forgive me this my virtue" to Gertrude (III, iv), one of its meanings is heroic; i.e., here is an exceptional person who wills virtue, is clearly aware that such a value separates him from others, and whose life must be a battle with powers, internal and external, opposed to the achievement of that goodness. "Virtue itself of vice must pardon beg" is Hamlet's perception of the paradox of his "rotten" world and it is the sort of situation most heroes confront.

The same century which saw the heroes of Corneille face conflicts of will, honor and duty also saw heroism characterized as a mad endeavor of knight-errantry in a work which suggested that attempts to redress the world's wrongs, or to avenge the oppressed and injured, were ridiculous. This aspiring hero does *not* perceive reality truly: to him flocks of sheep are armies, wind-mills giants, and galley-slaves gentlemen. Don Quixote is surely a satire on the exaggerated chivalric romances of Cervantes' time, but it exists in a limbo between the lost (classical) world and the cautious, contemporary world. As such, it is an ironic vision of an idealistic-hero mocked by and in a materialistic world. Sancho Panza, the shrewd, ignorant rustic, is a forerunner of the common man of common sense who will later sneer at heroes and heroism, and usurp for himself the hero's power of wise and honest perception, claiming that it is the heroes who are deluded or mad.

The historical moment of the decay of the hero is nearly impossible to fix across many different languages and literatures, but significant omens like Sancho appear on the horizon. Defoe's *Robinson Crusoe* presents the hero as a practical, ordinary man whose energy and self-possession overcome all difficulties; his prudence is assuredly successful, but not the stuff of high heroism. Heretofore, the hero lived in an alliance with his heroic destiny; his life was the deliberate affirmation of the existence forced on him by fate. He made destiny his own by embodying it in his action and experience. Carlyle observed that the hero's very suffering and misery come "out of his greatness." Certainly heroism was not a function of caution, sheer circumstance, common sense or random luck; the hero's deeds were strictly beyond the bounds of the common sphere. He flourished in the ideal of Renaissance humanists like Corneille, then declined as the aristocracy lost power and a commercial, middle-class culture supplanted it.

Despite attempts to find new kinds of heroes – the man of good heart, like Tom Jones, the tragic victims of fate, like Hardy's Tess – and despite romantic historical novels like those by Walter Scott, which struggled to create a world in which heroism was still possible, it could no longer flourish. Not in a modern world which like Sancho began to pride itself on its superiority to superior heroes: a world which distrusted large public gestures; which replaced individual initiative and judgment with group effort, even as the alienation and loneliness of group life increased; which saw glory and fame as mere vanity and presumption, and noble altruism as a neurotic compulsion. A world like this would see in Antigone not woman as moral hero, but an instance of unhealthy martyrdom or a proof of the Freudian theory that one-third of the female psyche is composed of masochism. If prudence is the enemy of heroism, so too does strictly scientific reason tend to destroy the hero's stature. Moral judgment can be annihilated by psychoanalytic theory or even by philosophers who refuse to discuss "meaningless" questions.

The old aristocratic ideal of heroism gradually became absurd in a bourgeoise society. Middle-class mentality was antithetical to true heroism: it cared not for honesty, courage, defiance, exploits of passion on the grand scale, but for money, power and comfort. A dialectic of pragmatic versus heroic ideals, of common versus exceptional man, occurred at many points — between hypocrisy and honesty, conformity and dissent, indecision and commitment, safety and risk, complicity and defiance, vulgarity and grandeur, and never the twain met.

The nineteenth century in England was full of such curious contrasts. Carlyle's 1840 classic work, which he dubbed "in search and study of Heroes," must be set against what has been called the eclipse of the hero in Victorian fiction.[7] When romanticism turned bourgeois, the impulse which led Byron to open his masterpiece, *Don Juan*, by lamenting the lack of heroes became somewhat irrelevant:

> I want a hero: an uncommon want,
> When every year and month sends forth a new one,
> Till, after cloying the gazettes with cant,
> The age discovers he is not the true one.

Dickens, Thackeray, and Trollope all hymned the praises of the humble and the mundane existence. Indeed, George Eliot, preoccupied with the small tragedies of average souls, compares Maggie and Tom's struggles in *The Mill on the Floss* to Homeric struggles in "the days of Hecuba, and of Hector" (Book V, Ch. II). Rather like Bryon, Carlyle observed that "Heroes have gone out; Quacks have come in," and attempted his own remedy within his study of the hero as reshaper of society. To Carlyle's heroic philosophy of history we might compare the fact of the great religious doubts of the century, when belief was shaken by Darwin's theories. With decreasing belief in a future life and immortality, without the sure convictions that inspired the heroism of saints and the bravery of martyrs, heroic action could not be justified by appeal to absolutes. There were only personally held values of respect for truth and concern for justice. The hero was even more alone, bereft of supernatural support: a modern Antigone, for example, cannot claim the gods' sanction for her heroic defiance. As Camus was to put it in *The Plague*, "can one be a saint without God? — that's the problem."

In the twentieth century, ambivalence about the hero is the norm, not only among different writers, but even within a single writer and/or a single work, where the author's attitude toward heroism may vary from skeptical disavowal to ringing acceptance. In E. M. Forster's philosophical novel, *The Longest Journey*, one character opines that "the chief characteristics of a hero are infinite disregard for the feelings of others, plus general inability to understand them." On another occasion, however, the narrator says "There comes a moment — God knows when — at which we can say, 'I will experience no longer, I will create. I will be an experience.' But to do this we must be both acute and heroic. For it is not easy, after accepting six cups of tea, to throw the seventh in the face of the hostess."

Yet even in modern unbelieving times, there are pockets of resistance to the death of the heroic ideal. In much the same spirit as the Camus character who wished to preserve "saintly" ideals without a belief in God, Miguel de Unamuno declared in *The Tragic Sense of Life*: "If it is nothingness that awaits us, let us make an injustice of it; let us fight against destiny even though without hope of victory; let us fight against it quixotically." In the preceding century, Carlyle had asserted the necessity of valuing heroic ideals: "I say great men are still admirable; I say there is, at bottom, nothing else admirable!" Wallace Stevens echoes that sentiment when he asks "Unless we believe in the hero what is there/To believe?" He concludes his poem on the hero:

> Each false thing ends. The bouquet of summer
> Turns blue and on its empty table
> It is stale and the water is discolored.
> True autumn stands then in the doorway.
> After the hero, the familiar
> Man makes the hero artificial.
> But was the summer false? The hero?
> How did we come to think that autumn
> Was the veritable season, that familiar
> Man was the veritable man? So
> Summer, jangling the savagest diamonds and
> Dressed in its azure-doubled crimsons,

May truly bear its heroic fortunes
For the large, the solitary figure.

"The large, the solitary figure" of the classic hero did not so much disappear as reappear in a different guise — the creative artist. Carlyle declared the Hero as Man of Letters to be "a product of these new ages; . . . one of the main forces of Heroism for all future ages; . . . [he] must be regarded as our most important person. . . . What he teaches, the whole world will do and make." In keeping with Carlyle's observation that "all sorts of Heroes are intrinsically of the same material," there was not as much distance between classical heroes of action and the new heroes of creativity as might at first have been supposed.[8] Stevens has noted that "The soldier is poor without the poet's lines, . . . /And war for war, each has its gallant kind" (*Notes Toward a Supreme Fiction*). The aesthetic unity of a work of art is akin to the personal unity that results from the creation of a superior self. The work of art accomplished is to its creator what the created individuality, the achieved sovereignty of the self, was to the older classic hero. Heroism, like art, is life in concentrated form, more intense than is usually available. Heroism is integrated behavior, art is synthesized experience; both epitomize a wholeness, a unity rarely present in raw reality.

Plato, who saw the artist as dangerously disruptive, ended by banishing poets from his ideal community. When mimetic theories of art gave way to theories of art as revelation of another sort of reality, the artist rose in philosophical stature, promoted metaphysically, as it were, from mere imitator of an illusory reality, thrice-removed from truth, to an important phenomenon. The idea of art as self-sufficient profession, hence the emergence of the artist-as-hero genre in literature, became current in the eighteenth century.[9] Goethe pioneered the use of artist-heroes in his fictional and dramatic works, but with mass man in the ascendency, the artist was a societal outcast. He returned the compliment, despising the common man as much as the later mistrusted and scorned him. "The great man is a public misfortune," the Chinese proverb notes, and Oscar Wilde was to remark that the public "is wonderfully tolerant. It forgives everything but genius" (The Critic as Artist").

Mid-nineteenth-century art-worship was accompanied by contempt for the common man; Philistine-baiting writers perceived the acute separation of mass culture from higher art forms. Nor was the art-for-art's-sake movement a narrowly national phenomenon. The dissociation of artist from society, of hero from common man, was as true for Baudelaire as for Oscar Wilde.

The inevitable split between the social being and the creative person had occurred.

Descended from their nineteenth-century anti-bourgeois ancestors, the artist and intellectual heroes of modern fiction displayed that maladjustment to society that results from a position at once different and alienated. The twentieth-century recipe for the artist-hero in inevitable conflict with his environment was the Joycean formula of silence, exile and cunning. Indeed, one feels one has come full circle, back to the hero as demi-god, when the artist-hero aspires to be a form of secular godhead; e.g., Joyce's Stephen Daedalus — literally "the cunning worker" who escaped his labyrinth — welcomes life like a genuine hero, and goes forth "to forge in the smithy of my soul the uncreated conscience of my race" (*Portrait of the Artist as a Young Man*). The use of words like *forging* and *conscience* unites the aesthetic and ethical dimensions of the artist's role as hero.

As for society, the classic hero and the artist hero stand in approximately the same relationship of opposition to it as to the common man. The values of the latter — property, wealth, comfort, security — are irreconcilable with the higher *ethos* of the artist-hero, who epitomized fearless use of the probing intellect, concentration of will, spontaneity, constructive power, solitary and detached existence. The artist-hero's quest requires that he be free of society's entrapments and independent of alignments to any community which might compromise his integrity or his quest for self-transcendence. In his benignly sardonic essay, 'The Duty of Society to the Artist,' E. M. Forster explores the irreconcilable differences between the artist and the society-bound average man, whom Forster is pleased to name Mr. Bumble. The artist confounds Mr. Bumble by declaring that his aims are "to experiment," to "extend human sensitiveness," to "instruct and inspire," and to create "something which will be understood when this society of ours is forgotten." "I know," he tells Mr. Bumble, "I don't fit in. And it's part of my duty not to fit in."

The artist-hero, like the classic hero of action, sees what the ordinary man cannot or will not see. Heroes are not deflected by what Edith Wharton called "the merciful veil which intervenes between intention and action" (*The House of Mirth*). Rather are they prophet figures, possessed of a faculty for seeing "the dawn before the rest of the world," as Oscar Wilde put it in 'The Critic as Artist.' Hero and artist, in what Carlyle termed their "wild wrestling naked with the truth of things," acknowledge whatever darkness and evil they discover in that "real" world allegedly beloved of the average man, yet so seldom acknowledged by him in its more unpleasant aspects. The hero-artist is "under the noble necessity of being true" (Carlyle, *H&HW*), and this heroic honesty makes him dangerously disconcerting to the common man, who often practices evasive hypocrisy with self and others. "See deep enough," Carlyle counseled, then added that the hero's obligation is to sheer truth, no matter how disturbing. Many writers have experienced Carlyle's "great Reality [which] stands glaring there" and felt impelled to "answer it, or perish miserably." Such is the clear-eyed honesty of one of W. H. Auden's lines: "This great society is going smash;/They cannot fool us with how fast they go" ('Woods'), and of Dostoevsky's passage from *Notes from Underground*:

> You see, if it were not a palace, but a henhouse, I might creep into it to avoid getting wet, and yet I would not call the henhouse a palace out of gratitude to it for keeping me dry. You laugh and say that in such circumstances a henhouse is as good as a mansion. Yes, I answer, if one had to live simply to keep out of the rain....

Heroes must have the inner strength and the vision to declare "I will put up with any mockery rather than pretend that I am satisfied when I am hungry" (*Notes from Underground*).

Perhaps *the* most dramatic extension of the concept of the hero is woman as hero,[10] for seldom is complete humanity evident in literary women characters, let alone something so singular as heroism. The hero must have a sustaining sense of his destiny and a basic autonomy upon which to rely as he makes his difficult quest. Given the many internal psychological and external societal pressures which mitigate against self-realization, women have rarely had such qualifications. Their relationship to heroism has nearly always been secondhand; they follow the male hero's destiny and remain an event in *his* life, not their own. Even extraordinary women, unable to extract themselves from an existence subservient to men, have faired badly. Queen Dido, founder of a great city, loses all for love of Aeneas; Medea, a semi-goddess of royal blood with supernatural powers, finds that once she links her life to Jason's, she is diminished — betrayed by him after all of her assistance and demoted to "barbarian" as well. Henry James, who along with Ibsen has been credited with discovering the heroic dimension of women in literature,[11] observed truly if unflatteringly: "Most women did with themselves nothing at all; they waited, in attitudes more or less gracefully passive, for a man to come that way and *furnish them with a destiny*" (*Portrait of a Lady*, italics mine).

Most *men*, it is true, have dared nothing beyond society's expectations, and most have lived constrained lives. Yet the lot of women makes even the average man's existence seem daring by comparison. Women have much more severely limited opportunities for the personal transfiguration necessary to heroism. Heroic qualities, however antithetical to the common life, are in even greater opposition to *woman's* traditional life of domesticity. Schopenhauer assures us of a war between genius and motherhood; Louise Bogan's poem, 'Women,' charges that "women have no wilderness in them," that "they wait, when they should turn to journeys."

Moreover, the chasm between purpose and perception, between action and consciousness, is experienced very differently by the woman as hero. The male hero's problem, that of joining wisdom to behavior, and his often fatal error of acting unwisely or arrogantly, is nearly the exact reverse of the typical dilemma of woman as hero. In her, consciousness tends to outpace the possibilities of action. She may grow to be both discontent and insightful, then find she possesses no existential arena in which to act upon her perceptions and convictions. Like some Cassandra who is never believed but always right, she is frequently doomed to being both correct and powerless.

Woman's task as hero involves her in a radical conflict of values, for the heroic stance does not supplement her traditional, gender-assigned characteristics, as it does with males. (Take, for example, the matter of assertive action necessary for heroism; even the average man is allowed assertiveness, the male hero is only more so.) Woman as hero finds herself in conflict with her womanly destiny, which is not to compromise her stereotypic passivity. She must not allow herself to be confined within femininity if she would be a hero; yet she must strive to avoid being "unfeminine" in her heroic ambition. Attempting to fulfill that impossible task, she discovers that being a hero (unlike the male situation) may cancel out her rightful claim to sexual identity.

The two heroic women I discuss below solve these dilemmas in very different ways, one as a classic, active/ethical hero, the other as a modern, creative/aesthetic hero. I have deliberately chosen to explore two works which span vast differences of time, culture, language, and sex of the writer, in order to throw into the sharpest possible relief the polarities of darkness and light, tragedy and comedy, pessimism and optimism, which the heroic vision of the human condition may entail.

Excepting historical figures later appropriated by literature (such as Joan of Arc), the first and perhaps finest example of female heroism was envisioned by a male playwright some twenty-four hundred years ago, when Sophocles created his maiden of the daring deed, Antigone. The play has metaphysical, ethical, epistemological and axiological dimensions, and its complex antinomies

—family allegiance versus political duty, religious versus secular responsibility, private versus public morality — are all expressed through a clash between two individuals who represent different ways of knowing and acting. Creon is pragmatic and materialistic, a political animal; his is the way of common sense. Antigone is intuitive, inner-directed and uncompromisingly loyal. Initially, the play suggests that Creon is the voice of reason and stability in the state of Thebes, Antigone the irrational fanatic who would disrupt the social order with her rebellious assertion of human and religious values. Until, of course, events prove otherwise.

The play opens with a discussion between Antigone and her sister Ismene of Creon's edict that the body of their brother Polynices should remain dishonored and unburied. The sisters' exchange establishes Ismene's pusillanimous toleration of tyranny, which in turn throws Antigone's integrity and devotion to principles into sharp contrast. Like most heroes, Antigone will have to act alone. The common sense values of her average sister are quite unequal to the exceptional task, and the average citizens of Thebes, although they believe Antigone is right to bury Polynices, have not the courage to support her defiance of the king. "All here would own that they thought it well, were not their lips sealed by fear. . . . they curb their tongues for thee," Antigone informs Creon. He tries to intimidate her into submission with the suggestion (which is not only false, but cannot work, because Antigone's is the heroic, not the common value system) that she is alone in her "peculiar" conviction. Moreover, the situation is illustrative of another truism about heroic action: despite his defensive scorn or lack of overt support for the hero, the common man needs heroic behavior in the world, needs someone to act out the very ethical principles which he fears to uphold. The hero is repository of the best principles of which mankind is capable.

Antigone's heroic behavior derives not just from her deeply felt emotional obligation to bury her brother, but from considerable intellectual ability, which sustains her in her exchanges with the Chorus and Creon. Her mental powers match her moral strength, making her a hero whose integrity is a fusion of right thinking and right doing, not accidental bravery. In her forensic confrontations with Creon, Antigone's arguments show her to be a considerable moral philosopher, interested in a higher ethical order than Creon's arbitrary political prohibitions. She is also something of an epistemologist in the rough who puts Creon's certainties about what the gods require and the sophistic distinctions by which he means to exclude Polynices from honorable burial into proper perspective. Her challenge is epistemologically incisive — "who knows?" Her acknowledgement of the limits of knowledge, her implicit recognition that the complex relationship between the knower and the known cannot be dismissed by Creon's arbitrary categorizing and blanket generalities is impressive.

But if she knows when to challenge irrelevant distinctions, she also knows when to invoke crucial ones. When Creon asks her if she is indeed the one who has dared to transgress the publically proclaimed law, her reply goes straight to the heart of the matter: "it was not Zeus that had published me that edict." She thus distinguishes the Law from arbitrary laws, the eternal moral code from the fleeting political prohibition, religious ordinance from secular whim. She sees that *rules* and *justice* are not synonymous; that there are conditions under which rebellion is not an evil but a necessity; that human society — to be human in any meaningful sense of that word — must rest not upon blind obedience to the imposed rules of a fallible leader, but upon a moral order in which what takes precedence is the discharge of obligations *to other human beings*, not to an abstract entity called "the state."

Moreover, she declares: "But if I am to die before my time, I count that a gain: for when anyone lives as I do compassed about with evils, can such a one find ought but gain in death?" This moves beyond the classic wisdom that the unexamined life is not worth living to the judgment that the morally corrupt life is not worth living. Antigone will oppose, not conspire in, the corruption of the world, even on pain of death. Creon's promise of doom Antigone calls "trifling grief," for to suffer the dishonor of her brother's unburied corpse is to her the *spiritual* death which true heroes must resist. In her severely beautiful moral world, virtue is indeed its own reward, and heroism is the risk of death for one's principles. The punishment and the honor both belong exclusively to her. Thus she disavows Ismene's tardy offer of partnership in blame for a deed in whose execution she (i.e., Ismene) had no part. It is, once again, starkly perfect justice, too strong a brew for average tastes.

So acute is Antigone's perception of her situation and its consequences that she can tell Creon: "if my present deeds are foolish in thy sight, it may be that a foolish judge arraigns my folly." If it is hubristic to judge one's judges, it is also absolutely necessary, lest absolute power like Creon's corrupt absolutely. Her stubborn persistence in her high moral purpose should not be seen as a tragic flaw: *that* defect resides in Creon, who ironically enough projects it onto Antigone (as does the Chorus upon occasion).

As true hero, Antigone is obliged not to abandon her convictions, whatever the cost. Defiant and determined, she insists that she has obeyed the higher laws of heaven. When she is led away to the entombment which Creon inflicts upon her, she utters a last prophetic curse: "if the sin is with my judges, I could wish them no fuller measure of evil than they, on their part, mete wrongfully to me." This too is a threat of perfectly balanced, eye-for-eye justice — one which turns out exactly as Antigone had predicted, for her punishment was not righteous, and the gods do indeed exact punishment from Creon, completing Antigone's justification.

When Antigone kills herself in the sunless, subterranean prison in which Creon has buried her alive, the ruthless purity of her moral life is sustained. For her act of self-murder completely and totally makes her what the Chorus had earlier called "mistress of thine own fate." She has mastered fortune with a freely chosen act, which deprives Creon of the decisive role in her death, mocking his "power" over her fate and outwitting his tyranny by dying *her* way, not his. Dying by her own hand before he can arrive to release her cheats Creon out of a pseudo-moral gesture, a handy hypocrite undoing of his initial error. Creon's moral fence-mending comes too little and too late. Through her suicide Antigone becomes the agent of the consummation she prophesied for Creon. Had she not taken her own life before Creon could prevent it, neither Haemon nor Eurydice would have committed suicide, and Creon would have been spared the terrible, crushing punishment of enduring the deaths of his wife and son, both of whom died hating him for his folly. It is, finally and ironically, Antigone who had the "power" to "determine" Creon's life. The restless divine plan has been fulfilled through a relentless moral agent, this mere young girl whose combined intellectual insight and moral passion caused her to do her duty — in other words, to fulfill her heroic destiny — as she saw it.

Some twenty-four hundred years later, Virginia Woolf responded with unusual intensity to the image of Antigone, declaring that she was impressed in Sophocles' play "by heroism itself, by fidelity itself" ('On Not Knowing Greek,' *The First Common Reader*). The power this play had over her imagination is evidenced in the fact of her frequent allusions to it or its heroine. For example, in her last feminist pamphlet, *Three Guineas*, and in the novel, *The Years*, which serves as a fictional companion piece to the factual work, the same line (in classical Greek) from *Antigone* is quoted — the splendid moment when Antigone informs Creon "Tis not my nature to join in hating, but in loving." Woolf asserts: "Lame as the English rendering is, Antigone's five words are worth all the sermons of all the archbishops" (*Three Guineas*).

"Consider the character of Creon," Woolf begins. "Consider Creon's claim to absolute rule over his subjects. That is a far more instructive analysis of tyranny than any our politicians can offer us." As for discerning true loyalties which we must honor from the false which we must oppose, Woolf advises: "Consider Antigone's distinction between the laws and the Law. That is a far more profound statement of the duties of the individual to society than any our sociologists can offer us."

Given her insight into the relevance of Sophocles' play and her admiration for Antigone, Woolf's own version of woman as genuine human being and hero is most interesting to contemplate: it is Orlando, protagonist of Woolf's like-named comic fantasy.

In Woolf's value world, the ultimate achievement is creative consciousness. She transforms the Sophoclean emphasis on ethics, where woman as hero is a moral hero, into an aesthetic realm where woman as hero is a creative artist. This transmutation reflects both the general literary tendencies which I have indicated above, and Woolf's personal predilection for stressing the aesthetic dimension of reality.

Orlando merges fiction, biography, literary history, comedy and fantasy as it tells its incredible story of a human being named Orlando who begins as a male in the Elizabethan era and survives three and one half centuries – changing sex along the way – to end as a thirty-six-year-old woman at midnight on "Thursday, the eleventh of October, Nineteen Hundred and Twenty-eight." Behind this fantastic conceit, however, is a serious philosophic purpose, for Woolf's Orlando, like Hobbes' Leviathan, is a conglomerate image in which are lodged various philosophic truths about human nature and society. For the philosophic mind in literature, one of the most potent metaphors has to do with the changing of the forms of objects. The metaphysics of Woolf's *Orlando* is metamorphoses: of physical objects, sexual gender, literary styles, approaches to literary criticism, consciousness itself – all of which are in constant transition from one form to another throughout the novel. Not even James Joyce believed more completely than Woolf that language is the raw material out of which we create the reality of the world. In *Orlando*, Woolf works out her conviction that idea is form: human personality, the art it creates, the societal institutions built and destroyed in historical time, are all articulations of an idea. Reality is a literary text; truth is a function of style; the self is a created work of art – such intricate propositions are at the heart of Woolf's brilliant romance.

Orlando is a mock "Biography" twice over, for if it tells the story of the life of the poet, it also tells the continuing saga of his/her poem. We first meet Orlando as a sly parody of the classic male hero: an Elizabethan soldier-adventurer whose deeds match his name, which contains the essence of the medieval French and Italian romances of the hero. Yet if he is the image of masculine derring-do, Orlando is also a philosopher-writer; when he is not penning literary compositions, he is contemplating the relationship of art to beauty. Specifically, he ponders *green* as it occurs in nature and as a word which can be made to conform to the poetic demands of meter and rhyme. He sees the poet as a special being who can conjure up "ogres, satyrs," but who also understands "the depths." "Afflicted," as Woolf comically puts it, with a love of literature, Orlando composes, before reaching the age of twenty-five, "some forty-seven plays, histories, romances, poems; some in prose, some in verse; some in French, some in Italian; all romantic, and all long." Having sampled the active and the artistic life, Orlando decides "there was a glory about a man who had written a book ... which outshone all the glories of blood and state." *His* image of a hero, therefore, is neither soldier nor statesman (classic hero figures), but an Elizabethan he once glimpsed in the flesh named Shakespeare.

Having established an early tension between art and reality, Woolf cunningly names Orlando's major poetical work "The Oak Tree," and repeatedly pictures him sitting under the physical object in nature after which his subjective mental creation is named. The title, in addition to marrying nature to poetry, suggests almost in a punning way the novel's themes of continuity and ancestry (i.e., a family tree, one's roots). The poem endures, as Orlando endures, all the radical changes — social, sexual and cultural — which Woolf's plot inflicts upon it. The manuscript is a palpable paradox which represents stability persisting through change, permanence in a transitory world. Orlando never neglects to carry it about with him; it is a stubborn emblem of the analogous development of selfhood and literary skill, going on simultaneously in time.

At the apogee of Orlando's masculine identity, he achieves a Dukedom and promptly falls into a "profound slumber" of seven days, from which he wakes utterly naked and a woman. The symbolic significances of his sex change are several, one of the most important being the implication that the fully realized human personality is androgynous, an idea Woolf had certainly encountered in Plato's *Symposium*. The literal metamorphosis from male to female suggests that the creative self is a psychological hermaphrodite. That is to say, any female poet has had, from the beginning, a fundamentally masculine aspect in her character; her imaginative life has been shaped by male socio-cultural realities. Orlando's first-male-then-female nature suggests that a male poet must awaken from its deep slumber the hidden female side of his nature and give it proper expression. The power of creative consciousness derives from an androgynous completeness beyond gender; the artist is the universal voice of humanity precisely because he is all things; male and female, knower and known, person and poem.

However comic and fantastic the mode in which Woolf chose to convey these ideas, the same sort of human dynamic operates in the Sophoclean play Woolf so admired. Like Orlando, Antigone is an androgynous composite of "womanly" qualities and qualities (intelligence, defiance and courage) usually deemed "masculine." Nor is it mere coincidence that the ultimate figure of wisdom in Sophocles' play is the blind prophet Tiresias, that androgynous being in whom, as in Orlando, both sexes meet. At the apex of heroic conflict and struggle is an androgynous level where augmented, conglomerate powers may be exercised and unification achieved.

The female Orlando discovers that while her anatomy may have altered radically, her talents and aspirations remain the same. She has retained the love of nature which characterized her as an Elizabethan boy, and she still is prone to philosophic reflections upon beauty, reality and truth. The tattered manuscript of her poem serves as a perpetual reminder "of the glory of poetry, and the great lines of Marlowe, Shakespeare, Ben Jonson, Milton." She realizes that she is nurturing "a spirit capable of resistance" — a phrase strikingly reminiscent of Antigone, and she vows to write "what I enjoy writing," echoing an earlier resolution, made when she was male, to "write, from this day forward, to please myself."

Orlando's Antigone-like spirit of resistance is put to the supreme test in the eighteenth and nineteenth centuries. Just as Antigone found herself in opposition to Creon and the laws of the land, Orlando finds herself in conflict with the misogyny of the greatest writers of the Augustan age — writers like Lord Chesterfield, Pope and Addison. Their denigration of women, combined with self-assured reasonableness, is almost as ironic a phenomenon as the "reasonable" King Creon. When Orlando accidentally spills some tea on Mr. Alexander Pope, his vengeful response is "the rough draught of a famous line in the 'Characters of Women'" — Woolf alludes here to the opening of the second *Moral Essay*, in which Pope declares "Most women have no Characters at all." Like laws and arbitrary edicts, literature may be the vehicle of oppression. If, as Woolf puts it, "the rapier is denied" a man, he will not long hesitate "to run her [i.e., a woman] through the body with his pen."

In the nineteenth century, Orlando's rebellion against the spirit of the age becomes even more acute, for in the repressive Victorian era women are no more tolerated "to range at large" (the phrase is Creon's and Woolf quoted it it her feminist writings) than they were in antiquity. Trapped in a historical period which glorifies only the narrowest of roles for women, Orlando, comically enough, can write only overdecorated, maudlin and insipid Ladies Verse, her florid poetry mimicking the sentimental realities of her claustrophobic world. Nevertheless, in spite of all, "she wrote. She wrote. She wrote." Woolf's blunt rhetoric captures Orlando's stubborn commitment to her craft. The manuscript of "The Oak Tree," dated 1586 in her former "boyish hand," is by now "sea-stained, blood-stained, travel-stained," but it endures as a fitting talisman of her struggles to perfect her art. Like the actual oak tree, the poem grows and develops into its maturity, surviving more than three centuries of drastic change, even as its creator.

Orlando manages to surmount all obstalces and finish her poem. It is published; she achieves fame; she wins a prize. After all this is accomplished, Orlando achieves the integrity of identity which comes to all heroes; she calls it "a single self, a real self." With it comes a heightened, comprehensive consciousness, "as if her mind had become a fluid that flowed round things and enclosed them completely." She makes a pilgrimage to the ancient oak tree which inspired her poem and considers burying an autographed copy of the first edition in the ground which covers the tangled roots of the great tree. Although she rejects the ritual as silly and sentimental, the mere consideration of such a ceremony once again underscores the symbolic mingling of eternal nature with immortal literature, the synonymy of "Oak Tree" as verbal construct with oak tree as object-in-nature. The integrated, self-actualizing hero Woolf has invented heals the gap between inner and outer, subjective and objective worlds. It is as Orlando had wondered at one point: "Life? Literature? One to be made into the other?"

As nature and art come together in her poem, so Orlando's disparate selves merge, and she has a final experience of unity and integration. The novel closes on this visionary note, a humorous epiphany in which Orlando experiences "ecstasy!" at the stroke of midnight in the present time (1928). She sees spring up into the sky overhead "a single wild bird . . . the wild goose." Serious and comic ramifications of that image surely apply: the bird symbolizes the unseizable truth and beauty of art, and the soaring flight of the artistic imagination, reiterating Orlando's earlier analogy of "wild bird's feathers" with the dreaming of "wild dreams." On the comic level, the image informs readers that Orlando's wild goose chase of some three and one-half centuries has at last come to fruition. The allegory of the poet's progress is complete. Woolf's comic vision of the creative powers of the self, of artist as hero, ends as comedy should — with the joyous achievement of Orlando's personal and artistic destiny.

For all its brilliant satire and learned irony, beneath Woolf's fantasy is a totally serious valuation of the authentic human being and of the literary endeavor. Woolf's protagonist is a hero who does not compromise his/her convictions and who remains steadfastly loyal to a difficult purpose until it is realized, despite ridicule and opposition (in the form of critical attacks). In both Antigone and Orlando we see total commitment to what each perceive as worthy and true. As Antigone never was diverted from her noble purpose of burying Polynices, Orlando never abandons his/her struggle to become a poet. As a male, Orlando did not doubt that being an artist was superior to other modes of existence: he conceived of literature in high philosophic terms as "the Bride and Bedfellow of Truth." In womanly form, Orlando persists in seeing poetry as "a dialogue with herself about this Beauty and Truth." Borrowing her metaphors from Plato, Woolf uses light as her symbol of knowledge and truth: Orlando perceives genius as the "most august, most lucid of beams," for example, or as "the lighthouse in its workings" or as "the only light that burns forever. But for you [i.e., poetic genius] the human pilgrimage would be performed in utter darkness." In a coupling of aesthetic and ethical values which transfigures each (and suggests the spiritual values that pervade the *Antigone*), Woolf has Orlando compare "art and religion," calling them, Platonically enough, "the reflections" of things ordinarily hidden from our knowledge by darkness and ignorance.

Woolf's ethic of art-as-religion is most strikingly revealed in a scene where Orlando is revising her manuscript of "The Oak Tree." There, the analogy of poetry to morality is overt; Woolf's terminology is deliberately ethical and religious. What is called Orlando's "faith" and her "religious ardour" is in fact her desire for literary excellence. There is consciousness of something "good" and an effort to realize it; conversely, the imperfections of Orlando's spiritual state are indistinguishable from the imperfections of her poem. Her *sins* — again the religious terminology — are such technical shortcomings as the excessive sibilance of the opening stanzas of her poem. The letter "S" is playfully called "the serpent in the Poet's Eden," and if "S" equals "these sinful reptiles" in Woolf's comic algebra, then the present participle is "the Devil himself."[12] Keeping to her theological ambiance, Woolf informs us that "the first duty of the poet" is to resist the "temptations" of bad art. "The poet's, then, is the highest office," Orlando concludes. "We must shape our words till they are the thinnest integument for our thoughts. Thoughts are divine."

Antigone's sense of spiritual responsibility was certainly more tragic, but Orlando's is no less keenly felt for being rendered in a comic context. Woolf's allegiance is to art for truth's sake. Her androgynous woman is heroic because her conception of the function of the literary artist is heroic. This conception, incidentally, is a constant of her fiction — it appears with great lyric power in *The Waves* — and her nonfiction. In her *Writer's Diary*, she speaks of "the creative power" as a force which "brings the whole universe to order," and of experiencing "the exalted sense of being above time and death which comes from a writing mood." Words like *exalted* should leave little doubt of the depth of Woolf's commitment to aesthetic values. Even the humor and playfulness of her farcical biography cannot hide the profound respect for creative consciousness which infuses the work.

Woolf's fictional chracter is convinced that artist-heroes can be decisive in the world of historical events; in *Orlando* it is asserted that "a silly song of Shakespeare's has done more for the poor and the wicked than all the preachers and philanthropists in the world." Woolf declared in her *Diary*: "It seems to me more and more clear that the only honest people are the artists. ... [not] social reformers and philanthropists ... [who] harbour so many discreditable desires under the guise of loving their kind." That 1919 remark, at the beginning of her writing career, may be compared to a remark made twenty years later, very near the end of her life: "I feel that by writing I am doing what is far more necessary than anything else." [13]

Not the least aspect of the artist's "honesty" is his heroic ability to distinguish between the apparently real and the really real, to discern the truth behind surface forms. Because Woolf sees aesthetic activity as a search for truth and significance, creativity becomes an ethical act. In *Orlando*, Woolf calls the poet "Atlantic and lion in one," for someone who "can destroy illusions is both beast and flood." If life is a dishonest or illusory dream for most people, the truth which wakens one is a truth brought by the poet as hero. If the performance of a moral deed can transform reality, so can the truthful fictions of the artist. Heroic beings like Antigone (through action) and Orlando (through language) shape not just their own destiny, but the destiny of others: both are agents of enlightenment, in the tragic and the comic mode.

If the dark side of woman as hero is the noble death of Antigone, the bright side is Orlando's triumphant survival. In Sophocles' play, the consequences of events are overwhelmingly fatalistic, the world view deeply pessimistic. By contrast, in Woolf's work, comedy, wit and fantasy combine to produce glimmerings of Platonic Idealism which sound a distinctly optimistic note. Woolf's work implies that victory and fulfillment occur when the dualisms, antagonisms and contradictions of mind and matter are eradicated in the experience of unity. The figure of Orlando is Woolf's exaggerated embodiment of such comprehensive wholeness. The novel's vision of fulfillment is *qualified* by Woolf, who was a relativist and too great an artist for oversimplification, but it is not *contradicted* by the satiric suggestion of the wild goose chase on which the work closes. The questing imagination, the wild goose chase after artistic excellence goes on. If we are willing, like heroes, to continue the struggle despite life's complexity and delusions, we may discover meaning, even that faint flash of "ecstacy!" which visits Orlando in her final moments of glory. Without being shallow in its optimism, Woolf's work proposes the creative power of art as morality, meaning, even salvation. As surely as Antigone's ethics, Orlando's aesthetics is heroic behavior in a world which could use examples of inspiring human conduct.

Transformation of the self and the world is the goal of the hero of action, of thought, or of creativity, be that hero male or female. The heroic act of warrior, intellectual, or artist is always, ultimately, dynamic and life-affirming. It preserves those life-enhancing values which are in danger of extinction. Even when it is not totally realized, the attempt, the process itself, is a demonstration of courage and imagination. The hero can turn defeat itself into victory, because the values for which he fought — justice, honesty, whatever it may have been — survive him. Rilke reminds us: "Consider: the Hero continues, even his fall/was a pretext for further existence, an ultimate birth." [14]

NOTES

[1] W. H. Auden, *The Enchafed Flood or the Romantic Iconography of the Sea* (London: Faber and Faber, 1951), p. 83.

[2] See Northrop Frye, *Anatomy of Criticism: Four Essays* (New York: Atheneum, 1965).

[3] The W. B. Yeats line is from 'Among School Children'; the Stevens quote from 'Examination of the Hero in a Time of War.' To eliminate myriad footnotes for the various works cited in this essay, I shall identify, parenthetically in the text, the work and its author (the exception being translated works). When no source is indicated, the quotation is understood to be taken from the work identified in the preceding notation.

[4] 'Guy Faux: The Same Subject Continued,' a review from the *Examiner*, November 18, 1821.

[5] The translation cited throughout is that of Sir Richard C. Jebb, the one with which Virginia Woolf worked.

[6] See *Heroic Love: Studies in Sidney and Spenser* by Mark Rose (Cambridge: Harvard University Press, 1968).
[7] Mario Praz, *The Hero in Eclipse in Victorian Fiction*, trans. Angus Davidson (London: Oxford University Press, 1956).
[8] Auden believes heroic authority "over the average" is of three kinds: "aesthetic, ethical, and religious" (*The Enchafed Flood*, p. 83).
[9] Maurice Beebe, *Ivory Towers and Sacred Founts: The Artist as Hero in Fiction from Goethe to Joyce* (New York: New York University Press, 1964).
[10] In a recent book, Ellen Moers coined the awkward word *heroinism* "because I could find nothing else in English to serve for the feminine of the heroic principle" (*Literary Women: The Great Writers* (Garden City, N. Y.: Doubleday & Company, 1976)). I have been using the word *hero* in a genderfree manner – as, for example, *mankind* – to signify the hero of either sex, and use it now to indicate both the aesthetic and active aspects of woman as hero.
[11] Carolyn Heilbrun, 'The Woman as Hero,' *Texas Quarterly* 8 (1965), 134.
[12] Cf. Woolf's September 7, 1924, entry in the *Writer's Diary*: "It is a disgrace that I write nothing, or if I write, write sloppily, using nothing but present participles."
[13] 'A Sketch of the Past,' in *Moments of Being: Unpublished Autobiographical Writings*, ed. Jeanne Schulkind (New York: Harcourt Brace Jovanovich, 1976), p. 73.
[14] Rainer Maria Rilke, *Duino Elegies*, trans. Stephen Spender and J. B. Leishman (New York: Norton Lirbary, 1963), 'The First Elegy,' lines 40–41, p. 23.

ROBERT MAGLIOLA

PERMUTATION AND MEANING: A HEIDEGGERIAN *TROISIÈME VOIE*

Let us introduce the issue by way of a practicum. Examples from the ambiguous writing of Cervantes, Nietzsche, or Kafka are usually adduced here (the controversy we are entering has a long history), but our practicum, though serving the same purpose, will be a different and therefore possibly a more refreshing one. Our "case in point" will be Flannery O'Connor's story 'The River.'[1] O'Connor's compacted plot recounts less than two days in the life of a child named Harry, who is about "four or five" years of age. His mother and father party in their apartment, and send him off to baby-sitters. As the story opens, a Christian woman named Mrs. Connin is about to assume charge of Harry for the day. On the way to her home, she announces she will take Harry and her own family to a healing service, where the Reverend Bevel Summers will preside. When she asks Harry his own name, the boy lies – declaring his name is Bevel. Mrs. Connin is astonished that the boy's name is the same as the preacher's. At her home, she introduces Harry to Christianity. Immediately afterward, her own children maltreat the boy. Then the whole entourage attends the healing service, conducted at water's edge. The preacher seems to identify the earthly river and the "river of Jesus' Blood." Throughout, a heckler named Mr. Paradise jeers at the proceedings. At Mrs. Connin's behest, the preacher then baptizes Harry – immersing him in the river water. When the boy returns home, he finds another party in progress. The next morning he goes to the river, intending "not to fool with preachers any more but to Baptize himself" (p. 50). Mr. Paradise, who lives near the river's edge, sees the boy headed for the water. Hoping to offer Harry a peppermint stick, he follows. The boy wades far into the water, and is soon sucked under. As the story closes, Mr. Paradise emerges from the water, his attempt at rescue thwarted. Little Harry has drowned.

Flannery O'Connor's 'The River' permits variant interpretations, two of which are clearly contradictory. 'The River' can be read as a vindication of secularism, and it can be read as a testament to Christian redemption. I propose to use this situation as a model for a whole congeries of hermeneutical problems. Do the norms of validity permit variant interpretations? Or must one and only one exegesis be adjudicated as "correct"? If one argues for multiple validity, does such a position lapse into relativism, or is there a middle ground between absolutism and relativity? What if the interpretations are not only variant but contradictory — a problematic that applies to 'The River'? And finally, what if variant interpretations unfold diachronically, so a given literary epoch interprets a work in one way, and a later epoch in a different way? Does the literary work itself change?[2] The first part of my essay offers substance from which the above questions can be evolved. The much longer second part attempts to bring forth solutions, and these are founded on section 32 (entitled 'Understanding and Interpretation') of Martin Heidegger's *Being and Time*.[3] The solutions proposed appear in the last chapter of my newly published book, *Phenomenology and Literature*,[4] but in the present essay enjoy the advantages of emendation. The critical feedback which follows upon publication of a book has apprised me of the arguments which have needed elaboration or clarification.

I. PRAXIS: VARIANT INTERPRETATIONS OF 'THE RIVER'

1. *A Secularist Interpretation*

In a society that is absurd and malicious, the only hope is to trust in the radically human — to trust in human instincts and the virtues they evoke. Grandiose belief in "another world" which puts all things aright is a dangerous illusion. Religion distorts the personality, and comports death. Thus the story accompanies Mrs. Connin with the imagery of death: she is a "skeleton" (p. 34) and has a "skeleton's appearance." Christianity has lubricated Mrs. Connin's pride, so she engages in the most egregious habit of the religionist: she sits in righteous judgment over the worth of others. She criticizes her husband's lack of faith, and disdains the alcoholic mother of Harry. Well-indoctrinated in Christianity, Mrs. Connin's children give Harry a cold welcome: "They looked at him silently, not smiling" (p. 34). In order to embarrass the boy, they seduce him into releasing the bottom board of the pig sty, so the pig escapes. Consistently, Mrs. Connin and her family exhibit uncharitableness — a sad comment on the religion they represent. As the Connins set out for the healing service, the death motif resurfaces: the whole contingent looks "like the skeleton of an old boat" (p. 37). At the service, the believers are grotesques, bearing in their bodies the sign of inner deformity. An old woman "with flapping arms whose head wobbled as if it might fall off any second" wades into the river, and moves "in a blind circle" (p. 41): she is the archetypal Christian. Indeed, the name of the preacher itself is symbolic. "Bevel" as a noun means "slant," and the clergyman is "slanted," biased, warped because of religiosity. And Harry, in assimilating the name, assimilates the self-destructive impulse. As for Harry's own baptism, it is repugnant to his instinctive nature — his body resists immersion, and he rises from the water with eyes "closed and half-closed" (p. 44). The surprising end to 'The River' is a specimen of black humor without compare. The river which is supposed to be the "River of Life" draws Harry to gruesome death. Significantly, the one heroic gesture of the story is executed by Mr. Paradise, the secularist who skoffs at the healing service. His name is of course ironical: the only true paradise is a kind of grim existentialism. In sharp contrast to the rarified imagery associated with religion, Mr. Paradise appears as primordial earth: he is a "humped stone" (p. 42) and an "old boulder" (p. 50). The old man struggles in desperation to save the boy, but all for nought. Cadaverous Christianity has triumphed.

2. *A Christian Interpretation*

Harry had lived in a household were "everything was a joke" (p. 43). After baptism "you'll count" (p. 43), the preacher promises the boy. With its message of redemptive love, Christianity brings meaning to an otherwise absurd world. Harry is destined to be an "old sheep" (p. 32) of the Lord, so Mrs. Connin introduces him to Christ. She reads from *The Life of Jesus Christ for Readers Under Twelve*, and displays the illustrations. One shows Christ "the carpenter driving a crowd of pigs out of a man" (p. 38). Mrs. Connin is herself a disagreeable person, but the process of exorcism lasts a lifetime. The Kingdom of Christ does incessant battle with the pettiness of earth — Christ does not achieve absolute victory until the Parousia. Even in the elect of the world, good and evil stand side by side.

As for Harry, he is from the start driven by grace to seek the kingdom. When Harry reports his name is Bevel, the name of the preacher, he expresses an innate desire to identify with the Christian message. It is significant that the narrative of the story itself affirms this subterfuge, and thereafter designates the boy by his new name. There is symbolic import in Harry's words, "Will he [the preacher] heal me? ... I'm hungry" (pp. 33, 34). After Mrs. Connin shows the boy *The Life of Jesus*, he steals the book, but even this should be seen as an attempt to assimilate the Kingdom. The image of the sun likewise functions as a symbol of Harry's spiritual desire. On the way to the service, the child wants to "snatch the sun" (p. 39). Indeed, throughout the story the sun appears and disappears: when Christ's love is triumphant, the sun shines; when human resistance thwarts the movement of grace, the sun is hidden in mist. Before reaching the river, Harry passes through dark woods and the sun vanishes. The woods are a "strange country" (p. 39) where the boy has difficulty keeping afoot. The episode symbolizes the dark night of temptation – a combination of Dante's dark Wood and Bunyan's Slough of Despond. Harry even meets the Devil, "two frozen green-gold eyes enclosed in the darkness of a tree hole." When the boy emerges from the forest, the "sun . . . set like a diamond," is awaiting him.

Then the preacher begins his sermon, proclaiming the "River of Life, made out of Jesus' Blood" (p. 40). Here the color red, the color of the precious Blood, assumes importance as a motif. "Red light reflected from the river" (p. 79) bathes the preacher's face, and later his face "burned redder" (p. 41). He baptizes Harry, announcing "You won't be the same again. . . . You'll count" (p. 43). The preacher's name is "Bevel," which bears the lexical significations "incline" or "ascent." The preacher builds the "incline" whereby Harry mounts to salvation; paradoxically, as St. Paul says, by immersion in the saving waters he rises to new life. The next morning, Harry goes back to the river. The salvific sun, "pale yellow and hot and high" (p. 49), follows him. At this point the symbolism of the story permits only one reading – the waters are not an earthly river, but the flow of Christ's grace. In the beginning Harry's body refuses to submerge, indicating the resistance human life offers to salvation. At the end, the "River of Life" prevails, catching the boy "like a long gentle hand, and pulling him swiftly forward and down" (p. 50).

Salvation foils the mysterious Mr. Paradise, who can only be an inverse paradise – Satan himself. Clues to the old man's identity had begun much earlier, when he had derided the preacher at the healing service. Staring for a moment at Mr. Paradise, Harry had instinctively recoiled: "Bevel [Harry] stared at him once and then moved into the folds of Mrs. Connin's coat and hid himself" (p. 42). The preacher had glanced at Paradise and then cried to the crowd, "Believe Jesus or the Devil. . . . Testify to one or the other!" (p. 42). At the end of the story, when the old man tries to balk Providence, he is called "a giant pig." Earlier, the illustration in *The Life of Christ* had pictured Christ "driving a crowd of pigs [devils] out of a man." The equation is self-evident: Mr. Paradise is really the Satanic force, prowling to destroy souls. In sum, the tale 'The River' is a kind of *Märchen*, replete with symbols of God and the Devil, goodness and evil. And in it, Christianity wrests an innocent soul from destruction, and absorbs that soul into the life of grace.

II. THEORY: AN INQUIRY INTO THE SIMULTANEOUS VALIDITY OF VARIANT INTERPRETATIONS, AND RELATED ISSUES

Is 'The River' a call for religious extrapolation, that is, the *Aufhebung* of mundane life into the supernatural? Or is the story a summons to retrenchment, so that man should confide in human nature alone? Is the skeletal imagery just *prima facie* portraiture — as a Christian reading would argue, or does this imagery connote more sinister meanings? Are the "foreign greengold eyes" in the tree hole a symbol of Satan, or are they a simple description of an owl — as a secularist reading could maintain? Christian and secularist interpreters can go on and on, arguing their respective cases. One alternative in such a situation is to insist that only one of the "cases" can be "correct," that is to say, "true." And this is so even if the debate is prolonged, or seemingly insoluble. There is a correct meaning even if it be forever unknown.[5] Another alternative is to argue that the reader can accept both of the conflicting interpretations as equally correct, and that he can accept many others as well. Indeed, this alternative is Heidegger's special claim.[6]

We shall now work our way through section 32 of *Being and Time*,[7] adjusting its more universal applications so they apply to literature as such. The start of the hermeneutical process, begins Heidegger, is the projection of *Dasein* (the human person) toward the possibilities "laid open" by a text. That is, the "aspects"[8] of a given text are open to the sight of the interpreter, and each of these aspects has the *potential* to participate in meaning — as long as the interpreter goes on to "meet" the aspect in question appropriately. The aspects, then, present "possibilities" for engagement in meaning. Significantly, "these possibilities, as disclosed, exert their own counter-thrust [*Rückschlag*] upon the human person" (p. 188). This counter-thrust collaborates in the mutual implication of interpreter and literary work, so that both belong to the same ontological field (that is, they share the same being).[9] In the first stage of hermeneutical activity, the critic is *at-one-with* a text. (We shall treat the words "a text" and "a work" synonymously, and we shall mean by them a composite of verbal signs,[10] and relations among verbal signs, identified by a society as a literary entity, distinguishable from other literary entities and all other entities.)

Next Heidegger says that the "development of the understanding [*Verstehen*]" is "interpretation [*Auslegung*]" proper. Though he will shortly describe the development more clearly, it suffices now to point out that by "understanding" he means the unitary ensemble wherein critic and text are one; and by "interpretation" he means a phenomenological description [11] of this understanding. Heidegger attaches an admonition: "Nor is interpretation the acquiring of information about what is understood; it is rather the working-out of possibilities projected in understanding" (pp. 188, 189). By all of this, he intends to place the origin (of what becomes hermeneutical activity) at the level of understanding (which is the level of primal Being),[12] and not at the level of "assertion" (*Aussage*) or even interpretation. To rephrase it, Heidegger maintains that philosophy should become meditative "Thought about Being" — philosophy should take as its main concern how entities are the same, not how entities are different from each other. Insofar as entities are the same, they share the same being — they are, indeed, one and the same! Ultimately, all entities are alike in one respect, all are "grounded" in a "one and the same." And this "one and the same" bears none of the triviality associated with a term such as "lowest common denominator": rather, the "one and the same" is a *vis primitiva activa*,[13] an "active primordial power," a dynamic. This dynamic is the origin [14] of individualities, of particularities, if you will. The *vis primitiva activa* "originates" meaning; or, to employ even more abstruse Heideggerian terminology, *Deutung* is the "impulsion which (working through the engagement of interpreter and aspect) will become meaning [*Sinn*]."[15] Interpretation, what we called "phenomenological description," is literally a "laying-out" of *Deutung* so the latter is apparent. Assertion, in and of itself, is propositional thinking: it deals with "information," and therefore with individual Things-in-being (*Seienden*) instead of Being (*Sein*).

For Heidegger, hermeneutics is ontology (description of Being, the matrix of all beings) and not just ontic study (description of individual entities which particularize themselves within the matrix). Assertion, as we shall see, divides the more holistic configuration which is interpretation into mutually distinct and isolable "subject" and "object," and then utters "truth-statements" which are measured by how well they "match" subject and object.[16] Heidegger laments the epistemological tradition which insists that philosophizing, from beginning through end, must operate on the level of assertion. Assertion is often necessary, says Heidegger, but should be at the service of interpretation and understanding. Otherwise, thinking effectively cuts itself off from the source of meaning.

Heidegger next attempts to ascertain what is common to all interpretation. Interpretation is the second level of hermeneutical activity, and is descriptive (thus, phenomenological) rather than propositional (assertional):

In interpreting, we do not, so to speak, throw a "signification" over some naked thing which is present-at-hand, we do not stick a value on it; but when something within-the-world is encountered as such, the thing in question already has an involvement which is disclosed in our understanding of the world, and this involvement is one which gets laid out by the interpretation. (pp. 190, 191)

In his section 32, Heidegger's greatest contribution is a phenomenology of a phenomenology (i.e., a concrete description of the second level, which is itself a description of understanding). He finds that interpretative activity manifests three functions: the "As-question" (what we may call the "interpretative question"), the "As-which" (or "textual aspect"), and the "As-structure" (or "interpretation" proper, which — as we shall see — is equivalent to "meaning").[17] Regarding the "interpretative question," Heidegger simply means that an interpreter is never neutral, but always approaches a text with an implicit or explicit question, so that the answer given is shaped by the question asked:

As the appropriation of understanding, the interpretation operates in Being towards a totality of involvements which is already understood — a being which understands. When something is understood but is still veiled, it becomes unveiled by an act of appropriation, and this is always done under the guidance of a point of view, which fixes that with regard to which what is understood is to be interpreted. (p. 191)

The next function of interpretative operation is the "textual aspect" which the text proffers in answer to the question: "That which is disclosed in understanding — that which is understood — is always accessible in such a way that its 'as which' can be made to stand out explicitly" (p. 189). Any given interpretative question should select and illuminate its affiliated textual aspect (if one exists),[18] an aspect which is "there" in the text and which is appropriate to the question.[19] In a narrative such as O'Connor's 'The River,' the plot and its mythemes, the images, prose rhythm, phonemes (and even their interstitial silences), and so on, all constitute aspects. The third function of hermeneutical knowing is the "As-structure," the taking of "something-as-something" (p. 189). The As-structure is the Articulation (*Artikulation*: literally, "exercising of the joints"), or description, of the "joining together" of interpretative question and textual aspect. The As-structure is the interpretation (or Articulation) proper: "That which is understood gets Articulated when the entity to be understood is brought close interpretatively by taking as our clue the 'something as something'; and this Articulation lies before [*liegt vor*] our making any thematic assertion about it" (p. 190). The Christian reader may ask religious questions of 'The River,' and if the story displays affiliated aspects, an As-structure or interpretation solidifies — the text AS a Christian document. The secularist may ask "humanistic" questions, and if appropriate textual aspects present themselves — the text appears AS a secular document. If there are aspects relevant to each set of questions, they are real aspects — parts of the real text. But what about the case of equivocal verbal signs, for example, signs which contain two or more denotative or connotative significations? A case in point is the noun "Bevel" in 'The River.' The noun "bevel" is lexically defined as a "slant" or "incline." While the denotative significations of these two may be more or less the same, their connotative significations are very different. The secularist interpreter can choose the signification "slant" (negative qualities), and the Christian interpreter the signification "incline" (positive qualities). My point here is that the verbal sign "bevel" has two real aspects (and many more, of course, with which we are not now concerned), and both of these are part of the text: one aspect is the "typical phonic form" of "bevel" plus the signification "slant" (with its negative qualities), and another aspect is the "typical phonic form" of "bevel" plus the signification "incline" (with its positive qualities).

To recapitulate, then, understanding is a prereflective at-oneness of critics and text; interpretation is the phenomenological description of understanding, and consists of an As-structure which is the articulated "hold" an interpretative question has on a textual aspect, and vice versa;[20] and finally, assertion is the logical language which abstracts from interpretation, and classifies interpretation into concepts (it therefore breaks interpretation all the way down into subject and object).

We can proceed now to Heidegger's next observation, that of "fore-structure" (*Vor-Struktur*). When using the word "fore," he means (1) that a kind of structure [21] antedates encounter with text, and in part determines how the text will be understood; and (2) that the structure meshes with a text before the interpreter even knows this is the case.[22] Fore-structure is characterized by three kinds of fore-awareness, namely (1) fore-having (*Vorhabe*), (2) fore-sight (*Vorsicht*), and (3) fore-conception (*Vorgriff*). Fore-having equates with the first grasp a critic has on a problem at the level of understanding (and it occurs be-fore he consciously knows it has). Recalling that interpretation is grounded in understanding, Heidegger tells us "interpretation is grounded in something we have in advance — in a fore-having" (p. 191). Fore-seeing adequates to the first interpretative grasp that the second level has on the first, so that "This fore-sight 'takes the first cut' out of what has been taken into fore-having, and it does so with a view to a definite way in which this [the understanding] can be interpreted" (p. 191). Fore-sight occurs before the interpreter knows he has seen anything, and is shaped by the way of seeing his environment has encouraged. As you will notice, Heidegger's earlier treatment of the interpretative question anticipated much of what he says about fore-sight. Finally, there occurs in some instances a fore-conception, which is a set of prereflexive ideas that eventually become reflexive, and assume logical form. Fore-conception works upon the material of interpretation, and transmutes it into concepts. Heidegger says "Anything understood which is held in our fore-having and towards which we set our sights 'fore-sightedly,' becomes conceptualizable through the interpretation" (p. 191). To rephrase it, fore-conception begins the process whereby the level of assertion conceptualizes the descriptive level.

Heidegger next takes up the question of "meaning" (*Sinn*), and the related percept of the hermeneutical circle. Keeping in mind that Articulation for Heidegger refers to the activity of the second stratum of awareness, we can conclude from the following that meaning is precisely that dimension of understood Being which can be "described": "Meaning is that wherein the intelligibility [*Verständlichkeit*] of something maintains itself. That which can be Articulated in a disclosure by which we understand, we call 'meaning.' The concept of meaning embraces the formal existential framework of what necessarily belongs to that which an understanding interpretation Articulates" (p. 193). Meaning, in other words, is an As-structure — and this aperçu functions as the pivot for the whole of section 32. Whereas meaning for the early Husserl is appropriated from a pool of ideal significations, and incorporated into the *intending act*;[23] whereas meaning for Mikel Dufrenne (whose theory of meaning is the opposite of Husserl's) is situated squarely in the *intended object*;[24] for Heidegger, meaning is the *holistic formation* constituted together by interpretative question and proper textual aspect.

As for Heidegger's famous "hermeneutical circle," in this context it refers to the circular movement from understanding to interpretation and back again: "Any interpretation which is to contribute understanding must already have understood what is to be interpreted" (p. 194). Heidegger grants that the hermeneutical circle is indeed circular, and that "according to the most elementary rules of logic, this circle is a *circulus viciosus*." Why so? Because in both traditional logic and "scientific proof, we may not presuppose what it is our task to provide grounds for." In his iconoclastic way, Heidegger continues, " . . . if we see this circle as a vicious one, and look for ways of avoiding it, even if we just 'sense' it as an inevitable imperfection, then the art of understanding has been misunderstood from the ground up." The understanding, or first stratum of awareness, operates on the preobjective level; and interpretation as such Articulates the understanding: this relationship is what Heidegger means when in the quotation cited earlier above he says that "Any interpretation which is to contribute understanding must have already understood what is to be interpreted." Interpretation presupposes understanding, but through the very process of Articulating, interpretation provides grounds for understanding. That is, interpretation is vitalized by understanding alone, but in turn "circles back," so it reveals the "how" and "what" of understanding, and indeed, constitutes the exclusive agency whereby understanding manifests itself.

Heidegger concludes section 32 with a very important discussion of "authentic" (*eigentliche*) interpretation, and he of course takes great care to distinguish the latter from mere "fancy" (*Einfall*). Heidegger's treatment of "authenticity" comprises the same range of issues subsumed under the notion of "validity" in English, though he eschews the literal German word for "validity" (namely, *Geltung*), because the term *Geltung* conjures up for him the antiquated methodology of positivism (I, however, shall use the English words "validity" and "authenticity" interchangeably). To understand the Heideggerian norms of validity, one must begin again with the notion of fore-structure. Recall that according to the first sense of this term, fore-structure is the psychological apparatus the individual brings to the text: for example, he may think in terms of English language syntax, he may be a Freudian, he may be an Archetypalist. Heidegger argues that without fore-structure, interpretation of any kind is impossible. That is, unless a person has the wherewithal to understand a phenomenon, it can make no sense to him whatsoever. Wittgenstein says as much when he remarks, "If you went to Mars and men were spheres with sticks coming out, you wouldn't know what to look for."[25]

To use an analogy, unless a person has a vantage point, he cannot look at something else (because he is simply un-situated, or "not there"). The vantage point is at once enabling and blinding. It enables him to see some profiles of the something else, but not other profiles. The vantage point helps to "structure" the character of his "view," and, chances are, people standing next to him and looking in the same direction will have much the same experience of the something else. But those on the other side of the something else may "see" profiles that are very different, or even contrary. Ethics, linguistics, philosophy, and other elements of fore-structure are psychological vantage points, as it were.

So fore-structure is essential to interpretation. But if fore-structures which illuminate a text differ from one another, how can interpretations ever be "invalid"? *Da capo*, let it be said that Heidegger most emphatically does not deny the possibility of invalid interpretations. Again, the crucial stratum is the interpretative stratum, which falls midway between the preobjective awareness of the first stratum and the objective awareness of the third. The second stratum, in other words, does not dichotomize experience into subject and object as the assertive level does, but neither is it as unitary a phenomenon as the level of understanding. After all, as soon as one talks of an As-question issuing from the interpreter, and an As-which offered by the text, one is talking in terms of a dichotomy. But the second level, in its *essential* construct, does *remain true* to the experiential unity found in understanding: the interpretation strictly defined, that is, the As-structure, is precisely the mutual engagement of critic and text (and by engagement, we do not just mean the interface of critic and text, but the holistic formation which includes the relevant elements of critical fore-structure and the text). The interpretation is constituted simultaneously by an interpretative question and a textual aspect, and thereby bridges the dichotomy of subject and object. The As-structure, in short, is a unitary phenomenon.

Regarding authenticity, Heidegger says the following:

> To be sure, we genuinely take hold of this possibility [primordial knowing] only when, in our interpretation, we have understood that our first, last, and constant task is never to allow our fore-having, fore-sight, and fore-conception to be presented to us by fancies and popular conceptions, but rather to make the scientific theme secure by working out these fore-structures in terms of the things themselves (p. 195).

Notice, this stricture applies not only to the interpretative level, but to contact between all three strata of fore-structure on the one hand, and texts on the other. In fact, earlier Heidegger spoke of the same requirement in regard to conceptualization (fore-conception), warning the thinker not to "force the entity into concepts to which it is opposed in its manner of Being." Heidegger is saying that the particular As-questions appropriate to the phenomena at hand (and not inappropriate As-questions, "fancies and popular conceptions" which do not suit the phenomena) are essential for validity. Or, to approach the matter from the other direction, Heidegger is saying that As-questions must grasp textual aspects that are *really there* in the phenomena. When such contact occurs, a *valid* interpretation, and therefore a *correct* interpretation, materializes. Otherwise, there remains a merely fanciful, or pseudointerpretation. Since Being exhibits itself precisely in and through meaning, and meaning (as we have seen) is interpretation, a valid interpretation is by necessity *true*. To have several valid interpretations of a text, then, is to have several true interpretations — each of them determinate and self-identical (that is to say, each exegesis discloses a different aspect of a literary work which by nature has many aspects, and each combination of question and relevant aspect is a "truth"). The issue that next surfaces for us, of course, concerns criteriology. How does one determine what are apposite As-questions? How does one demonstrate that the textual aspects alleged are "really there"? Clearly, a literary critic must convince his peers of the appropriateness of his interrogation; clearly, he must "show" the presence of the textual aspects he claims; and clearly, he must demonstrate how his interpretative questions engage the aspects. I shall address the conditions which characterize the "making of a case" for an interpretation. And I shall defend Heidegger from accusations of relativism. First, though, some remarks on "authorial intent" are in order. And I wish also at some length to consider the matter of contradictory interpretations.

You will notice that throughout section 32 Heidegger ignores the relevance of authorial intentionality, that is, the significations [26] the author related to a text when he created it.[27] Hans-Georg Gadamer, Heidegger's disciple, goes on to strictly exclude the relevance of these significations. In *Truth and Method*, Gadamer says "The *mens auctoris* is not admissible as a yardstick for the meaning of a work of art."[28] And in another book, he repudiates the tradition of "objective hermeneutics," and its quest for the author's significations: " . . . all that Schleiermacher and Romanticism report to us on the subjective factors of comprehension seem to us unconvincing. When we understand a text we do not put ourselves in the place of another, and it is not a question of penetrating the spiritual activity of the author. . . . "[29] Since in his practical exegesis of literary writers and others,[30] Heidegger often does involve the author's "willed significations" for a word or passage, it seems safe to conclude that Heidegger's position is more inclusive than Gadamer's: a critic may ask an "authorial" As-question, but he *need not*. Much depends on the critic's purpose. A critic may try to approximate (through biographical or other means) the author's fore-structure, and a textual aspect which accommodates it, so the meaning that arises is similar to the author's meaning. Or a critic may investigate the medieval significations available to a medieval poet, say, and encounter the poem in that light. But again, the attached provision is that such a maneuver is by no means requisite. *In nuce*, Flannery O'Connor's own deep Christian commitment need not deter an atheist or anyone else from a secular reading of 'The River.' Biographical and historical critics incorporate their researched data into fore-structures, and bring these structures to bear on aspects of a work. Christian, Jungian, and other critics exercise their own distinctive options, dependent on other values and other kinds of research.

As this juncture I call to the reader's attention that the way I have "laid out" (*ausgelegt*) the above issues – the structure of hermeneutical activity, and so on – is itself Heideggerian and phenomenological. Obviously I have not tried to argue in the way a "logical atomist" does, or a "logical positivist" does. Rather, with the aid of Heideggerian As-questions, I have described what one "sees" if he collates the practical criticism of many reputable critics identified with many critical schools. He invariably sees fore-structures and textual aspects in collusion or attempted collusion. What we have called the authorial As-question, or an attempt to read a text in terms of the author's fore-structure, is a variable: sometimes it plays a role in a given critique and sometimes it does not.

As I have already indicated, I shall treat anon the functions of verification used by literary critics. Concerning my own principle of verification, in advancing a whole phenomenology of the hermeneutic act, I turn consistently to a traditional maxim of phenomenology: "Corroborative description is the only verification." In other words, phenomenology is ultimately a communal activity: we check our own concrete experiences of what is common to the subject at hand – in this case, critical practice. With many other phenomenologists over a long period of time, I have experienced and then described interpretative activity. *Sic feliciter evenit*! The reader is invited to do likewise.

The reader may further ask what is the status of this very essay of mine? First, I call to your attention the fact that my essay manifests the "aspects" of an "assertive" paper (as Heidegger would say). It operates primarily on the level of assertion, so it is not just *Auslegung* (in this case, description of hermeneutical activity) but *Aussagen* ("explaining," or "the making of a case" for the embedded description). And I usually choose to "make my case" by the phenomenological means – I provide examples of representative critical activity, and explain them.[31] I ask the reader to consult the examples, and any others he pleases. Expository assertion, the kind of explaining of which my essay purports to be a kind, is by nature concerned with authorial intent, in this case, the significations Heidegger attached to section 32 of *Being and Time*. So I consistently ask the authorial As-question. But for that matter, my essay is far from true philosophizing, the deep "Meditation on Being," and hopes to perform just a humbler, ancillary service.

The contrast of secular and Christian readings of 'The River' exemplifies a special kind of multiple interpretation: the readings in question are not only variant but contradictory. In my book I state that a sequence from Heidegger's essay, 'What Are Poets For?' can provide the wherewithal to justify the simultaneous validity of contradictory interpretations [32] (indeed, I could have adduced many other similar passages from the late-phase Heidegger). Heidegger's turn in his later career toward Oriental thought is well-known (though as early as 1929 he already resorts to phraseology which is Orientalist, even if it may come by way of Oriental motifs long miscegenated with nineteenth-century German idealism: witness the passage in his *Vom Wesen des Grundes*: "The self can yield up the 'I' in order to achieve its true self").[33] Heidegger's 'A Dialogue on Language' originated in 1953–1954, on the occasion of a visit by Professor Tezuka from Japan, and the "Dialogue" demonstrates the affinities of Heideggerian and Japanese attitudes toward poetic language. Moreover, the renowned Professor Chang Chung-yuan, whose earlier work on Chinese philosophy, poetry, and art moved Heidegger to express his admiration, has written a new translation of the *Tao Tê Ching*, accompanied by commentaries which draw extensive comparisons to Heideggerian thought.[34]

Let us begin with the quotation cited in my book from Heidegger's 'What Are Poets For?':

If Being is what is unique to beings, by what can Being still be surpassed? Only by itself, only by its own, and indeed by expressly entering into its own. Then Being would be the unique which wholly surpasses itself (the *transcendens* pure and simple). But this surpassing, this transcending does not go up and over into something else: it comes up to its own self and back into the very nature of its truth. Being itself traverses this going over and is itself its dimension.

When we think on this, we experience within Being itself that there lies in it something "more" belonging to it, and thus the possibility that there too, where Being is thought of as venture, something more daring may prevail than even Being itself, so far as we commonly conceive Being in terms of particular beings. Being, as itself, spans its own province, which is marked off (*temnein, tempus*) by Being's being present in the word. Language is the precinct (*templum*), that is, the house of Being.[35]

A really substantive grasp of the above passage can only be had in the light of Heidegger's treatise, *The Question of Being*,[36] which with his other late treatise *Identity and Difference*[37] sum up Heidegger's last appraisal of Being. And the reader is urged to read these two treatises (both quite short). The passage quoted above first attests to the "surplus" which characterizes Being (indeed, Heidegger's phraseology throughout the two paragraphs recalls the medieval notion of *bonum diffusivum sibi*). Being, as the French say, "se jaillisse." Yet we know individual beings – on the level treated by explaining – are often contradictory. Yet "Being is what is unique to beings," so each being shares in the transcendent Being (the *transcendens* pure and simple) which is at one and the same time immanent in the individual being.

The "surpassing" which is overarching Being still "enters into its own." Thus as early as *Sein und Zeit*[38] Heidegger repudiates inquiries into Being which are *tiefsinnig* ("deep"); that is, "deep" in the sense that they "deduce" or "speculate" about Being because they fail to recognize Being is "at-work" and "to-be-experienced." But the Being which enters "into its own" also "surpasses," with the enormous consequence that the "more" which is universal Being subsumes all contradiction into itself. What are logical contradictions on the level of explaining are organized into a holistic formation (what Heidegger earlier called "the one and the same") at the primal level of understanding. As a result, contradicting elements (as unique entities) can be simultaneously valid – since they are emanations, if you will, of the one Being. This paradox, furthermore, finds its privileged locus "in the word." Being "ventures" language, and language becomes the "house of Being."

In the face of interpretative questions which produce contradictory readings, Anglo-American critics often praise "ambiguity." These critics are in fact affirming the mystery of which Heidegger speaks – when two or more readings argue solid but irreconcilable cases, these critics rest in awe before the spectacle, confident that at a deeper level of experience the interpretations bespeak the *one and the same*. I say "deeper level of experience" advisedly, because – as Professor Chang Chung-yuan testifies – and as even I and some of my colleagues can testify, this oneness can be *experienced*. And within this experience, the interpreter circles from the phenomenal contradictories to phenomenological oneness, and back again. And the "aesthetic" (that which involves beauty) tensions as well as ontological tensions which arise from all this interplay are dazzling indeed! On the phenomenal level, contradicting interpretations both lure and elude each other, like coquettes fixated in hypnotic dance. Yet at the phenomenological center, this dance – which Roman Ingarden so beautifully calls "Opalescent Multiplicity"[39] – dissolves in the *one and the same*. As Heidegger would put it, there is a kind of "homecoming." The ontic becomes ontological. The existents become Being.

As demonstrated at length in my book,[40] Heidegger's theoretical writing on literary texts gives short shrift to the relevance of the author's personality. Poetic language summons the "World" of *Dasein*, a world where the primordial powers of Sky, Earth, Divinities, and Mortals intersect; and this World is "mutually implicated" with the realm of Things-as-things.[41] In and through the simultaneous tension and intimacy of World and Thing, language "presences" Being. But why is poetic language, which Heidegger so often deprives of its authorial filiations, somehow privileged? And so gloriously privileged that it is for all practical purposes a secular sacrament — "a symbol which makes present what it symbolizes"? The answer lies in the Heideggerian understanding of poetic language (*Dichtung*) — its genesis and nature. For Heidegger, poetic language (and meditative philosophy is a kind of poetic language and vice versa) is the most human of all activities; poetic language is at once bodily and spiritual, concrete and intellectual — poetic language brings us home to Being and Being home to us. Nor does Heidegger normally invoke the romantic image of the seer, who transmits his inspired vision of Being. Rather, Heidegger says "the artist remains inconsequential as compared with the work, almost like a passageway that destroys itself in the creative process for the work to emerge."[42] Heideggerian thought sees poetic language as quintessentially human, which is to say — quintessentially communal[43] (remember, Heidegger is not the isolated existentialist, but the ontologist — even at times, I dare say, the cultural anthropologist). Being "be-ings" itself in the human being, since human being "gathers" the World and Things. And the quintessential display of communal human being, and therefore of Being, is poetic language.

Heidegger's attitude toward contradictories approximates the essence of Zen described by the famous Orientalist, R. H. Blyth: "We live supposedly in a world of opposites, of white against black, of here versus there. But beneath this level of opposition lies a sea of tranquillity in which all things are complementary rather than contradictory." And Blyth quotes the ancient Buddhist dictum:

> On the sea of death and life,
> The diver's boat is freighted
>   With "Is" and "Is not";
> But if the bottom is broken through,
>   "Is" and "Is not" disappear.[44]

A kindred example that comes to my mind at once is from Seng Ts'an, third Ch'an patriarch in China (ca. 600 A.D.):

> At the least thought of "Is" and "Is not,"
> There is chaos and the Mind is lost . . .
> In its essence the Great Way is all-embracing.[45]

Many other passages in both Taoist and Buddhist traditions make it clear that so monist a vision is not intended to undermine the particularities and even the contradictions of the phenomenal world; indeed, anyone familiar with Japanese art knows that Zen, for example, celebrates uniqueness, but the uniqueness is "measured out" in its relation to the *oneness*. And this is precisely Heidegger's idea too.[46]

But perhaps some further clarification is in order. Though Taoism and Buddhism both preach the complementarity of opposites, so that each event, say, has a "happy" quality and a "sad" quality, and these should be harmonized so the human being achieves stasis — it is not this principle as such which concerns us here. For this principle would have as its analogy the phenomenon of multiple (or "variant," or "disparate" literary interpretations), where textual event may, conceivably, profile various "complementary" aspects — an event may be simultaneously erotic in a Freudian way, politically aggressive in a Marxist way, and still esthetically "holistic" in an American "New Critical" way. Many of the "complementary" oppositions can be what are technically called contraries, as "happy" quality and "sad" quality attributed to the same textual event can function as logical contraries.

But I have been maintaining that Heidegger does much more than approximate the law of complementary contraries, as found in Eastern philosophy and indeed some Western philosophy. I have been maintaining that he approximates a more radical Taoist and Buddhist principle, called by Lao Tzu "the unification of affirmation and negation,"[47] and illustrated by Lao Tzu's phrase, "Great white is as if it is black."[48] That is, not only contraries (e.g., one literary event sad in one way and happy in another way) but also contradictories (one literary event sad and nonsad in the same way) can be simultaneously valid. Perhaps a rationale of this can be that for Being, all time is ultimately *one and the same*, all experience ultimately *one and the same*, so *sub specie aeternitatis* all real contraries reduce into real contradictories, and all real contradictories into oneness. Yet again, up from the oneness "se jaillissent" the glorious multiplicities. (Notice that I use the word "real" above advisedly, because not all contradictories are simultaneously valid — in many cases, one interpretation is simply true and its contradictory is false — but more of this later.)

My discovery, subsequent to the writing of my book, that the eminent Professor Chang had already noted Heidegger's reformation (in the Western world) of the law of contradiction, has given me a corroboration that is most appreciated. Of the many relevant Heideggerian passages cited by Chang, I quote here a line from Heidegger's 'The Origin of the Work of Art': "This rift carries the opponents into the source of their unity by virtue of their common ground."[49] And Chang adds in his own words, "The source of the unity of opposites maintained by Heidegger is what Taoists call the Tao."[50] If one recalls that thinking (as consciousness or *pour-soi* in European philosophy) is the contradictory of Being (or *être-en-soi* according to European philosophy), then Chang is correct in recognizing the iconoclasm of Heidegger's reinterpretation: "Different things, thinking and Being, are here thought of as the same. . . . Thinking and Being belong together in the same and by virtue of this same."[51] Recall that conventionally, thinking and Being are considered not just contraries, but contradictories — thinking is the contradictory of nonthinking, and nonthinking is Being. In the cited passage, in other words, Heidegger is affirming the identity of these contradictories. For much further evidence from Chang, see the footnoted references.[52]

Now let us face the issue of validity. If one reviews the panoramic history of literary criticism (if he does a "phenomenology" of the situation, if you will, seeking out the concrete essential structures), the following becomes readily apparent. Privately, the critic adjudicates for himself the validity of his own interpretation (his norms may resemble or differ markedly from those of his contemporaries). Publicly, if he wants to convince others his interpretation is valid, the critic must accept the norms of his audience, and try to show his interpretation suits these norms; or he must convert his audience to new norms, and demonstrate his interpretation suits the latter. Even if a critic maintains the "author's meaning" is the only valid meaning, the critic must obviously convince his peers that such a norm is justified, and indeed, that his interpretation really coincides with the author's meaning. The role of the critical audience leads at once, of course, to the concomitant issue of intersubjectivity. That is, norms which are publicly operable are shared among a community of subjects (eighteenth-century neoclassicists, for example, were a body of subjects who agreed on many hermeneutical norms; so are American "New Critics" today, or Marxists, or Freudians, and so on).

A further description of literary history reveals that the meanings which various groups accept as valid can differ because of two different reasons. One reason is that the "work" or "text," which we have already defined as a literary composite of verbal signs and relations among verbal signs, is in fact a different work to different audiences. Recall that the meanings of a work are engagements of interpretative questions and textual aspects. If we suspend for the moment the factor of variable interpretative questions, and deal only with textual aspects, we find the following. Textual aspects are appearances of verbal signs and relations among verbal signs, and these appearances are really "in and of the text" (see Note 19). Now if the verbal signs of a "work" change, it follows that the work itself changes. And some of the aspects or appearances of the work, since they are "in and of the work," likewise change. If we recall that meanings are engagements of interpretative questions and textual aspects, the conclusion is inescapable: meanings can differ because textual aspects have changed. Sometimes these changes are synchronic — for example, different dialectical groups attach different significations (lexical values) to a given word sound.

Most changes, however, are diachronic. A famous instance is provided by René Wellek in his *Theory of Literature*.[53] In the seventeenth century, Andrew Marvell's phrase "vegetable love" signified what today could be called "vegetative love." That is, the signification, or lexical value of "vegetable" was about the same as today's "vegetative." And according to the scholar Louis Teeter (quoted by Wellek), the usual qualitative value (something like "connotation") imparted by the expression "vegetable love" at the time was "life-giving principle." But in the twentieth century, "vegetable" signifies "edible plant," and this signification opens the way to a new qualitative value: "vegetable love" connotes "torpidity," or "slow and stifled love" (love as an "erotic cabbage," as Teeter says). In short, the word sound "vegetable" has taken on a new signification, so that we can indeed have a new verbal sign (and to that extent, a new "work" or "text"). In terms of the concept "textual aspect" or "textual appearance" (and there are all kinds of textual aspects or textual appearances), the word sound "vegetable" signifying "vegetative" is an "aspect"; the word sound "vegetable" signifying "edible plant" is another aspect. The verbal signs "vegetable love" (the word sounds plus the significations vegetative and love) connoting the qualitative value "life-giving principle" is an "aspect." The verbal signs "vegetable love" (here, the word sounds plus the significations edible plant and love) connoting the qualitative value "torpidity" is an "aspect."

In the story 'The River,' perhaps at some future time the significations of "skeleton" will change, and with these mutations the qualitative values contingent upon "skeleton" will change, and these permutations of aspect will to such an extent change the literary work *per se*. In any case, it is crucial to recognize that changes in verbal sign are effected by the language-group which constitutes a work's audience at any given time. Signification, and qualitative values dependent on signification, are conferred by the language-community; individual interpreters and even whole "schools" of criticism naturally accept the lexical values, multiple as they may be, which a *langue* confers on word sounds so that verbal signs can occur. In this sense, and to this extent, the reader is merely a mechanism whereby *langue* vivifies word sounds — Freudian, Marxist, Christian — one and all will normally accept that the word sound indicated by "triangle" bears the lexical values attributed to it by the *langue* which is English.

We can even submit the notion of *langue* to further examination, by putting our conclusions to the test of the "extreme case." Here we can assume the original significations of a text have been lost completely. In other words, the *langue* has been lost. In terms of the twentieth-century interpreters and their culture, the text may then make no sense or perhaps some sense — the latter through chance alone. The marks on the page happen to make sense in a language known by the interpreters, even though their language is different from the language of the author (and the author's culture). Let us turn the screw another spiral. Through a fortuitous happening, the text makes complete sense in the second language. Surely one cannot appeal to the author's sense now! Yet by chance perhaps an exquisite poem has arisen. Whence come the significations? They can only come from the language of the *interpreter's* culture.

An entirely different matter is the change in meaning which arises because interpretative questions differ. In the latter case, the work and its aspects remain constant, that is, a given language culture at a given time makes available one or several significations for a word sound. Since a literary work is the totality of verbal signs (word sounds plus significations) made possible by the circumambient culture, different As-questions can contact different aspects of the one work. Such changes are initiated on the side of the individual interrogator, but the meaning as a whole changes, since meaning is the mutual engagement of As-question and relevant textual aspect. Notice that in such cases the literary work itself has not changed — only meanings have changed. In sum, the meanings which various groups accept as valid can differ because of two reasons: the work itself can change, so it is indeed a different work; or, while the work remains the same, the interpretative questions — assuming they are valid — can differ (the latter is often a synchronic phenomenon).

In his famous work *Validity in Interpretation*,[54] E. D. Hirsch argues that formulations of the above kind are relativistic. Though Hirsch has modified his position somewhat in a second book, *The Aims of Interpretation*,[55] his first book remains the classic attack on Heideggerian thought as relativist. And Hirsch's arguments in that first book are still upheld by many theorists. So I turn to what can be, I think, a Heideggerian *riposte*. In *Validity in Interpretation*, Hirsch discusses Wellek's example out of Marvell (see above). (Before we begin, however, I interject a provisional statement. In that the significations of "vegetable" have changed, historically, a Heideggerian can say that to such an extent the work has changed. So a comparison of "valid meanings" is here in a way inappropriate. It is really a matter of two different literary passages, each participating in several possible meanings.) However, since the work as a whole (the poem, "To His Coy Mistress") has not changed (i.e., most of its words retain the same significations they had in the seventeenth century), we can make the practical choice of considering it the one and the same work which existed in the seventeenth century. (Actually, a more suitable ground on which to argue all this would be a modern text that bears contradicting significations — with each signification alive in the *langue* today: but the example out of Marvell has been sanctified by usage — Wellek's, Hirsch's, and mine — so I shall stay with it.) We proceed to hear Hirsch's case. Hirsch advances the following argument. Wellek's very thesis, that the modern interpretation is also valid, assumes the distinction between the author's "sense" (what we have called "signification") and subsequent "senses" (significations, again). In order to avoid a relativism, a chaotic flux of meanings, it is absolutely essential to distinguish between an author's "willed" significations and other possible significations. Thus the necessary distinction, says Hirsch, between "meaning" (which is Hirsch's term for the author's willed signification) and "significance"[56] (Hirsch's term for the relation of authorial meaning to the reader's other meanings, emotions, etc.).

Heidegger would answer that he too sees distinctions among significations (how could he not?). However, he refuses to attribute an exclusivity to the author's willed significations. When one does a concrete phenomenology of any hermeneutical experience, one plainly sees that what Hirsch calls "meaning" and what Hirsch calls "significance" both arise from a live contact of an As-question and textual aspect. *In situ*, there is absolutely no difference in their functional nexus. Thus the same validity obtains for both. Hirsch's As-question can be: Did the author intend such and such a qualitative value for this verbal sign? A critic identified with the New Hermeneutics may ask: Can an archetypalist, say, intend such and such a qualitative value for this verbal sign? And, I might add, much excitement in exegesis arises from the dazzling interplays of As-questions and textual aspects, as various combinations complement and contradict each other, converge upon and tug away from each other.

Remember that Heidegger proposes that there are a plural number of aspects in texts, but that each of these aspects is determinate and self-identical (whether the author "willed" them there or not). To invert Hirsch, the work — in the grasp of various As-questions — means *many things* in particular. Hirsch's perception of a text is remarkably flat: a text cannot have more than one facet (our analogy for "aspect"). The author's chosen signification excludes all others. Thus *Validity in Interpretation* announces categorically: "It may be asserted as a general rule that whenever a reader confronts two interpretations which impose different emphases on similar meaning components, at least one of the interpretations must be wrong."[57] But why must this be the case? The often-used analogy with a diamond is appropriate here. Our interpretative glance at the diamond can contact one or more facets (aspects), and each is really part of the diamond. But let us take a verbal example – the word "cleave," which has two opposing definitions. "Cleave" can mean "to separate" or "to adhere."[58] Let us say a given text describes a God who descends in blinding theophany, and utters to his hushed disciples, "Men and women, cleave!" Let us assume furthermore that the verbal context is of no help. But, and this is of utmost importance, let us take it as given that all authorial evidence points to one interpretation as the author's own – that he willed, let us say, "cleave" to mean "separate." Does this authorial data make the alternative interpretation, that to "cleave" is to "adhere," less determinate and self-identical? Both significations (to "separate" and to "adhere") are held firm by the syntax and lexicon of the *langue*, which gives them shareability and particularity. What further determinacy is needed? For a Heideggerian, Hirsch's argument remains ineffectual.

That many interpretative questions can be axiologically sound does not by any means deny the importance of historical criticism – and the reactions of some past interlocutors of mine seem to require of me an addendum of this kind. Heidegger, especially in his revaluation of the past history of metaphysics, has repeatedly asked "the historical question" – to put it another way, he has asked what significations were attached to verbal signs by given historical periods. And he has also asked "authorial questions" – what significations a given author intended for given word sounds. Heidegger has often stressed the importance of "sedimentation," whereby word sounds accumulate significances historically.[59] And much depends, as I have previously suggested, on the kind of language under study – be it explanation, for example, which involves interlocution of several kinds between authors and listeners; or poetic language, which in a deeper and richer sense communicates Being without regard for the historical author.[60] But my point here is that, in the case of poetic language, or literary language, modern interpretative questions, and modern significations, can be just as valid, and indeed fruitful, as historical or authorial readings. Hirsch's rejoinder is that such a proposition reduces to pure relativism. In answer, I postulate there is a constancy adduced by Heideggerian theory – the constancy imparted by intersubjectivism.

We can begin with an example. In Henry James's novella, *The Turn of the*

Screw, a governess struggles to protect her two young charges against diabolical ghosts. It so happens that nineteenth-century critics brought a traditional Christian fore-structure to the text, and saw the governess *as* an integrated and wholesome personality, fighting the war of a Christian heroine against Satan. Freudian critics of the twentieth century have seen the governess *as* a neurotic personality, perverting the children through her malign fantasies. If we were to perform a phenomenology of critical dialogue, we would find each critic trying to "make a case" for his own interpretation (of course, this does not preclude learning from others as well). Each critic would try to "convince" the others. But "convincing" can take place only to the extent that the critics share values in common. In the above example, Freudians could convince both those already sympathetic to Freudianism, and those converted to Freudian insights by the vigor of the Freudian argument in this case. Of course, the "making of the case" would require much more than beliefs held in common. The Freudian interpreter would have to show that, in terms of Freudianism, the precise questions they have asked are appropriate. The interpreter would also have to show that the textual aspects he espies are "really there."

That, for example, in the story 'The River,' Bevel exhibits the behavior of a young child who is rejected by his parents, and who, rebounding from this hurt, is subconsciously driven to suicide — a motive he must consciously disguise as holy and beneficent, viz., as self-baptism. In this case, the Freudian critic would be showing an engagement of As-question and textual aspect: Bevel as delusive and suicidal. Obviously, the traits of repression, self-delusion, and so on, are seen as "really there" in the events of the story only by readers who believe repression and self-delusion of this kind can possibly occur, either in reality or in fiction. I add that important second phrase, "in fiction," because a reader may accept repression and self-delusion as possible ways of presenting behavior in fiction — even if they are not (for the reader) a correct way of explaining behavior in the real world.

But another reader — say a devout Fundamentalist Christian — may pose other questions of 'The River,' and "see" other textual aspects "really there" in the work. Because of the fore-structure this interpreter brings to the text, he does not see repression and self-delusion. Yet he can "make a case" for his Christian audience that the meaning, or As-structure, he advances is appropriate: As-questions meet textual aspects according to norms that are intersubjective among his peers. Even if he is told that the historical author of 'The River' intended a Freudian interpretation (which, incidentally, is probably not the case, but for our purposes makes no difference), he can with perfect legitimacy answer that consultation of authorial intent is unnatural and inappropriate for poetic language. After all, the Christian critic, from a formal perspective, is doing here no more and no less than other critics, including critics asking the "authorial question." For example, the critic asking the authorial question attaches a Freudian qualitative value to a verbal sign because the author did. The Christian critic attaches a Christian qualitative value to the same verbal sign because the Bible does.

The Christian critic is asking questions vital to his group; he is showing to this group's satisfaction that textual aspects answer to these questions. Surely, on the side of the work and its aspects, he must adhere to couplings of signification and word sound permitted by the language-community's *langue*. (If he and his group do not, they are bespeaking a new and different work.) But at any given time in history, several significations and even more qualitative values are available through culture and its *langue* to a given word sound, and *a fortiori*, to the work as a whole. The literary work, as we have said, is an organon comprising all of these synchronic significations and qualitative values. Practically speaking, all interpretations contact only some facets of a multifaceted work. The Christian critic in the above case is "finding" in the work significations and qualitative values the Freudian is not finding. The Freudian is finding other significations and qualitative values. But the collective language of our culture (which includes, indeed, many subcultures) comprises both sets of lexical and qualitative values, the Freudian and the Christian (and many, many more).

A critique that has been leveled at all of my account above is that it would, in effect, validate, say, both an astrological (Ptolemaic) and astronomical description of the heavens, though the whole civilized world surely "knows" the astronomical description is the "correct" one. The conventional answer which the European hermeneutical tradition would give to this argument is that hermeneutics concerns the realm of *Geist*, or human spirit, and not physical matter. Thus, examples drawn from science do not apply to aesthetics, the liberal arts, social sciences, and other human activities and constructs. Scientific laws, and the certainty they provide, apply to matter alone; a great fallacy of the empirical tradition has been to apply physical laws to human spiritual activity.[61] Attacking empiricism even more earnestly, existentialists and others can even argue here that astrology presents a "truth" of its own kind, just as valid in its own way as the astronomical truth. The heavens can and indeed have "appeared" to many perceivers in a way which Ptolemaic description satisfies — just as ocean water has a molecular structure, say, which empirical science discovers and claims to be "true," and also an array of other characteristics — archetypal values of motherhood, death, and so on, which the *sciences humaines* discover in ocean water and claim to be "true" (and claim to be more important, in terms of the human psyche, than mere "physical facts" about "matter").

The fact remains, however, that the astrologers, in their heyday, claimed to be asserting specifically scientific truth. My own stand on these matters is perhaps more radical than that of others who oppose an overextended empiricism. It seems to me that truth of all kinds, including scientific as well as spiritual truth, is adjudicated by the group. Medieval scientists judged astrological theories true, and by their norms the theories were true. Today's empirical scientists consider the conclusions of astronomy true, and these scientists are believed by all those who accept the normative constants identified with empiricism — namely, at this point in history, the "scientific method" and its corollaries.[62] But obviously astronomical "certainties" are certain only for those who accept the scientific method. That virtually all civilized people, at this point in history, accept the scientific method's suitability for the investigation of material objects is in a way deceptive. Just as the majority of people once accepted astrology as a valid science, and then succeeding generations lost that conviction, so too are there indications that the scientific method will cede in time to other "scientific" norms which will in turn become "intersubjective constants" for a relative chronological period. But, as we shall see, all this does not despoil the empirical scientist, for example, of the right to declare the astrologer dead wrong, and a fool at that. Likewise, it does not despoil the Hindu swami, and thousands of his devotees, of the equal right to deny the efficacy of the scientific method, even in relation to physical matter.

Relativism is the absence of truth conditions. But in the Heideggerian system, the collective *langue* vouches for the presence of determinate significations and qualitative values in the work. And each critical school accepts as a prerequisite the engagement of As-questions and textual aspects, and establishes norms held in common to "test" whether the engagement is achieved. Thus Heidegger establishes a universal truth condition for the function of validity (namely, the mechanism of mutual implication, as already described), and upholds as a truth condition the necessity of group norms whereby this function can be evaluated (though he obviously affirms a plurality of norms, and leaves the determination of these various sets of norms to the various critical subcultures which appoint them).

But the further objection can be raised: perhaps Heidegger's formulation is not solipsistic relativism, but it does suggest a kind of peer-group relativism (similar to Dilthey's, and others). My answer is that neither is this the case. Group relativism would "found" its truths on the beliefs of the group pure and simple. But Heidegger says, "That which is 'shared' is our *Being towards* what has been pointed out — a Being in which we see it in common. One must keep in mind that this Being-towards is Being-in-the-world, and that from out of this very world what has been pointed out gets encountered."[63] The validity of a meaning requires implication of As-question and As-which, but whenever this implication really occurs, Being is manifested. Validity, and the truth that consecrates validity, is at bottom founded not on subcultures or "critical schools," but on Being; truth is only mediated through subcultures and critical schools. Different critical schools can reflect different facets of Being.

Nor does Heidegger imply that one must accept every critical school that emerges in history. One may be convinced a given school, with its own distinguishing norms and so on, does not reflect Being. It may just produce fancy. And what about a perverted author who produces textual aspects which bespeak hatred? A reader may (and indeed, for my part, I think should decide) that the proper As-questions for such aspects would be questions which display the plot, qualitative values, and so on, of the work *as* evil and perverted. Then the As-structure or meaning of the work appears *properly* — Being ultimately is "the way things really are," and the interpretation here would show evil to be evil. So again, Heidegger is not advocating an "anything goes" criticism. But he is saying that if we wish we can simultaneously affirm some contradictory schools and interpretations, and that we can do so without ontological embarrassment.

Starting from a base in Flannery O'Connor's own marvelous work, we have attempted to describe the structure of literary experience (including interpreter and interpreted). We have described the nature of meaning, the function of validity, and the principles of literary transformation. The overriding influence, of course, has been a Heideggerian theory of Being. Heidegger's perception is rooted in what the Scholastics called *admiratio*, and what Flannery O'Connor herself calls so often a "sense of Mystery."[64] *Admiratio* pauses in awe before the multiple unity of reality. If, in O'Connor's words, fiction pushes "its own limits towards the limits of mystery,"[65] the horizon of this mystery is that Reality can mean several things at the same time, and still be radically one. For us, the most crucial turn in the argument is the last — that the poetic language of a Flannery O'Connor, and of others like her, utters *sans pareil* the mystery which is Being. Thus we can close with the so appropriate words of Martin Heidegger —

When thought's courage stems from
the bidding of Being, then
destiny's language thrives.[66]

NOTES

[1] Flannery O'Connor, 'The River,' in *A Good Man Is Hard To Find* (New York: Doubleday, Image Books, 1970); cited by page references within the text.
[2] In recent American critical debate, the above issues (or most of them) have been skillfully treated by Monroe Beardsley, Denis Dutton, Joseph Margolis, E. D. Hirsch, and several others, and many of these scholars have disagreed radically with each other. My contribution, I hope, will be to scrutinize these issues in a manner which is of recent European origin. Such a task will involve the suspension of what are some current Anglo-American presuppositions, and a reconstitution of the problematic in phenomenological – and specifically Heideggerian terms.
[3] Martin Heidegger, *Being and Time*, trans. John Macquarrie and Edward Robinson (New York: Harper and Row, 1962); cited by page references in the text. The translation is from the seventh German edition, *Sein und Zeit* (Tübingen: Neomarius Verlag); the first German edition appeared in 1927.
[4] Robert R. Magliola, *Phenomenology and Literature* (West Lafayette, Indiana: Purdue University Press, 1977).
[5] On the other hand, in his 'Plausibility and Aesthetic Interpretation' 7 (1977), 327–40, Denis Dutton argues that interpretation of aesthetic phenomena should operate in terms of a special kind of plausibility, rather than "truth" understood in the commonly accepted scientific sense. The Heideggerian approach to such topics as validity, plausibility, "true" interpretation, "correct" interpretation, and the norms for determining them, will emerge later in my presentation.
[6] In *Phenomenology and Literature*, I demonstrate passim that Heidegger's critical practice, and even some parts of his theoretical work, seem to imply a heuristic absolutism rather than pluralism. But in regard to theory of multiple interpretation, I think his usefulness depends on another current in his thought, and it is this second and stronger current that I examine in detail.
[7] Heidegger, *Being and Time*, pp. 188–95.
[8] A definition of "aspect," and various insights into its functioning, will appear in several places in my presentation.
[9] For definitions of "mutual implication," see *Phenomenology and Literature*, pp. 14, 61, 69, 70, 72, 76, and passim. "Implication" here bears the signification of "enfoldment," the signification of its Latin etymological root.
[10] By verbal sign, we mean (and in my opinion Heidegger can concur) what Roman Ingarden calls a "word sound" or "typical phonic form," plus the significations (lexical values) available to this form by way of a culture's *langue*. Typical phonic form is to be distinguished, of course, from written or phonic material, such as quality of voice or shape of print, which can be individually new and different with each implementation. Thus, to use an example cited by Husserl, the word sound *Hund* transcends any

individual articulation of it, and the primary German significations available to this word sound are "a dog" and "a truck used in mining." See Roman Ingarden, *The Literary Work of Art*, trans. George G. Grabowicz (Evanston: Northwestern University Press, 1973), pp. 34–56, for a discussion of word sound.

[11] For Heidegger, the word "phenomenology" means "that which shows itself in itself." Phenomenological description delineates what shows forth concretely, *in* experience.

[12] I implore the reader's patience: indications of what Heidegger means by the elusive term "Being" are forthcoming later in my paper.

[13] The phrases "the one and the same" and *vis primitiva activa* are applied to Being by Heidegger himself. See Heidegger, 'Hölderlin and the Essence of Poetry,' in his *Existence and Being*, trans. with introd. by Werner Brock (Chicago: Henry Regnery, 1965), p. 278; and Heidegger, 'What Are Poets For?' in *Poetry, Language, Thought*, trans. with introd. by Albert Hofstadter (New York: Harper and Row, 1971), p. 100.

[14] Jacques Derrida, the brilliant and chilling contemporary French philosopher, would decry Heidegger's quest for ontological origin, and even Heidegger's "logocentricism." Though there is no space to treat the matter here, my next book (now underway, and entitled *Beyond Derrida: The Recovery of Poetic Presence*) will argue at length that Heidegger, in his own way, "deconstructs" metaphysical language, but does so in order to suggest a *via negativa* similar to that of traditional mysticism. Heidegger's *Dif-ferenz* provides fullness where Derrida's *Différance* "closes" ontology.

[15] I owe this apposite definition to my colleague, and internationally recognized phenomenologist, Professor Calvin Schrag.

[16] For a comparison of "correspondence theory," which "matches," and "commemorative truth," which "brings forth," see *Phenomenology and Lieterature*, pp. 65, 66.

[17] In order to avoid what may seem to be the jargonistic timbre of the actual Heideggerian terms, all of which feature the prepositional "as," I offer the alternate terms provided within the parentheses. Hereafter, where possible I will substitute the alternate terminology. But it is important to recognize that Heidegger features the prepositional "as" for the sake of precision and emphasis: we shall see that the interpretative question takes the text *as* something; the textual aspect, for its part, is taken *as* something; and the interpretation proper is textual aspect *as* something. Heidegger, like Derrida after him, invents new terms because conventional terminology reflects a world view he repudiates. Moreover, Heideggerian terms are no more "jargonistic" than the phraseology of American "New Criticism," for example, or even "Analytic" philosophy – just less familiar to an Anglo-American audience.

[18] As we shall see, if a relevant aspect is lacking, the interpretation is invalid.

[19] I do not use Husserlian nomenclature – *noema* instead of "As-which" or "textual aspect," because Husserl's *noema* (at least in his later philosophy) may "be distinguished from the real object" (see Aron Gurwitsch, the great Husserlian specialist, in 'On the Intentionality of Consciousness,' *Phenomenology*, ed. J. Kockelmans, [Garden City: Doubleday, 1967], p. 128). Heidegger's As-which, on the other hand, is a facet of the real object, the text. Nor is my surrogate term for the Heideggerian As-which, namely, "textual aspect," to be confounded with Husserlian "aspect": the latter is usually adjudicated unreal (see R. Ingarden, *The Literary Work of Art*, sections 40 and 42, where Ingarden speaks as a faithful Husserlian). As for Husserlian *noesis*, it is unlike

Heideggerian "As-question," since Husserl's *noesis* bestows meaning (see *Phenomenology and Literature*, pp. 98–101) and Heidegger's As-question does not.

[20] Heidegger's As-structure reminds one of Husserl's old maxim – intentionality is at one and the same time the grasp and the grip which grasps it.

[21] Take care not to confuse the terms "fore-structure" and "As-structure."

[22] Perhaps not enough Heideggerian phenomenologists have realized that Husserl, with his notion of *habitus*, approximates some of what Heidegger means by fore-structure. For example, Husserl says that each experience causes a *habitus*, or "new abiding property" which further determines the ego; and the ego, thus habituated, goes on to its next experience in a different (i.e., proportionately modified) way. See Husserl, *Cartesian Meditations*, trans. Dorion Cairns (The Hague: Martinus Nijhoff, 1960), pp. 66, 67, for his discussion of *habitus*. However, when dealing with formal interpretative activity, I think Heidegger would validate some fore-structures that Husserl would consider "presuppositions" in the technical sense, and thus "bracket out."

[23] See *Phenomenology and Literature*, pp. 97–104.

[24] See ibid., pp. 146, 148, 151, 156, 157, 160, 161, and passim.

[25] Ludwig Wittgenstein, 'Lectures on Aesthetics,' in *Philosophy of Art and Aesthetics from Plato to Wittgenstein*, ed. Frank Tillman and Steven Cahn (New York: Harper and Row, 1969), p. 517.

[26] In my essay, I use the term "signification" in lieu of the more common term "sense," because of what has been a knotty problem for translators. Husserl's and Ingarden's word *Sinn* has been customarily translated "sense," but Heidegger's word *Sinn* is translated "meaning," and Heideggerian "meaning" is very different from the Husserlian or Ingardenian notion of "sense." When I mean an idea and/or image operative in a culture's *langue*, I resort to the word "signification" – not to be at all confused, by the way, with E. D. Hirsch's term "significance."

[27] Contrast Roman Ingarden, who speaks of the "bestowal of meaning" by the author's "sense-giving acts." See *Phenomenology and Literature*, pp. 110, 115, 116.

[28] Hans-Georg Gadamer, *Truth and Method*, trans. Garrett Barden and John Cumming from the second German edition (New York: Seabury Press, 1975), p. xix.

[29] Gadamer, *Le Problème de la conscience historique* (Louvain; Publications universitaires de Louvain, and Editions Béatrice-Nauwelaerts, 1963), p. 75.

[30] See *Phenomenology and Literature*, pp. 73–78. Notice, however, that Heidegger uses As-questions even when involving the author's "willed significations." Heidegger is by no means "objective" in his own practical literary criticism.

[31] Consult also ibid., pp. 185, 186, for the examples I provide from Nathaniel Hawthorne's short story, "My Kinsman, Major Molineux."

[32] Ibid., pp. 186, 187.

[33] Cited from Heidegger's *Vom Wesen des Grundes* (1929) by Theophil Spoerri in 'Style of Distance, Style of Nearness,' in *Essays in Stylistic Analysis*, ed. Howard Babb (New York: Harcourt, Brace, Jovanovich, 1972), p. 78.

[34] Heidegger 'Dialogue on Language,' in *On the Way to Language*, trans. Peter D. Hertz (New York: Harper and Row, 1971), pp. 1–54. Professor Chang Chung-yuan's book is *Tao: A New Way of Thinking: A Translation of the Tao Tê Ching with an Introduction and Commentaries* (New York: Harper and Row, Perennial Library, 1977).

[35] Heidegger, *Poetry, Language, Thought*, pp. 131, 132.

[36] Martin Heidegger, *The Question of Being*, trans. with introd. by William Kluback and Jean T. Wilde (New Haven: College and University Presses, 1958).
[37] Martin Heidegger, *Identity and Difference*, trans. with introd. by Joan Stambaugh (New York: Harper and Row, 1969).
[38] See *Being and Time*, pp. 193, 194.
[39] Ingarden, *The Literary Work of Art*, p. 142 (though it must be kept in mind, of course, that the Ingardenian norms for "opalescence" are somewhat different from Heidegger's).
[40] See *Phenomenology and Literature*, pp. 67, 73, and passim.
[41] For the convergence of *"das Gevierte"* to generate human *Dasein*, and for the *"Differenz"* between *Welt* and *Seienden*, see ibid., pp. 69–72, and its composite of quotations from Heidegger.
[42] Heidegger, *Poetry, Language, Thought*, p. 40.
[43] For the communal nature of *Dasein*, see Heidegger's 'Remembrance of the Poet' and 'Hölderlin and the Essence of Poetry,' in *Existence and Being*.
[44] R. H. Blyth, *Games Zen Masters Play* (New York: New American Library, 1976), p. 14.
[45] *The Teachings of the Compassionate Buddha*, ed. E. A. Burtt (New York: New American Library, Mentor Books, 1955), p. 228.
[46] William Barrett, to his great credit, and long before the idea seemed due, recognized the possibility of correlating Heideggerian and Oriental thought. See Barrett, *Irrational Man* (1958; New York: Doubleday, Anchor Books, 1962), p. 234 and passim.
[47] Chang, *Tao: A New Way of Thinking*, p. 108.
[48] Ibid., p. 106.
[49] Heidegger, *Poetry, Language, Thought*, p. 63.
[50] Chang, *Tao*, p. 150.
[51] Heidegger, *Identity and Difference*, p. 27.
[52] For more passages where Heidegger affirms the simultaneous validity of contradictories, see the quotations from Heidegger in Chang, *Tao*, pp. 14, 61, 108, 124, and passim.
[53] René Wellek and Austin Warren, *Theory of Literature*, 3d. ed. (New York: Harcourt, Brace and World, 1956), pp. 177, 178.
[54] E. D. Hirsch, *Validity in Interpretation* (New Haven: Yale University Press, 1967).
[55] Hirsch, *The Aims of Interpretation* (Chicago: University of Chicago Press, 1976).
[56] Take care not to confuse "signification" and Hirsch's term "significance."
[57] Hirsch, *Validity in Interpretation*, p. 230.
[58] *Webster's Seventh New Collegiate Dictionary* (Springfield: G. and C. Merriam, 1963), p. 154.
[59] See Heidegger, *Being and Time*, pp. 377–80, and *On the Way to Language*, p. 34.
[60] But see my discussion of an inconsistency in Heidegger's practical criticism of literary works, *Phenomenology and Literature*, pp. 77, 78.
[61] Denis Dutton, in his own way, also argues effectively that norms for "scientific truth" should not function as models for the adjudication of literary interpretations (see my Note 5, above).
[62] But see Dutton's article, pp. 331, 332, his footnote 2. Some scientists are coming to regard the "scientific method" as "plausible," not "certain."
[63] Heidegger, *Being and Time*, p. 197.

[64] Flannery O'Connor, *Mystery and Manners*, ed. Sally and Robert Fitzgerald (London: Faber and Faber, 1972), p. 153.
[65] O'Connor, ibid., p. 41.
[66] Heidegger, *Poetry, Language, Thought*, p. 5.

EUGENE F. KAELIN

CRITICISM OF ROBERT MAGLIOLA'S PAPER

From the foregoing it is clear that Professor Magliola has found in Martin Heidegger's hermeneutical phenomenology a way to ground his own critical pluralism. As a teacher of comparative literature, perhaps, he can be nothing but a pluralist. He meets and enters into dialogue with students every day who possess different "fore-structures," each of whom demands an equal right to interpret a text as he or she sees and feels it. The value of his approach is its centering on the critical consciousness as it struggles to make its interpretation. Since as Heidegger puts it in *Being and Time*, "Dasein ist je meines," no explanation of anything at all can be put in any other terms. Indeed, *Jemeinigkeit* is the first existential (*Existenzial*) noted by Heidegger as proper to Dasein's constitution.

But this same consideration applies to three other existentials of Heidegger's existential analysis, all of which are accepted by Professor Magliola as valid, since he uses them in explaining the principles of interpretation: affectivity, understanding, and discourse. Interpretation is only the way in which *Dasein*'s prior understanding of its being-in-the-world gets worked out. This prior understanding is, of course, preontological; and it is this preontological comprehension by human consciousness of its involvement in a world into which it is thrown that gets interpreted through the existentials of *Being and Time*. Not unimportantly, the concept of meaning itself (*Sinn*) is also an existential, understood, to be sure, by virtue of *Dasein*'s projection of itself upon its own possibilities. Heidegger is explicit on this point: "Sinn ist ein Existenzial des Daseins, nicht eine Eigenschaft, die am Seienden haftet, 'hinter' ihm liegt oder als 'Zwischenreich' irgendwo schwebt."[1] The question arises, then, why Professor Magliola treats some of Heidegger's existentials as speculative and ideological (and therefore not phenomenological), yet retains others to lay out his program for textual analysis. Perhaps he has not taken Heidegger at his word, that the so-called "distinction" between the earlier and the later Heidegger is not indicative of two distinct inquiries.

The question of the meaning of Being can only be posed by human consciousness. Again, section 32 of *Being and Time* is quite explicit:

Und wenn wir nach dem Sinn von Sein fragen, dann wird die Untersuchung nicht tiefsinnig und ergrübelt nichts, was hinter dem Sein steht, sondern fragt nach ihm selbst, sofern es in die Verständlichkeit des Daseins hereinsteht. Der Sinn von Sin kann nie in Gegensatz gebracht werden zum Seienden oder zum Sein als tragenden "Grund" des Seienden, weil "Grund" nur als Sinn zugänglich wird, und sei er selbst der Abgrund der Sinnlosigkeit.[2]

In ontological inquiry one questions entities about their being, i.e., about their position in the totality of involvements that constitute *Dasein*'s worlding world. Being as the ground of beings can be nothing more than a meaning, which remains a function of *Dasein*'s projection of self-understanding.

Why, then, the apparent change in the Heideggerean phenomenology? After *Being and Time* Heidegger's own fore-structure undergoes a change. *Dasein*'s disclosedness becomes more explicity a region, and Being becomes "that which regions." The innate ambiguity of the German '*da*,' meaning both here and there, had already been used for a description of *Dasein*'s spatiality. It is the basis of the hither and the yon, in which human consciousness meets the entities of its concern. And Being presences itself in the area of human conern, and nowhere else. Circumspective concern produces the understanding for any act of human consciousness living the conditions of its life-world, and for this reason Heidegger begins section 32 with the manner in which the understanding of circumspective concern gets worked out in an interpretation.

Like anything else uncovered in the totality of involvements a text gets interpreted in terms of *Dasein*'s projection of possibilities from the implicit knowledge contained in its fore-structure to explicit knowledge of its own being. But ontologically considered, for every being-disclosing (*Dasein*'s projecting itself understandingly upon its own possibilities) there is a being-disclosed, be it the tools within the totality of its involvements or the things of the natural environment. This is precisely why a self and a world are mutually implicative. If the being of *Dasein* is its careful concern, the meaning of its being is its temporality (for the Genevans, a "mode" of consciousness). And the "presencing" of Being may be understood only as a ground, the regioning of *Dasein*'s world, in which entities are made present.

In poems, then, the regioning of *Dasein*'s being is made manifest, but, as interpreted by Professor Magliola, only within the representational stratum of the work. In poems, we are presented with a world within which appeal is made to the fourfold: the earth and the sky, mortals and divinities. Each of these, being metaphors, are symbols for human meanings, the ultimate locus of human involvements. The earth is that upon which we live, and the sky, its circumferential horizon. Things can make their appearance only within these regions; and when they do, the mortals are disclosed by virtue of their relations to the things of their concern. Only the divinities remain aloof, since we lack their names. To name them would be to make them present, at least as a meaning. So poets, questioning the meaning of their existence, initiate the dialogue with the philosophical hermeneuts among us concerning the ultimate meaning of them all — that of the God who is always proximate, as the ground of all beings, yet remains in reserve, since the ground disappears beyond the figures of the things. This disappearance of the ground "behind" the figures is of an epistemological order and not the ontological. There are, then, two ways in the Heideggerean system of thought, which is always hermeneutical, for thinking being: either in ontology itself, in which our discourse is brought to the level of spoken language; or in poetry, where language, which houses Being, is allowed to let the presence of Being shine forth.

There is no escaping the conclusion: the meaning of any being may shine forth (the root meaning of "phenomenon," it must be recalled) only to the consciousness that expresses its concern. Whether we use the existentials of *Being and Time* or the symbols of his later works, Heidegger wants to avoid the abyss of meaninglessness ("der Abgrund der Sinnlosigkeit") that occurs every time our concern expresses itself for things in abstraction from their ground, i.e., apart from the totality of involvements. Professor Magliola's preference for the hermeneutics of the later Heidegger, I fear, expresses a fore-structure that contains a preference for the mystical.

Let us merely recall that to question the meaning of beings is to question the beings themselves. In fundamental ontology, *à la* Heidegger, *Dasein* questions its own being, and in doing so reveals the structures of its own existentiality. *Dasein*'s disclosedness (affectivity, understanding, and authentic discourse) reveals only two things: its own being-disclosing and the being of the entities disclosed. But in our own understanding of our relationships to our worlds at this prejudicative level of "understanding," Being remains the ground, the regioning wherein the entities appear, and is itself grasped, therefore, only as a meaning resulting from the projection of ontic possibilities. The poetic trick is to produce the image of the worlding world, in which Be-ing is allowed to presence itself as that which regions in *Dasein*'s self-projection. It is in this way that the poet enacts the unconcealment of beings, and the hermeneut commemorates the enactment.

Section 32 of *Being and Time*, I am suggesting, should be read as descriptive of both creative and interpretative understanding. Before being written, the poem is a projected significance in the life of the poet; once written, it is a means of dialogue in which the hermeneut projects the significance of that poem as one of his own possibilities. In both cases, the projects are existential, or, if you prefer, existential-ontological. In a very real sense, therefore, we discover who and what we are by writing and reading poems. That, it seems to me, is sufficient a result to expect from any human enterprise. To look for the meaning of Being in general as the significance of every poem is to erect an hypothesis that is too powerful and to lose oneself in the abyss of meaninglessness, if for no other reason than the one mentioned above: Being presences itself in a poem not by virtue of the representation alone, but between the "earth" of the sensuous vehicle (the words) and the representational depth references these words may reveal or conceal, the "world" of the expression. The Heideggerean hermeneut questions the significance of words in terms of their signifieds, and tends to ignore the expressiveness of the signifiers, the earth of the poem itself. We can only glimpse Be-ing as a meaningful ground when we let beings be, and whether we do this by becoming shepherds of Being (as in the earlier Heidegger) or whether we meet Being by traveling through the house of language, the result will be the same: we shall be constructing an abode in which to live, to move, and to enjoy our being. It is this situation, ultimately, that yields its aspects to any as-question we may entertain.

We may, therefore, like Professor Magliola, use this hermeneutical model to justify any interpretation of a given literary text; or we may, like the structuralists, find that the living context of language restricts the number of aspects that might be revealed therein. In choosing between the one or the other attitude, there is still the tutor text to be used as one's guide. And that text contains both signifiers and signifieds. It was for this reason, it will be remembered, that in 'Origin of The Work of Art' Heidegger maintained that every work of art sets up a tension between earth and world, between the organized sensuous medium of sound and whatever level of depth signification these sounds of the poem may intend.

Until this back reference of representational depth to the expressing surface is made, Professor Magliola will have failed to show the concreteness of his hermeneutical analysis. For, without feeling the dynamic tension between expressing surface and expressed depth the hermeneut will have no foreconception on which to base an analysis of the work. In practice, then, Professor Magliola's application of the Heideggerean theory of meaning fails to fulfill the second criterion he himself has stipulated for a method to be accepted as "phenomenological." Poets, of course, know better, since they deal concernfully every day with the tensions generated between signifying word and signified meanings.[3]

NOTES

[1] Martin Heidegger, *Sein und Zeit* (Tübingen: Max Niemeyer Verlag, 1957), p. 151.
[2] Ibid., p. 152.
[3] Cf. my 'Notes toward an Understanding of Heidegger's Aesthetics,' in *Phenomenology and Existentialism*, ed. Lee and Mandelbaum (Baltimore: Johns Hopkins University Press, 1967), pp. 59–92.

VEDA COBB-STEVENS

MYTHOS AND *LOGOS* IN PLATO'S *PHAEDO*

Aristotle remarked in the first book of the *Metaphysics* that "the lover of myth is in a sense a lover of wisdom" (982b18). In Plato's *Phaedo*, the converse of that proposition is shown to hold as well. Socrates, who characterized himself not as a *sophos* but as a *philosophos*, proves to be, despite his disclaimer at the beginning of the dialogue, a *mythologikos*, a fabricator of myths. Given that Socrates, the lover of wisdom, is also a lover of myth, two questions immediately arise: (1) *How* exactly is philosophy related to myth? and (2) *Why* should the philosopher need to appeal to it? Aristotle's explanation of his own thesis is rather straightforward. Philosophy begins in wonder, he says, and since myth is composed of wonders, the lover of myth is in a sense a lover of wisdom. Yet Aristotle's comments are not meant as a real attempt to explore the relation between philosophy and myth, for in his view " . . . it is not worthwhile to consider seriously the subtleties of the mythologists" (*Metaphysics*, 1000a18). In the *Phaedo*, however, Socrates appears as a philosopher who appeals to myth, not just casually as one might quote a line of Homer, but thoroughly and consistently, within the context of philosophical inquiry itself. As Friedländer has noted, it seems that one method is not enough.[1]

As a first approach to understanding the relation between philosophy and myth, we shall begin from the side of philosophy and examine Socrates' defense of the philosophic life (63b–69e). In many places throughout the dialogue, philosophical argumentation is characterized as *logon didonai*, giving an account of whatever object is the focus of inquiry. Socrates begins his defense by saying that he wants to "give an account" (*logon apodounai*, 63e9) of his philosophical mode of life, thus indicating that the method which accounts for other things must also account for itself. He points out that philosophy is a search for wisdom, and that wisdom involves a knowledge of such things as the just, the good and the beautiful, none of which can be perceived by the senses. Hence, to obtain wisdom, a person must approach each object ". . . as far as possible, with the unaided intellect . . . [and pursue the truth] by applying his pure and unadulterated thought to the pure and unadulterated object, cutting himself off as much as possible form his eyes and ears and virtually all the rest of his body . . ." (65e7–66a5).[2] However, it is not simply a question of recognizing that the perceptions of the senses are deceptive and resolving thenceforth to avoid them. On the contrary, even when the soul has decided to investigate by itself the things themselves, the body continues to intrude with its needs, desires, and fears. Thus, if we are ever to know anything purely, we must be completely rid of the body (66d8–e1). Now, death is by definition the release and separation of the soul from the body. Since, as we have seen, philosophy is nothing but the constant practice of separating the soul from the body, it follows that philosophy is the practice of dying. So, Socrates concludes, given that a philosopher has spent his whole life in preparation for death, it would be unreasonable that he should not welcome it when it arrives (67d4–69e5).

The defense might have ended here, had Cebes not raised an objection. A philosopher would be justified in his confidence, Cebes maintains, if it can be shown (1) that the soul actually continues to exist after death and (2) that it is possessed of some intelligence (70b3–4), for that complete wisdom for which the philosopher is striving would otherwise be impossible to obtain. At this point, Socrates begins to "give an account" of these two requirements.

First of all, by the argument from the cyclic recurrence of opposite processes, Socrates infers that those who are living must continue to exist in the world of the dead. Because opposites are generated from opposites and living is opposite to dying, then the dead must come from the living and the living from the dead, since the cycle must continue if all becoming is not to come to a standstill (70c4–72e2). Second, by the argument from recollection, we are shown that the soul is possessed of intelligence when separated from the body. Since we have knowledge of absolute equality, justice, and beauty, a knowledge which could not have been derived from the world of becoming through sense perception, our souls must have existed in a state of knowledge before our birth (72e3–77a5).

At this conclusion, Simmias remarks that although he is impressed by the argument, he must agree with Cebes: the real issue is whether the soul continues to exist and to be possessed of intelligence *after death*. Socrates suggests that one need only combine the second argument with the first, since in effect, given the cyclic recurrence of opposites, the time before birth *is* the time after death. However, Socrates does not stop here, but goes on to pinpoint the sources of Simmias' and Cebes' discomfort: "You are afraid, as children are, that when the soul emerges from the body the wind may really puff it away and scatter it, especially when a person does not die on a calm day but with a gale blowing" (79d5–e1). Cebes responds: ". . . don't suppose that it is we that are afraid. Probably even in us there is a little boy who has these childish terrors. Try to persuade him not to be afraid of death as though it were a bogey" (77e5–7). Socrates agrees and advises them to "say a magic spell over him" (*epadein*) every day until the child's fears have been soothed and charmed away (77e8–9). When Cebes objects that Socrates, the only capable charmer of such things, will soon be departing, Socrates replies that, actually, the best charmer one can find is oneself.

What is the meaning of Socrates' advice to charm and sing incantations to this child in us? In the *Charmides*, we find Socrates offering Charmides a remedy for a headache, consisting of a "kind of leaf" which must be accompanied by a charm. The charm, it seems, is by far the more important of the two, for ". . . without the charm, the leaf would be of no avail" (155e).[3] But ultimately, we discover, the true charm is philosophy, since one ought not to attempt to cure the body without attending to the soul. "And the cure of the soul," Socrates says, "has to be effected by the use of certain charms and these charms are fair words (*logoi*)" (157b). In the *Phaedo*, too, it would seem that the charms which will soothe the fearful child in us are nothing but the *logoi* themselves, the words spoken in a philosophical discussion and, more importantly, the conceptual distinctions which these words express. As Socrates said in his defense, the body is the source of many troubles and fears, and the philosopher, in keeping his soul off by itself, progressively purifies it from the influence of the body. The very attempts to prove that the soul is immortal are charms to soothe the fear that it might be mortal. But if it is the *logos* which has the role of charming or soothing the child in us, both by its *method* or *structure* (when it is directed to any philosophical problem) and by its *content* (when its subject is the fate of the soul itself), it would seem that myth, which is composed of concrete images taken from the world of sense perception, could have only a negative value in the philosophic life. And yet we find elements of myth throughout the dialogue, not only in passages where myth is "gathered together by itself" but also within the very fiber of the *logoi*.

I. *MYTHOS* WITHIN THE CONTEXT OF *LOGOS*

First of all, one can discern a use of myth which appears to be somewhat similar to the dialectical method later specified by Aristotle for the investigation of ethical matters. For the resolution of a given problem, he said, one should make a survey of opinions (*endoxa*) held both by the many and the wise, and use these as the basis of inquiry. "For every man has some contribution to make to the truth, and with this as a starting point, we must give some sort of proof about these matters. For by advancing from true but obscure judgments, we will arrive at clear ones, exchanging ever the usual confused statement for more real knowledge" (*Eudemian Ethics*, 1216b30–35).

At least twice in the *Phaedo*, Socrates initiates the resolution of a problem by means of a myth which is appropriated as being relevant to the issue at hand, but in need of either clarification or justification. The first such instance (which involves the clarification of a myth) occurs at the beginning of the dialogue and constitutes the factor which occasions Socrates' defense of his philosophical way of life. Socrates has just claimed that although the philosopher should not do violence to himself by suicide, he should be willing to follow the dying. When Cebes demands that this paradox be explained and the injunction against suicide be vindicated, Socrates begins by appealing to a myth. There is a secret legend, he says, "... that we men are in a sort of custody, and that a man must not release himself or run away, which appears a great mystery to me and not easy to see through" (62b3–6).[4] This tale, presumably of Orphic or Pythagorean origin, Socrates proceeds to examine and clarify, attempting to "see through" it by means of a familiar analogue. He asks Cebes: "If one of your possessions were to destroy itself without intimation from you that you wanted it to die, wouldn't you be angry with it and punish it, if you had any means of doing so?" (62c1–4) If we take a person's relation to one of his possessions as analogous to the gods' relation to humanity, then we can *understand* (*diidein*, 62b6), Socrates is claiming, why suicide is prohibited.

It would seem that the procedure here involves no great complexity. The myth tells us that we are possessions of the gods, in their care and custody, and must not release ourselves or run away. When Socrates, in constructing the clarifying analogue, speaks of a person's relation to one of his possessions, we assume that he is referring to that living possession, the slave, which is the only one capable of releasing itself or running away.[5] Thus the proportion to which he appeals in his clarification of the myth may be expressed as follows: as a master (a) is to a slave (b), so is a god (A) to a human being (B). The structure of the analogy is such that two distinct *parts* of (B), viz., (a) and (b), are used to clarify the relation between the *whole* of (B) and something else, (A). We are enabled to conceptualize the import of the myth more clearly by considering the familiar relationship between master and slave. Were the slave to run away, the master would be angry and justified in punishing him. Thus, were a person to commit suicide, the gods would be angry and their punishment just.

This account of Socrates' clarification of the myth is perfectly in order.

The only difficulty is that this is not what he has *said* (cf. 62c1–4 quoted above). It is Cebes who employs the analogy in its logically proper form. It is Cebes who, in taking up the line of discussion initiated by Socrates, emphasizes the element of "running away." Socrates had begun by claiming, Cebes points out, that the philosopher is not foolish in his confidence concerning death, but rather justified in his willingness to die. Appealing implicitly to the analogue of master and slave, Cebes argues that if the gods are the best masters there are, then a person would be foolish indeed to want to "run away" or release himself from them. The analogue may help us to understand why we should not commit suicide (because we would incur the wrath of our "masters"), but it makes even more incomprehensible the problem which occasioned the discussion in the first place: why a philosopher should be willing to die (62c9–e7). It seems that Socrates, rather than resolving the original paradox, has only succeeded in making it more puzzling. But Socrates is quick to point out Cebes' questionable presupposition, viz., that death is a release from the gods. On the contrary, he suggests, there is hope that after death the philosopher will be in even closer relation to these good masters (63b5–6). It must be admitted, however, that Cebes' presupposition is quite natural, first of all, in light of the original statement of the Custody Myth which indicated that we should not "run away" from the gods, and second, given the proper development of the clarifying analogue in terms of the prohibition of the slave's *running away* from his master.

The claim that the element of *running away* constitutes the proper point of the analogue's development may be justified by the following considerations. Since Socrates appeals to the Custody Myth as an aid to understanding the prohibition against suicide, he sees it, presumably, as *expressing* that very prohibition in its own terms. The myth suggests that death is a "release" or "running away," and that we are "possessions of the gods." Thus we must understand the import of this "running away" if we are to understand the myth and, in that way, understand the prohibition against suicide. Now when the myth speaks of "running away" it is no longer talking about running away in the literal sense, but about that mode of self-release known as suicide. Hence, in order to understand the transformation effected by the myth of an ordinary language expression, it is necessary to investigate the meaning of the expression in its original ordinary language context. Since the myth suggests also that we are "possessions" of the gods, the most obvious context in which to place the literal meaning of "running away" is that of a slave running away from the master who owns him. We would then be able to grasp the meaning of the myth more clearly: as a (human) slave must not run away from his (human) master, so a human being must not "run away" from the gods.

Socrates, however, does not develop the analogy in this way. Rather than saying that just as a slave ought not to run away (or release himself by escape), so no human being should "run away" (or release himself by suicide), he says that the possession ought not "kill itself" (*apokteinuoi*, 62c2). In the analogue as properly articulated, all the elements are known: the slave, the master, and the relation between them. In this way the *explanans* helps us to understand the relation between the elements of the *explanandum*. But Socrates does not simply take an analogue from a known domain (the slave running away from his master) to throw light upon its counterpart which extends into an unknown sphere (a human being committing suicide). Rather, he places the unknown itself into the supposedly clarifying analogue. Very much aware of what he is doing, he insists that the possession ought to be punished, *if there were any punishment* (62c3–4). Of course, there is no punishment, since the slave would be dead. If the myth supposedly expresses a prohibition against suicide in general, then the appeal to a particular type of suicide (that of the slave) does not significantly clarify the issue.

However, in terms of extending our awareness of the human situation, Socrates' treatment of the problem has accomplished a great deal. By saying that just as a slave ought not to "run away" from his master, so we ought not to "run away" from the gods, we would be entertaining an analogy which is structurally correct, but possibly misleading about the true situation. As Socrates suggests, it is probable that at death the philosopher will be in even closer relation to the gods. This makes us question the concept of death as release from *the gods* (the assumption guiding Cebes' reflections) and opens the way to the possibility that death is a release from *something else*, a possibility which is developed more fully in Socrates' defense and forms the basic presupposition of the four arguments for immortality.

More significantly, however, the analogue which uses "running away" as the connecting relation hides from us, under the mask of the extensive power of logical analogy, the fact that the analogy is precisely that: an extension beyond the immediately known. The slave's running away is something that the master can accommodate. He sends out a party to hunt him down, quite naturally planning to have him brought back and punished. Were the slave to *kill himself*, the master might be angry, but would be rather nonplussed as to how to punish him. The effect Socrates achieves by putting the *explanandum* (death) back into the *explanans* is to remind us that we are unavoidably looking at the problem from our own human perspective and that we must not be deceived by the extensive power of analogy into thinking that we have *seen* the relation between ourselves and the gods.

Thus what at first seemed to be a use of myth similar to Aristotle's appeal to *endoxa* turns out to be much more involved. Indeed, Socrates appropriates the myth and then submits it to a clarifying reflection, much as Aristotle appropriates and examines *endoxa*. But the reflection is not, as in Aristotle's case, straightforward. The insight to be gained is not presented outright, but is rather indicated obliquely. In this way, Socrates not only succeeds in bringing his interlocutors into philosophical discussion at the present moment, but offers them (and us) hints for further reflection when the problem is reexamined at a subsequent time.

In a later passage, when the first argument for immortality is offered, we find that Socrates once more uses a myth. Cebes, we remember, first wants assurance that the soul *exists* after death. Socrates appeals to "an old legend," again one of Orphic or Pythagorean origin, which holds that souls " ... *do* exist [in Hades], after leaving here, and that they return again to this world and come into being from the dead" (70c5–8). If what the legend says is true, then of course our souls must *exist* in the world of the dead, since they are to be born again. In this argument, Socrates uses a method somewhat like[6] the method of hypothesis which he describes (99b4–102a2) prior to the final argument for immortality. First of all, a given hypothesis (here, the Rebirth Myth) is considered and what harmonizes with it (or follows from it) is taken to be true. In the present case, if the legend is true, then what follows is that our souls must exist in the world of the dead. Second, the hypothesis is subjected to scrutiny and is itself justified. The justification here consists in establishing general principles from which the legend follows as a particular instance. The course of Socrates' argument deserves to be examined in detail.

There are two general principles which he establishes as being relevant to the justification of the legend, and hence, to the problem of the fate of the soul: (1) " ... that everything is generated in this way — opposite from opposite" (71a9–10) and (2) " ... that between each pair of opposites there are two processes of generation, one from the first to the second, and another from the second to the first" (71a12–b2). Socrates supports both these principles in precisely the same way, viz., by pointing out a number of particular instances and then generalizing inductively on the basis of them. First of all, he claims, we can see that whatever becomes bigger becomes so from having been previously smaller, that whatever becomes weaker becomes so from having been stronger, and likewise with the slower, the worse, and the more just, in that each comes into being out of its opposite. The first principle, then, is sufficiently established (71a9) and we can be assured that everything is generated by opposites coming from opposites. The second principle, also, is supported by various examples. By noting the processes of increasing and diminishing, being separated and being mingled, growing cold and growing hot, etc., we are forced to recognize that there is a *transition* from one opposite to another.

From these two general principles, Socrates proceeds to make specific applications which will justify the Rebirth Myth. We are to reflect upon being awake and sleeping, two opposites which are analogous to being alive and being dead. Socrates himself considers being awake and sleeping and asks Cebes to explore the other pair. He says: " . . . I say that waking comes from sleeping and sleeping from waking, and that the processes between them are going to sleep and waking up. . . . Now you tell me in the same way . . . about life and death" (71c11–d5). In response to Socrates' questions, Cebes agrees that (1) from the living the dead come into being, and again, from the dead, the living. On the basis of this, he agrees further that (2) our souls must exist in the next world. It would seem that the conclusion for which the argument was striving has now been reached. However, we note that Cebes' responses carry less than full conviction. He agrees to (1) and (2) by saying, "Evidently" (*Phainetai*, 71e1) and "So it seems" (*Eoiken*, 71e3). Socrates continues then to argue on the basis of the second general principle, that for any two opposites there is a process of becoming between them. In the case of coming to life and dying, Socrates notes, one process of becoming is quite clear, viz., the process of dying. Thus if nature is not to be lame in this point (71e9), then the other process must be present also. Socrates now concludes: "So we agree upon this too – that the living have come from the dead no less than the dead from the living. But I think we decided that if this was so, it was a sufficient proof that the souls of the dead must exist in some place from which they are reborn" (72a4–8).

What exactly has been accomplished by this argument? When we look back over the progression of the discussion, we note that the pair of opposites, being awake and sleeping, plays a rather peculiar role. Given Socrates' pattern of reasoning, we would expect that this pair of opposites, as well as the relevant opposite processes, would have been cited in *support* of the two general principles, for only one pair of opposites, being alive and being dead, must be justified deductively by means of the inductively established general principles. When we examine the nature of the examples given as the basis for the inductive generalizations, we discover that they are all taken from the world of possible experience. Hence, the general laws thus established should only be applicable to objects within this world which, although they are not presently *known*, are yet *knowable* (or experienceable) in principle. Now although being alive is a quality which certainly falls within the range of experience, its opposite, being dead, does not. The pair, being awake and sleeping, constitutes a means of mediation between all those supporting examples, both terms of which belong to the world of experience, and the one pair of opposites which is the focus of the argument, being alive and being dead. For although being awake is quite obviously experienceable, sleeping is at most "half-experienceable," that is, only through dreams. Through his quasi-illegitimate application of the two general principles to being awake and sleeping, Socrates can proceed to the completely illegitimate application of them to the case of being alive and being dead.

But although the structure of the argument is quite fallacious, Socrates has, as with the Custody Myth, greatly increased our understanding. It is quite true that, formally, Socrates proceeds again in a manner quite similar to Aristotle's in his examination and justification of *endoxa*. His pattern of argumentation involves, as we have seen, first recognizing that the myth would entail the conclusion desired (that the soul exist after death) and then establishing general principles from which the myth could be deduced. The general principles established, however, have a limited range of applicability and cannot justifiably be extended to a sphere which is nonexperienceable in principle, viz., that of death. Yet the recognition of the fallaciousness of Socrates' procedure provides the ground for an insight that will guide the reflections in the remainder of the dialogue. The argument fails because qualities from sense perception alone and the principles of their generation and destruction cannot possibly reach beyond the world of becoming. As the last three arguments indicate, it is only by means of a purely intellectual grasp of unchanging forms that we have any intimation at all of a dimension transcending the empirical, a dimension which just possibly could be the home of our souls after death.

So far, we have seen myth taken as the *beginning* point of a train of argumentation. Although a close examination of the discussions has revealed that there is more to Socrates' reasonings than would at first sight appear, nonetheless the general surface-structure of his argument follows one of two patterns: either the myth is clarified (as with the Custody Myth) or justified (as with the Rebirth Myth) by logical analysis and reflection. However, myth can be seen to play another role within the context of argumentation. At the end of Socrates' defense of the philosophic life, a myth is introduced which is seen, *after* the argumentation, to express much the same thing as the argument itself. Philosophy, we have been told, is a means of separating the soul from the body, " ... and truth is in reality a cleansing from all such [bodily] things, and temperance and justice and courage and wisdom itself are a means of purification" (69b8–c3).[7] On this basis, we can understand the practitioners of mystic rites who spoke with a hidden meaning when they said that " ... he who enters the next world uninitiated and unenlightened shall lie in the mire, but he who arrives there purified and enlightened shall dwell among the gods" (69c5–8). They spoke with a hidden meaning, but ironically, with a meaning hidden from themselves. After Socrates' discussion, we understand that the purified are those who have practiced philosophy, the separation of the soul from the body by means of intellectual inquiry and moral self-control. As H.-G. Gadamer has pointed out, the older Pythagorean notion of purity has, in Socrates, been replaced by a new one. "Purity for him is no longer to be identified with prescribed cultic rites of purification, to which the members of an order cling as their 'symbol' without the least bit of self-understanding. On the contrary, for Socrates 'purity' means a new awareness of oneself as found in the life of the philosopher who concentrates upon thinking."[8]

Thus we find that Socrates appeals to myth both at the beginning of a train of argumentation and at the end. But in either case, *mythos* is in some sense reduced to *logos*; it is not taken at face value, but is questioned, clarified, or justified before it is accepted. That the clarifications and justifications themselves are not to be taken at face value does not undermine the primacy of *logos*. For it is precisely a further reflection, not a further appeal to myth, which the difficulties in the analysis of the myth demand. But even though myth, as it occurs within the context of argument, appears as subordinate to *logos* and is subjected to analytical scrutiny, it is still open to question whether the Myth of the Earth, which constitutes an independent section of the dialogue (108e4–115a8), has the same status.

II. *MYTHOS* OUTSIDE THE CONTEXT OF *LOGOS*

The main body of the Myth of the Earth is devoted to depicting the nature of the earth as it "really is." We human beings, Socrates says, live around the sea, breathe the air, and think we are living on the true earth, whereas we are in much the same position in relation to the true earth as sea creatures are in relation to us. If through sluggishness and weakness, those living underneath the sea never strive upward to break the surface of the water and view the earth where we live, they would continue to think that the murky vision which they have underwater yields a correct picture of reality. However, the "reality" to be found under the sea falls short of what is present in our world: " ... in the sea everything is corroded by the brine, and there is no vegetation worth mentioning, and scarcely any degree of perfect formation ... " (110a3–5). But we human beings are in the same position in relation to the true earth. For we live in hollows of the earth into which air has collected, just as sea creatures live in hollows filled with water. And just as those living under the sea would think that the water was heaven when they saw the sun and stars shining through, so we, too, think we see the real heaven, when we see the sun and stars shining through the air. The real earth, however, when seen from above is like a twelve-patch leather ball of various colors, which are brighter and purer than those we know. On the true surface of the earth there are flowers and fruits more perfect than we have ever seen, as well as precious stones of which the jewels that we know are only corrupted fragments. On the real earth, everything is " ... clearly to be seen, ... so that to see it is a sight for happy spectators" (111a2–4).[9]

John Burnet and others spend a great deal of time trying to demonstrate that Socrates is here presenting a kind of masked natural science. It is held that the passage involves an attempt to reconcile various pre-Socratic scientific traditions, most notably the Pythagorean and the Ionic. Such an account of the true surface of the earth where everything is clearly to be seen is, Burnet claims, " ... an astronomer's vision of blessedness."[10] I think Hackforth is correct in rejecting this interpretation. The role of the Myth of the Earth, Hackforth claims, is " ... to contrast the world of ordinary experience with another, better world which is literally nowhere, but which we can conceive as a place (τόπος). ... "[11] The true significance of the myth is not cosmological, but rather more broadly ontological. In support of this thesis, we recall that beginning with Socrates' defense of the philosophic life and continuing all through the discussion of immortality, a certain distinction was hinted at, presupposed or explicitly used, viz., the ontological distinction between an invisible, unchanging, perfect world of the forms and the visible, changing, imperfect world of becoming. Our task in life, Socrates has said, is a progressive purification of our souls from the influence of the body and the world of becoming in general, such that we turn away from those things which are merely imperfect copies (or "corrupted fragments") of the perfect forms and turn toward those forms themselves.

The myth, however, does not stop with sketching a picture of the two orders of reality, but goes on to delineate an eschatology. Socrates describes the inner structure of the earth, that abode of souls who have died in a state of impurity. All the hollows in the earth, he says, are connected by systems of channels " ... through which, from one basin to another, there flows a great volume of water ... and of fire too, great rivers of fire, and many of liquid mud, some clearer, some more turbid, like the rivers in Sicily that flow mud before the lava comes, and the lava stream itself" (111d2–e2). Furthermore, "all this movement to and from is caused by an oscillation inside the earth ... " (111e4–5). Burnet notes that "the whole description shows that a sort of pulsation, like the systole and diastotle of the heart, is intended. The theory is in fact an instance of the analogy between the microcosm and macrocosm and depends on Empedocles' view of a close connection between respiration and the circulation of the blood."[12] Burnet's assessment seems to be confirmed when we read a few lines later that these rivers move in and out with the wind " ... just as when we breathe we exhale and inhale the breath in a continuous stream ... " (112b6–7).

However, it is not the case that the body is taken as a starting point for reflection upon the cosmos. It is not, as Aristotle would say, an instance of beginning with that which is familiar and reasoning analogically to obtain insight into matters that are more obscure. Rather than considerations of bodily existence representing the inner structure of the earth, the inner structure of the earth represents bodily existence. The Myth of the Earth mentions four underground rivers: Acheron, the river of pain; Styx, the river of hate; Cocytos, the river of wailing; and Pyriphlegethon, the river of burning fire. These rivers carry souls who had died in a state of impurity through the earth, into Tartaros and back again, until, if possible, they are finally purified. If the *pulsations* of these rivers represent the life of the body, their *names* indicate even more pointedly the suffering derived from living a life devoted to that part of us which, as Socrates said earlier, " ... fills us with loves and desires and fears and all sorts of fancies and a great deal of nonsense ... " (66c2–4). It is a life of excess, without form or measure, cut off from true being and "wallowing in utter ignorance" (82e4–5). As Socrates said in the *Apology*, "the difficulty is not so much to escape death, the real difficulty is to escape from doing wrong, which is far more fleet of foot" (39a ff.).[13] Thus, the important consideration is care for the soul in this life and if the soul is immortal, then for all time (*Phaedo*, 107c1–4). At this point, I would disagree with Hackforth's claim that the ontological dimension of the myth is somewhat at odds with the eschatological element. He states: "For the purpose of eschatology, of contrasting earthly misery with heavenly bliss, this ignorance, this mistaking one world for another is irrelevant."[14] On the contrary, the decisive factor determining the misery or bliss of the soul is precisely its relation to being. As Socrates pointed out earlier in the *Phaedo*, the greatest evil and misery to be derived from a life devoted to the body is not becoming ill or squandering money through uncontrolled desires, but is the ontological error which this life entails (82d9–83e3). It is by its relation to being that the soul, first of all, orders its life as far as possible according to the ideals of the forms, and second, purifies itself so as to escape the punishment of "lying in the mire" in Hades.

III. THE RELATION BETWEEN *MYTHOS* AND *LOGOS* IN THE PHILOSOPHIC LIFE

From the present standpoint, we have a better grasp of the status of *mythos* in relation to *logos*. We have seen that myth plays at least two roles in the *Phaedo*. On the one hand, it is woven into the very texture of the *logos* itself, either from the beginning of an argument or at the end. That which is "spoken with a hidden meaning" is examined, clarified, or justified by philosophic reflection. In effect, *mythos* is resolved into *logos*. On the other hand, the Myth of the Earth at the end of the dialogue is not a simple legend, expressible in two or three propositions, which is appropriated and reflected upon. Rather it is an extended piece of discourse which expresses, in its own terms, the entire epistemological, ontological, and moral outlook of the dialogue. In this case, the *logos* is resolved into myth.

But even though the role of myth in relation to philosophic inquiry has been clarified, the second question with which this paper began has not yet been answered. *Why*, to reverse Aristotle's remark, is the lover of wisdom also a lover of myth? With regard to the first function of myth, i.e., as offering in an unclarified fashion insights worthy of philosophic reflection, no great difficulty is encountered. In being a lover of myth in this sense, the philosopher has not stepped outside his own sphere, but has brought other things within his purview. Furthermore, even if the reflection upon myth is in some sense logically inadequate (as an analysis of Socrates' treatment of the Custody Myth and Rebirth Myth has shown), this very inadequacy is a veiled demand for *philosophical* inquiry, not for further appeal to myth. The present question, however, is not so easily answered regarding the second role of myth. If philosophy, conceived as *logon didonai*, is sufficient unto itself (as Socrates' defense of the philosophic life seems to indicate), why should its results need to be transposed into mythic form?

Socrates' comment at the close of the Myth of the Earth deserves examination. He says: "No sensible man would think it proper to rely on things of this kind being just as I've described.... [But] such things he must sing like a healing charm (*epadein*) to himself, and that is why I have lingered so long over the story" (114d1–7).[15] Previously, it was the *logos* which was to act as a healing charm (cf. p. 393 above). By reflecting philosophically and giving an account of our beliefs, we could bring the soul away by itself, separate it as much as possible from the body, and thereby soothe the child in us who is afraid of death. In this light, the soothing power of the Myth of the Earth might appear to be a regression to the concerns of real childhood, where the fretful infant is calmed by simple stories.

It must not be forgotten, however, that the myth which Socrates spins out is not a "simple story," but a *philosophic* myth, which expresses the basic outlook of the entire dialogue. That this transposition of philosophical insight into mythical terms is a necessary step in the process of allaying our fears of death can be seen by considering that insight itself. The framework of the dialogue is constructed on the basis of the ontological distinction between the world of becoming and the world of being. Within this general cadre, a similar distinction (that of body and soul) is found in the ontological structure of human nature. By transforming this philosophical insight of *logos* into mythic terms, Socrates is perfectly consistent with a point made many times throughout the discussion. We are to separate the soul from the body *as much as possible*; but this separation can never be achieved *completely* during our life time. Since we are composed of both body and soul, the language of each must be used to address the fear of death. We must use language expressing concrete perceptual images, as well as that which is purely abstract and conceptual. However, since the soul must retain primacy and control over the body, the basic *structure* of the philosophical insight gained by *logos* must guide and order the content of the myth. In this way, the myth can speak directly to the fear which has its origin in the body, but speak with the "hidden meaning" and calming effect of its logical structure. Thus the myth, instead of inducing the soul to regress to the level of the body, speaks to the body in a language structured by the insights of the soul. Indeed, immediately after speaking of the myth as a source of comfort, Socrates continues and reaffirms his earlier contention that confidence in the face of death must have its justification in philosophy. "There is one way, then, in which a man can be free from all anxiety about the fate of his soul – if in life he has abandoned bodily pleasures and adornments ... and has devoted himself to the pleasures of acquiring knowledge ... " (114d8–e4).

NOTES

[1] Paul Friedländer, *Plato*, vol. 3, trans. Hans Meyherhoff (Princeton: Princeton University Press, 1969), p. 41. Cf. also Olympiodorus: διὰ δύο ἐπιχειρημάτων, ἑνὸς μὲν μυθικοῦ καὶ Ὀρφικοῦ, ἑτέρου δὲ διαλεκτικοῦ καὶ φιλοσόφου (*in loc. Phaed.*).

[2] Unless indicated otherwise, all passages quoted from the *Phaedo* are taken from Hugh Tredennick's translation, in *Plato: the Collected Dialogues*, ed. Edith Hamilton and Huntington Cairns (New York: Pantheon Books, 1961); hereafter cited as Hamilton and Cairns.

[3] Passages quoted from the *Charmides* are taken from Benjamin Jowett's translation which appears in Hamilton and Cairns.

[4] Translated by W. H. D. Rouse, *Great Dialogue of Plato* (New York: New American Library, 1956), p. 465; hereafter cited as Rouse.

[5] Rouse, in fact, mentions the slave explicitly in his translation of 62c1–4: "... if one of your possessions, your slave, should kill himself ... " (p. 465). For a parallel use of *ktēma*, cf. Euripides' *Medea*, l. 49.

[6] The method of hypothesis entails the positing of the *eidos* rather than the positing of a myth.

[7] Rouse, p. 472.

[8] H.-G. Gadamer, 'Die Unsterblichkeitsbeweise in Platons *Phaidon*' in *Wirklichkeit und Reflexion: Walter Schulz zum 60 Geburtstag* (Pfulligen: Neske), p. 148; translated by P. Christopher Smith in *Dialogue and Dialectic: Eight Hermeneutical Studies on Plato* by H.-G. Gadamer (New Haven: Yale University Press, 1980), pp. 24–25.

[9] Rouse, p. 515.

[10] John Burnett, *Plato's Phaedo* (Oxford: Clarendon Press, 1911), p. 133; hereafter cited as Burnet.

[11] R. Hackforth, *Plato's Phaedo* (Cambridge: Cambridge University Press, 1955), p. 174; hereafter cited as Hackforth.

[12] Burnet, p. 135.

[13] Translated by Hugh Tredennick in Hamilton and Cairns.

[14] Hackforth, p. 174.

[15] Rouse, p. 518.

RAMONA CORMIER

SARTRE'S CONCEPTION OF THE READER-WRITER RELATIONSHIP

The genius of Jean-Paul Sartre has ranged over a variety of topics, which he has discussed in philosophic treatises and exhibited in novels, plays, and short stories. In these writings, Sartre has sought totality not only in the creation of his essence but also in the rendering of human possibility. Yet twentieth-century trends toward specialization have generally led critics to deal with his conceptions in a fragmentary manner by focusing either on his philosophical or literary writings. On the other hand, I will endeavor to integrate ideas in these seemingly diverse writings in an effort to understand Sartre's contention that he has been impotent as a writer [1] even though he claims that the prose writer writes to provoke guilt in the reader.[2] I will formulate my argument by showing that Sartre's conception of the reader as inventing the writer is ontologically analogous to the love relationship and that the reflection of the ontological assumptions underlying these relationships in his novels, short stories, and plays might very well give Sartre the impression that he has been an impotent author.

I

Sartre describes the love relationship in the following passage from *Being and Nothingness*:

Why does the lover want to be loved? If Love were in fact a pure desire for physical possession, it could in many cases be easily satisfied. Proust's hero, for example, who installs his mistress in his home, who can see her and possess her at any hour of the day, who has been able to make her completely dependent on him economically, ought to be free from worry. Yet we know that he is, on the contrary, continually gnawed by anxiety. Through her consciousness Albertine escapes Marcel even when he is at her side, and that is why he knows relief only when he gazes on her while she sleeps. It is certain that the lover wishes to capture a "consciousness." But why does he wish it? And how?
    The notion of "ownership" by which love is so often explained, is not actually primary. Why should I want to appropriate the Other if it were not precisely that the Other makes me be? But this implies precisely a certain mode of appropriation; it is the Other's freedom as such that we want to get hold of ... the lover does not desire to possess the beloved as one possesses a thing; he demands a special type of appropriation. He wants to possess a freedom as freedom. ... He wants to be loved by a freedom but demands that this freedom as freedom should no longer be free.[3]

Succinctly characterizing the plight of the lover as well as the Sartrean inferno in which hell is other people,[4] the love relationship demonstrates the ontological dependence of the lover on the loved one as well as the paradoxical nature of this dependence, which rests upon Sartre's contention that existence has an architecture delimiting the manner in which a conscious being acquires an essence. This architecture consists of consciousness, spontaneity seeking content derived from *êtres-en-soi*, entities possessing essence, and *êtres-pour-autrui*, other conscious beings. As long as consciousness, through its relationship to other conscious beings, is unaware of itself as an object this dynamic essence or content is freely acquired. However, when confronted by the look of the other, which may provoke either feelings of shame, guilt, fear, or similar emotive reactions, consciousness loses its freedom and becomes aware of itself as the object of another consciousness. Sartre illustrates this flight of freedom with the well-known example of the voyeur caught in the act. At that moment the voyeur ceases to exist as pure spontaneity and can only regain his autonomy by turning the cause of his objectivity into an object. The conflict exhibited in the voyeur's predicament characterizes all human relations, for Sartre maintains that "While I attempt to free myself from the hold of the Other, the Other is trying to free himself from mine; while I seek to enslave the Other, the Other seeks to enslave me.... Conflict is the original meaning of being-for-others."[5] Thus consciousness in its need of the Other and by virtue of the sadistic-masochistic nature of its relationship to the Other is doomed to a hell void of sympathy or love.

The lover then needs the loved one to make him be. Yet, in the act of loving, the lover "does not want to possess an automaton, and if we want to humiliate him, we need only try to persuade him that the beloved's passion is the result of psychological determinism. The lover will then feel that both his love and his being are cheapened."[6] On the other hand, if the lover consents to being an object he does not want to be manipulated like the pieces on a chess board. Rather he "wants to be the object in which the Other's freedom consents to lose itself, the object in which the Other consents to find his being and his *raison d'être*...."[7] But these desires remain unfilled, for love is, according to Sartre, triply destructive: "in the first place it is, in essence, a deception and a reference to infinity since to love is to wish to be loved, hence to wish that the Other wish that I love him."[8] Consciousness' autonomy is at stake here, since the satisfaction of this desire depends upon the freedom of another. However, even if this desire were satisfied, the lover's status as a free being would still be in jeopardy, for "in the second place the Other's awakening is always possible; at any moment he can make me appear as an object – hence the lover's perpetual insecurity."[9] Furthermore, were this difficulty overcome the lovers could not realize their desires, for "in the third place love is an absolute which is perpetually *made relative* by others. One would have to be alone in the world with the beloved in order for love to preserve its character as an absolute axis of reference...."[10] The lover then wishes the impossible. He wishes to love a free being but instead his love enslaves the loved one by turning her into an object or the loved one responds by turning the lover into an object. Either one or the other is a sadist or a masochist.

An analogous situation arises in the writer-reader relationship. Sartre describes the writer's activity as "action by disclosure.... The 'engaged' writer knows that words are action. He knows that to reveal is to change and that one can reveal only by planning to change."[11] One generally assumes that the act of disclosing through prose writing is complete with the publication of the work. But the writer is, for Sartre, as dependent upon the reader as the lover is on the loved one since the writer cannot read what he writes. While reading his prose, the writer confronts his consciousness, "*his* knowledge, *his* will, *his* plans, in short, himself."[12] In other words, the writer's prose cannot be an object for his consciousness. Prose becomes an object only when

reading is composed of a host of hypotheses, of dreams followed by awakenings, of hopes and deceptions. Readers are always ahead of the sentence they are reading in a merely probable future which partly collapses and partly comes together in proportion as they progress, which withdraws from one page to the next and forms the moving horizon of the literary object. Without waiting, without a future, without ignorance, there is no objectivity.[13]

Consequently the writer needs the reader if his work is to be. This beingness of the work is its reinvention by the reader. During the act of reinvention, the substance of the work becomes the reader's subjectivity while the writer serves as the reader's guide.[14] As guide the writer, like the lover, assumes the freedom of the reader, who upon choosing to read this work rather than that one, accepts responsibility for the work chosen. But this responsibility as well as the reader's reinvention of the prose work is subject to the unique essence or lived experience of each reader.

Furthermore, Sartre claims that since the writer writes for free men he "has only one subject – freedom."[15] In elucidating the freedom, which is the proper subject matter of prose, Sartre distinguishes eternal and concrete freedom. He seems to associate the former with ontological freedom, the conception he analyzes at length in *Being and Nothingness*, and the latter with a freedom or liberation associated with institutions, customs, or certain forms of oppression and conflict. Accordingly, it is concrete rather than eternal freedom which should engage the writer, for "One cannot write without a public and without a myth — without a *certain* public which historical circumstances have made, without a *certain* myth of literature which depends to a very great extent upon the demand of this public."[16] In a word, the writer chooses his public by writing about a specific alienation in a particular situation to give "society *a guilty conscience.*"[17] But this guilt is one in which the writer, by his anatagonism toward the conservative forces that are maintaining the balance he intends to upset, seeks to enslave the reader.[18] This enslavement, however, is distinct from that involved in the love relationship because the writer, once he releases his work, finds that its invention by the reader is beyond his look. (The writer's situation is analogous to that of Estelle in *No Exit* who has no control over the interpretation her best friend gives to her infanticide.) Also the writer must enslave through a presentation that enables the reader to make an aesthetic withdrawal, to freely suspend his disbelief. Sartre assumes that this withdrawal of the reader, in which prose writing becomes an object and to which the reader has lent his subjectivity, might result in a guilty conscience. Yet, through this suspension, the reader may disassociate his lived experience from that depicted in the prose he is reading and fail to feel responsible for the alienation or oppression described in it.

Sartre further erodes the effect of the writer upon the reader by contending, in the lectures "A Plea for Intellectuals," that prose says something that is unsayable.[19] This nonknowledge, which is prose, is distinct from scientific knowledge conveyed through technical language, which transmits a maximum of information with a minimum of misinformation. The unsayable, on the other hand, conveys nonknowledge through the use of ordinary language, a medium about which there is no agreement. Yet, despite these handicaps, the writer, according to Sartre, communicates a nonsignifying silence enclosed by and produced by words.[20] This nonsignifying silence expresses the prose writer's being in the world as it is filtered through his consciousness. Similarly, the reader filters the writer's world through his lived experience. But here's the rub: the reader's invention of the writer expresses his situation, that is, what he values in terms of his life-style. Thus his comprehension of the prose work may not reflect the writer's intent.

II

Despite the barriers between the writer and the reader, the former does enter the consciousness of the latter. One need only examine one's reading experiences to recall sympathetic or antipathetic involvement with characters in novels, plays, etc., and the influence this involvement might have had on one's awareness of oneself or others. (Some aesthetic theories have been constructed with this relationship as central.)[21] Despite this empirical phenomenon, however, Sartre claims in an interview with Madeleine Chapsal that his writings have changed nothing — that ever since his youth he has experienced utter impotence as a writer.[22] Let us examine this contention in the context of his prose writings. There we observe that the freedoms expressed in these works are not concrete freedoms alienating one human being from another but are rather eternal freedoms pertaining to man's existential predicament, a condition about which nothing can be done. Man is not the cause of nor does he feel guilty about the contingency of existence, the conflict that defines his relationship with others, the distinction between himself and the objects of his consciousness, his lack of essence and so forth. Yet these are the topics of interest to the characters in Sartre's novels, short stories, and plays. Sartre presents these issues in the context of situations in which human beings are in the act of enslaving or being enslaved. For example, Pablo, Juan, and Tom in *The Wall* are enslaved by the values of the Belgian doctor and their executioners; Garcin in *No Exit* by Estelle and Inez and Estelle and Inez by Garcin; and Mathieu in *The Age of Reason* by Marcelle, his mistress of several years.

Sartre presents these conflicts in a stylistic framework in which characters appear solitary to the reader by virtue of their engrossment in their existential predicament. There is interaction among them as exhibited in their conversations, but these conversations reflect their solitary assessment of their predicament. For example, Mathieu, the ineffectual intellectual, continuously struggles with the issue of his freedom. He tells Marcelle that "If I didn't try to assume responsibility for my own existence it would be utterly absurd to go on existing."[23] Yet when he acts he seems unable to act with reason. He wishes to tell Marcelle, now pregnant, that he loves her but finds himself saying he does not love her. He seems unable to furnish reasons for his actions and appears irresponsible to the reader. On the other hand, Daniel, Mathieu's friend, wishes to be a homosexual as a stone is a stone but finds it impossible, for his experience and his vice can never coincide. He will always be a consciousness distinct from its content or objects. He resolves his problem by choosing the opposite of his desires: he marries the pregnant Marcelle. Moreover, Lucien and Roquentin, the "heroes" of *The Childhood of a Leader* and *Nausea* respectively, struggle directly not with their freedom but with issues pertaining to their existence and essence.

In a word, Sartre's characters exhibit concerns with that ontological freedom presumed in the writer-reader relationship and not with those social ills which might induce guilt in the reader. Engrossed in their existential predicament, these characters are objects to each other and also appear to be objects to the reader. They are, of course, objects to Sartre, their creator. But as objects their emotive life appears vacuous. Consequently, as a reader, I am, for example, unable to sympathize with any of them, but rather become involved in their philosophical vagaries. Furthermore, since all values are equal in the Sartrean inferno I perceive neither a victim who is unjustly treated nor one who errs and suffers as a result of his or her mistake. Instead I find myself in a world peopled by atomic entities whose interactions with others are less important than their quest of ontological freedom or their search for essence. My involvement is cerebral rather than emotional. I hesitantly generalize these particular comments since my responses may not be someone else's. After all, I am, by Sartre's own view, inventing him, i.e., filtering the world of his characters through my lived experiences.

Nevertheless, I contend that the aesthetic withdrawal or suspension of disbelief, which may occur in reading Sartre's work, is induced specifically by his conception of man, and since one cannot exhibit in prose writings compelling reasons why one should cerebrate at length about one's existential state, then Sartre may indeed by impotent as a writer, if he writes to induce guilt that might lead to social change. Furthermore, his claim that prose writing conveys the unsayable suggests that the critical analysis of literary works is futile and improper and reduces the critic to silence. I find his world of silent readers uncharacteristic of consciousness and suggest that Sartre's conception of himself as an impotent writer may be false, since his efforts at enslaving the reader set the scene for further activity in the lived world of other conscious beings. Hopefully this activity will seek to overcome the pessimism and isolation inherent in Sartre's conception of the human being as an anguished creature whose essence forever escapes him.

NOTES

[1] Jean-Paul Sartre, *Between Existentialism and Marxism*, trans. John Mathews (New York, 1974), p. 21.
[2] Jean-Paul Sartre, *Literature and Existentialism*, trans. Bernard Frechtman (New York, 1966), p. 81.
[3] Trans. Hazel Barnes (New York, 1956), pp. 366–67.
[4] Jean-Paul Sartre, *No Exit and Three Other Plays*, trans. Stuart Gilbert (New York, 1946), p. 47.

[5] *Being and Nothingness*, p. 364.
[6] Ibid., p. 367.
[7] Ibid., pp. 367–68.
[8] Ibid., p. 377.
[9] Ibid.
[10] Ibid. Sartre exhibits this last phenomenon in *No Exit* when the presence of Inez hampers the lovemaking of Garcin and Estelle.
[11] *Literature and Existentialism*, p. 23.
[12] Ibid., p. 42.
[13] Ibid., p. 41.
[14] Ibid., p. 45.
[15] Ibid., p. 64.
[16] Ibid., p. 150.
[17] Ibid., p. 81.
[18] Ibid.
[19] *Between Existentialism and Marxism*, p. 272.
[20] Ibid.
[21] The most original of these theories in English is that of Veron Lee.
[22] *Between Existentialism and Marxism*, p. 21.
[23] Jean-Paul Sartre, *The Age of Reason*, trans. Eric Sutton (New York, 1973), p. 15.

CHRISTOPH EYKMAN

"SOUVENIR" AND "IMAGINATION" IN THE WORKS OF ROUSSEAU AND NERVAL

Two mental processes, remembering and fantasizing, characterize much of the autobiographical and literary work of the eighteenth-century philosopher Jean-Jacques Rousseau and the nineteenth-century French romanticist Gerard de Nerval.[1] The influence of such works as Rousseau's *La Nouvelle Héloïse, Confessions, Les Rêveries d'un Promeneur Solitaire* on Nerval has already been studied in detail.[2] The way in which Nerval makes use of "souvenir" and "imagination" in his work, shows not only a definite affinity of mind and thought between himself and Rousseau but also the originality of his own poetic genius. This study, therefore, does not deal with the question of influence but compares the scope of two thematic complexes in the writings of the two authors.

As to the role of imagination within the totality of mental functions, we find a variety of definitions by French philosophers of the eighteenth century. Voltaire, in his *Dictionnaire Philosophique*, attributes to imagination nothing but the faculty to recompose the sensations received in the acts of sensory perception or to construct independently new fictitious inner perceptions out of the pregiven sensory material: "On ne fait aucune image, on les assemble, on les combine. Les extravagances des *Mille et une Nuits* et des *Contes des Fées* etc. etc., ne sont que des combinaisons."[3] Diderot, in the article "Génie" of the *Encyclopédie*, emphasizes the close interrelation between "imagination" and "sentiment," thereby showing a great deal of affinity with Rousseau and foreshadowing the emergence of romanticism: "... dans l'homme de génie, l'imagination va plus loin: il se rappelle des idées avec un sentiment plus vif qu'il ne les a reçues, parce qu'à ces idées mille autres se lient, plus propres a faire naître le sentiment."[4] He also insists on what today might be termed the "existential" quality of imagination, i.e., the fact that the prime source of imagination is the personality of the individual: "L'imagination prend des formes différentes, elles les emprunte des différentes qualités qui forment le caractère d l'âme. Quelques passions, la diversité des circonstances, certaines qualités d l'esprit, donnent un tour particulier a l'imagination; elle ne se rappelle pas avec sentiment toutes les êtres, parce qu'il n'y a pas toujours des rapports entre elle et les êtres."[5] Diderot took the first steps on a road which was eventually to lead to a new appraisal of imagination and ultimately to Fichte's idealism according to which *Einbildungskraft* produces all reality (*Grundlage der gesamten Wissenschaftslehre*, 1794) and to Schelling's thesis that the same *Einbildungskraft*, "hovering" (*schwebend*) between finitude and infinitude, produces ideas (as opposed to concepts (*Begriffe*)). Schelling equates it with *Vernunft* ("reason") which, as in Kant, assumes a higher level of mental activity than mere *Verstand* ("understanding," "intelligence").[6]

Rousseau envisages and experiences imagination in a way which clearly places him at a distance from both Fichte and Schelling, yet his "position" is also sufficiently removed from eighteenth-century rationalism so as to justify the role he has traditionally been assigned by literary historians: that of a pacemaker of European romanticism.

His image of man is dualistic. Human nature reveals a split between reason and emotion, nature and civilization, egoistic instincts and the aspiration to eternal verities or the demands of conscience.[7] The sciences and civilization, i.e., man's own achievements, have corrupted man who once had been good after he had left the hands of his Creator. In a natural presocial state, man had known peace and harmony. But that was in a golden age long since lost. The conflict between the "moi humain" and the "moi commun," i.e., the struggle between self and society has yet to be solved in a new society to be created, a society which will be both civilized and virtuous and which is to reestablish the lost state of "natural" primeval man.

More than anything else, it is the antagonism between reason and emotion in human nature which accounts for Rousseau's obsession with memory and imagination. Feeling emerges as an eminently creative faculty in the form of remembering and fantasizing. It overwhelms both reason and any rational sense of reality in the mind and work of the French philosopher.

In Rousseau's autobiographical writings the notion of happiness (*bonheur*) assumes a key function, as Albert Béguin points out in his book *L'Ame Romantique et le Rêve*: "Il faut bien prendre garde que tout s'oriente, chez lui, authour d'un désir immense de bonheur."[8] Rousseau defined happiness in terms of an inner satisfaction that is totally independent of the outer vicissitudes of life.[9] As the old Rousseau puts it in *Les Rêveries du Promeneur Solitaire*: ". . . la source du vrai bonheur est en nous. . . ."[10] To him, one of the sources of happiness consists in his ability to escape from a painful reality or to overcome a feeling of inner void by means of fantasizing: "Dépouillé par des mains cruelles de tous les biens de cette vie, l'espérance l'en dédommage dans l'avenir, l'imagination les lui rend dans l'instant même; d'heureuses fictions lui tiennent lieu d'un bonheur réel; et, que dis-je? lui seul est solidement heureux, puisque les biens terrestres peuvent à chaque instant échapper en mille manières à celui qui croit les tenir; mais rien ne peut ôter ceux de l'imagination à quiconque sait en jouir."[11]

Rousseau is a virtuoso in constructing visions of a chimerical world which reflect his own unfulfilled wishes and aspirations.[12] In his *Confessions* he recalls how, in the years of his adolescence, he imagined himself as "maréchal Rousseau," anticipating in his imagination a splendid military career which he would never have been able to attain in real life. Before entering Paris for the first time in his life, he daydreams of a city consisting only of magnificent palaces. His high expectations are bitterly disappointed when he finds himself in the dirty and poverty-stricken suburb of Saint-Marceau.

Toward the end of Rousseau's novel *La Nouvelle Héloïse*, Julie defends imagination and desire by attributing to them a higher emotional and aesthetic value than to reality or real possession:

Malheur à qui n'a plus rien à desirer! il perd pour ainsi dire tout ce qu'il possède. On jouit moins de ce qu'on obtient que de ce qu'on espère, et l'on n'est heureux qu'avant d'être heureux. En effet, l'homme avide et borne, fait pour tout vouloir et peu obtenir, a reçu du ciel une force consolante qui rapproche de lui tout ce qu'il désire, qui le soumet a son imagination, qui le lui rend présent et sensible, qui le lui livre en quelque sorte, et pour lui rendre cette imaginaire propriété plus douce, le modifie au gré de sa passion. Mais tout ce prestige disparoit devant l'objet même; rien n'embellit plus cet objet aux yeux du possesseur; on ne se figure point ce qu'on voit; l'imagination ne pare plus rien de ce qu'on possède, l'illusion cesse ou commence la jouissance. Le pays des chimères est en ce monde le seul digne d'être habité, et tel est le néant des choses humaines, qu'hors l'Etre existant par lui-même, il n'y a rien du beau que ce qui n'est pas.[13]

While Rousseau's mind is an easy prey of fantasies, he finds himself unable to experience and perceive physical reality as it *is*. He remains indifferent vis-à-vis the real features and qualities of the outer world: "Ma mauvaise tête ne peut s'assujettir aux choses. Elle ne saurait embellir, elle veut créer. Les objets réels s'y peignent tout au plus tels qu'ils sont; elle ne sait parer que les objets imaginaires. Si je veux peindre le printemps, il faut que je sois en hiver; si je veux décrire un beau paysage, il faut que je sois dans des murs; et j'ai dit cent fois que si j'étais mis à la Bastille, j'y ferais le tableau de la liberté."[14]

Reality might trigger the workings of Rousseau's creative imagination, but the real object as such is only of interest to him if it is "aimable," i.e., if it presents itself to him as an object of his inexhaustible urge to love. A neat borderline between "fiction" and "réalité" cannot be established in his mind. In the words of the old Rousseau: "... j'assimilais a mes fictions tous ces aimables objets et me trouvant enfin ramené par degrés a moi-même et à ce qui m'entourait, je ne pouvais marquer le point de séparation des fictions aux réalités."[15] External objects attract his attention only when the powers of his imagination are temporarily exhausted and when the mere act of sensory perception brings about a state of relaxation (*délassement*) of his mind that allows him to replenish the energies of his creative faculties.[16]

Although Rousseau frequently admits and describes his own very strong inclination toward fantasizing, he also shows a sense of sober self-criticism which makes him aware of the dangers of the escapist attitude inherent in this mental activity. In his educational novel *Emile* he definitely disapproves of any type of existence wavering between reality and imaginary worlds:

C'est l'imagination qui étend pour nous la mesure des possibles, soit en bien, soit en mal, et qui, par conséquent, excite et nourrit les désirs par l'espoir de les satisfaire. Mais l'objet qui paraissait d'abord sous la main fuit plus vite qu'on ne peut le poursuivre; quand on croit l'atteindre, il se transforme et se montre au loin devant nous. Ne voyant plus le pays déjà parcouru, nous le comptons pour rien; celui qui reste à parcourir s'agrandit, s'étend sans cesse. Ainsi l'on s'épuise sans arriver au terme; et plus nous gagnons sur la jouissance, plus le bonheur s'éloigne de nous.[17]

The ultimate source of Rousseau's creative imagination and of his ever unfulfilled longing for happiness is feeling. Thus Diderot might have used him as a prime example of his definition of imagination which emphasizes the emotional aspect (see above). The symbol of feeling in Rousseau's writings is the heart as the organ of feeling ("... son coeur, avide de bonheur et de joie ...").[18] "Je ne vis plus que par le coeur," he writes to the Marquis de Mirabeau in 1767.[19] The object of his striving for happiness remains unknown and undefinable. He feels "... dans les lueurs même de prospérité que quand j'aurais obtenu tout ce que je croyais chercher je n'y aurais point trouvé ce bonheur dont mon coeur était avide sans en savoir démêler l'objet."[20] The simultaneous emotional acts of "désirer et jouir"[21] produce fantasies and embellish the real world. Rousseau's fantasies are by no means rational constructions. In the same book in which he voices some skepticism as to the educational value of imagination, he also writes: "Ce sont les chimères qui ornent les objets réels; et si l'imagination n'ajoute un charme à ce qui nous frappe, le stérile plaisir qu'on y prend, se borne a l'organe, et laisse toujours le coeur froid."[22] The contrast between "organe" (mere sensory perception) and "coeur" returns elsewhere in Rousseau's writings in the form of "voir" versus "sentir"[23] or "sens" versus "coeur."[24]

Bringing to mind memories of pleasant events experienced in reality serves the same purpose in Rousseau's intellectual and emotional life as fantasizing: "... sans cesse occupé de mon bonheur passé, je le rappelle et le rumine, pour ainsi dire, au point d'en jouir derechef quand je veux."[25] Unpleasant situations in the real present are denied their full impact on Rousseau's mind by an act of plunging back into the happier memories of the real past while unpleasant memories are, of course, suppressed.[26] In turning to the real past, imagination does not create but recreate. Rousseau even claims that he experiences things and events fully only when looking at them *in retrospect*: "... je ne vois bien que ce que je me rappelle, et je n'ai de l'esprit que dans mes souvenirs."[27] To use an anachronistic comparison: his mind functions like the light-sensitive layer on a photographic film the impressions on which have to be developed *after* they were received.

In the ninth book of his *Confessions*, Rousseau describes the genesis of his novel *La Nouvelle Héloïse*. Dissatisfaction with reality (the intrigues of Therese's mother, his lack of genuine friends) make him seek happiness in a world designed by his emotionally charged imagination. He creates the fictional idol of Julie; later, he projects her radiant character onto the real Mme. d'Houdetot, who finally, having been thus transfigured, appears in those parts of the novel which were written after Rousseau made the acquaintance of the real woman. In a letter to A. M. de Saint-Germain from the year 1770, Rousseau states in retrospect: "L'amour que je concois, celui que j'ai pu sentir, s'enflamme à l'image illusoire de la perfection de l'objet aimé; et cette illusion même le porte a l'enthousiasme de la vertu; car cette idee entre toujours dans celle d'une femme parfaite."[28] Thus love in reality as in fiction is, in Rousseau's case, inspired by an imaginary transfiguration of a real person (l'objet aimé) into a perfect model of moral grace.

Happy childhood memories constitute yet another essential ingredient of Rousseau's novel. The interplay of "souvenir," "imagination," and "bonheur" accounts for much of the emotional substance of the work. Here is Rousseau's own testimony:

L'impossibilité d'atteindre aux êtres réels me jeta dans les pays des chimères, et ne voyant rein d'existant qui fut digne de mon délire, je le nourris dans un monde idéal, que mon imagination créatrice eut bientôt peuplé d'êtres selon mon coeur.[29]

Ce fut alors que la fantaisie me prit d'exprimer sur le papier quelques-unes des situations qu'elles m'offraient, et rappellant tout ce que j'avais senti dans ma jeunesse, de donner ainsi l'essor en quelque sorte au désir d'aimer, que je n'avais pu satisfaire, et dont je me sentais dévoré.[30]

It is quite obvious that the mere plot of this novel calls for "souvenir" as a major factor. Indeed, the two lovers, Saint-Preux and Julie, are forced to substitute memory and imagination for happiness and fulfillment which they taste only for an all too brief period during the early stages of their relationship. One example shall stand for many others: Julie sends Saint-Preux a locket containing her portrait. In one of his letters he describes to her the way in which he accepts the miniature portrait as a substitute for the real Julie. In turn, Julie transforms herself — by means of her imagination — into the art-object and savors in her fantasies what she would not dare to enjoy in reality. Imagination provides her the fulfillment which a virtuous life will deny her for ever: "Je m'imagine que tu tiens mon portrait, et suis si folle que je crois sentir l'impression des caresses que tu lui fais et des baisers que tu lui donnes: ma bouche croit les recevoir, mon tendre coeur croit les goûter. O douces illusions! ô chimères, dernières ressources des malheureux! Ah, s'il se peut, tenez-nous lieu de réalité!"[31]

The two lovers are well aware of the traps of imagination ("des pièges de l'imagination"),[32] as Julie once puts it, for it can so easily rekindle the flames which they and their mentors have taken such pains to extinguish. This is why Wolmar, Julie's all-too-reasonable husband, remarks about Saint-Preux: "Otez-lui la mémoire, il n'aura plus d'amour."[33] How paradoxical the dialectics of virtue, love, memory, and imagination can be is shown by a remark Saint-Preux makes in a letter to Julie in which he ponders his weakness (*foiblesse*): "Qu'elle abuse mon imagination, que cette erreur me soit douce encore, il suffit pour mon repos qu'elle ne puisse plus vous offenser, et la chimère qui m'égare à sa poursuite me sauve d'un danger réel."[34] Here imagination is no longer presented as a threat to virtue but as its bulwark.

In the works of Gérard de Nerval man is also portrayed as being torn between two opposing poles of his nature. Like Rousseau's dichotomy of nature versus civilization, Nerval's philosophy of man envisages time as a negative factor. But unlike Rousseau's, Nerval's thought is predominantly theological with mythological implications. Nerval interprets history as a continuous process of the rise and fall of religious beliefs. The divine powers which animate and inspire the world have ceased to be the prime mover of history, and the poet finds himself in the typically romantic situation of being stranded in a materialistic and god-forsaken world. In Nerval's writings, the golden age of a distant past is described not so much in terms of natural goodness but in terms of a timeless, perfect, and harmonious state of communication and interpenetration between God and man.

As a result of some cosmic catastrophe the nature of which remains mysterious, time came into existence and with it a long process of degeneration set in.[35] It is Nerval's ardent desire to recapture the lost paradise. This is the unique religious-emotional thrust of his private and often esoteric mythology.

If primeval harmony cannot be reestablished, it is, at least, accessible through memory charged with emotion and supplemented by imagination. Such memory extends far yeond the personal life of any individual into the prehistorical realm of mythology. Reality is still linked with that distant past, for it displays, like a palimpsest, the traces of a happier time. What is more, there still exists a timeless realm where everything man has ever loved and enjoyed is preserved and shielded against the annihilating effect of time.

Upon reading the second part of Goethe's *Faust*, Nerval was particularly impressed by the revival of the historical and mythical past through magic. He interprets this revival as an expression of Goethe's belief in a timeless immaterial realm where the past which seems to be doomed to disintegrate into nothingness and oblivion is rescued from annihilation and for ever preserved. "Pour lui [i.e., Goethe]," writes Nerval,

comme pour Dieu sans doute, rien ne finit, ou du moins rien ne se transforme que la matière, et les siècles écoulés se conservent tout entiers à l'état d'intelligences et d'ombres, dans une suite de régions concentriques, étendues à l'entour du monde matériel. Là, ces fantômes accomplissent encore ou rêvent d'accomplir les actions qui furent éclairées jadis par le soleil de la vie, et dans lesquelles elles ont prouvé l'individualité de leur âme immortelle. Il serait consolant de penser, en effet, que rien ne meurt de ce qui a frappé l'intelligence, et que l'éternité conserve dans son sein une sorte d'histoire universelle, visible par les yeux de l'âme, synchronisme divin, qui nous ferait participer un jour à la science de Celui qui voit d'un seul coup d'oeil tout l'avenir et tout le passé.[36]

Nerval, in this somewhat subjective interpretation, intentionally or unintentionally overlooks the fact that the above-mentioned "regions" in Goethe's text are the creation of the Devil, the product of magic, even though the Devil and his powers are, of course, tools in the hands of God.

Nerval believes that we can gain admittance to the same timeless realm of which Goethe speaks: in our dreams. For it is also the realm of the immaterial and eternal soul. Therefore, dreaming is remembering, and memory in that very special sense means a temporary (though illusionary) reconstitution of the lost paradise, thus fulfilling what Nerval calls "un désir secret de rétablir quelque chose de divine...."[37] However, the dreamer who manipulates legends and myths according to his own desires is, inevitably and regrettably, always flung back again into reality.

The common denominator of Nerval's and Rousseau's views consists in the trivial truth that reality and imagination are irreconcilable. In the work of both authors biographical and psychological factors blend with a particular historical situation to produce two images of man and his world which show a certain degree of affinity as well as divergence.

In Gerard de Nerval's writings the desire to transcend a reality falling short of the high and idealistic aspirations of romanticism takes on a special form of fantasizing. The image of a real beloved woman is transformed into that of a goddess. The term "goddess," though, is not applied to the woman in question in the traditional symbolic sense. Indeed, Nerval presents the real person to his readers as a reincarnation of a mythical deity of the distant past. Only the thus revealed "femme divine" merits the love of Nerval or his protagonists.[38] Once the real person stands divested of her mythical aura, the magic luster disappears. Nerval describes his love as "amour pour une étoile fugitive"[39] and uses the verb "chimérer"[40] to characterize his quest for the feminine ideal. A form of romantic antirealism appears to be behind this urge to idealize as the following lines from Nerval's story *Sylvie* show: "Il ne nous restait pour asile que cette tour d'ivoire des poètes, où nous montions toujours plus haut pour nous isoler de la foule. . . . Amour, hélas! des formes vagues, des teintes roses et bleues, des fantômes métaphysiques! Vue de près, la femme réelle révoltait notre ingénuité, il fallait qu'elle apparut reine ou déesse, et sourtout n'en pas approcher."[41] The parallel to Rousseau's attitude toward real things and persons is obvious. Rousseau remains indifferent toward the features of physical reality (except when seeking "délassement" after having strained and exhausted his imagination), Nerval despises reality as it is and pursues chimerical idols even if that means that desire and fulfillment do not coincide, as seems to be the case in Rousseau's mind. Nevertheless, Nerval, like Rousseau, assigns a higher value to desire and imaginary anticipation than to real possession: "La vie s'attache tout entier à une chimère irréalisable qu'on serait heureux de conserver à l'état de désir et d'aspiration, mais qui s'évanouit dès que l'on veut toucher l'idole."[42]

As in the autobiographical and literary work of Rousseau, there is, in Nerval, an important interrelation between the mental faculties of remembering and fantasizing. Throughout his *Voyage en Orient* we find literary "memories," i.e., a real experience involving the sights and the people of the Orient remind the author of certain passages or scenes from literary works he once read.[43] The Orient, however, does not only evoke literary but also historical reminiscences. History has left its traces in almost every place Nerval visits, and it is the historical "behind" the present and real Orient that fascinates the visitor.[44] Sometimes memory and imagination produce a peculiar experience of "déjà-vu." In the years preceding his voyage, Nerval establishes in his own mind – by means of his imagination and with the aid of pictures and descriptions – a vivid image of the city of Cairo. Once he actually sets foot upon the real city, he is amazed how congruent imagination and reality turn out to be: "Je l'avais vue tant de fois dans les rêves de la jeunesse, qu'il me semblait y avoir séjourné dans je ne sais quel temps; je reconstruisais mon Cafre d'autrefois au milieu des quartiers déserts ou des mosquées croulantes! Il me semblait que j'imprimais les pieds dans la trace de mes pas anciens."[45]

Nerval's obsession with the past often draws upon works of art (especially paintings) as a source of stimulation. Works of art are to him the only surviving witnesses of perished historical epochs. They preserve the ephemeral almost as well as those immaterial and timeless "regions concentriques et étendues" which he mentions in his preface to Goethe's *Faust* part 2. Art is thus "la coquille ou l'écaille splendide d'une créature anéantie."[46] It is also the most essential type of document for the historian, since "l'art renouait la chaîne des temps, conservait les idées, les examples et les enseignements du passé."[47] When Nerval enters the ancient city of Beyrouth, he remembers old paintings of the city which he has seen. Those memories serve as a link to the past which seems to have come to life once again transforming even the spectator Nerval into a historical figurine: "On a déjà lu cela dans les livres, on l'a admiré dans les tableaux, surtout dans ces vielles peintures italiennes qui se rapportent à l'époque de la puissance maritime des Vénitiens et des Génois; mais ce qui surprend aujourd'hui, c'est de le trouver encore si pareil à l'idée qu'on s'en est formée. On coudoie avec surprise cette foule bigarrée, qui semble dater de deux siècles, comme si l'esprit remontait les âges, comme si le passé splendide des temps écoulés s'était reformé pour un instant. . . . Me voilà transformé moi-même, observant et posant à la fois, figure decoupée d'une marine de Joseph Vernet."[48]

What motivates Nerval's search for the past? It is, first of all, his belief in divine providence as the prime mover of history. The study of the past thus reveals to us something about the direction in which future history is to move: ". . . le science et la divination du passé, qui leur découvre la loi providentielle de l'avenir et leur fait retrouver parmi les ténèbres et les décombres les cercles les plus anciens de la grande spirale du progrès et le point ou devront être soudés les cercles nouveauz qui la continueront dans l'avenir."[49]

But there is more to Nerval's obsession with the past and with "souvenir." He claims that his "memory" does not only extend over his personal life as well as over the realm of world history (as far as he acquainted himself with it) but also over the *mythical* past: "Ma pensée remontait au-delà: j'entrevoyais *comme* en un souvenir le premier pacte formé par les génies au moyen de talismans."[50] Here Nerval still hesitates to call his mythical vision "souvenir"; therefore he modifies and qualifies his statement by means of the word "comme." Elsewhere, however, for example, in his *Histoire du Calife Hakem*, he endows himself or his characters with mythical memory (i.e., memory not of myths *read* but of mythological existence "actually" experienced – which amounts of course in the final analysis to the same thing). A good example is Hakem's description of his sister: "Il y a chez elle quelque chose de céleste que je devine a travers les voiles de la chair ... par instants je crois ressaisir a travers les âges et les ténèbres les apparances de notre filiation secrète. Des scènes qui se passaient avant l'apparition des hommes sur la terre me reviennent *en mémoire*, et je me vois sous les rameaux d'or de l'Eden assis auprès d'elle et servi par les esprits obéissants."[51] Mythical memory means to Nerval and his protagonists in many instances an act of reaching back for the divine, an attempt to reestablish a lost relationship with God or the gods. This is also the prime motif of Nerval's imaginary transfigurations of "la femme réelle" into "la femme divine," which clearly distinguishes Nerval's "souvenir" from Rousseau's.[52]

Nerval believes in metempsychosis and traces the incarnations of a real woman (who is his contemporary) back through the ages to mythical times. As he puts it in *Aurélia*, memory reveals to him in his dreams the incarnations and the mythical identification of his own soul.[53] It also reveals the identity of all those women "que j'aimais transfigurées et radieuses."[54] Jenny Colon, the actress whom Nerval loved, once reveals to him in a dream this mythical identity: "Je suis la même que Marie, la même que ta mère, la même aussi que sous toutes les formes tu as toujours aimée."[55] The feminine idol can be mother, beloved woman, and saint (goddess) at the same time. Whether one of those three forms of love is prior and basic to the other two or whether all three converge into one complex emotional phenomenon without being derivatives of each other, is impossible to decide. Thus, Nerval loves not only the latest reincarnation of his idol but the one and only soul in *all* its temporal and ephemeral guises: "... une figure animique collective."[56] The real and ephemeral human being becomes transparent as the latest embodiment of a multiplicity of beings whose identity is of mythical quality. We now understand why the character Yousuf in Nerval's *Histoire du Calife Hakem* can say when he sees the image of his beloved in a hashish dream: "Mon âme se grandissait dans le passé et dans l'avenir; l'amour que j'exprimais, j'avais la conviction de l'avoir ressenti de toute éternité."[57]

Yet it is not only the real beloved woman who is transformed into a mythical person by Nerval's imaginary memory. Any real event experienced by the writer can be thus transfigured. In *Les Nuits d'Octobre* a character by name of Baratte falls asleep on a pile of red roses in the Paris market halls.

Nerval projects the myth of Silenus (for whom the bacchantes prepare a bed of flowers) onto the much more prosaic situation in which Baratte finds himself.[58] A duetto sung by a rustic young couple evokes the mythical "memory" of Daphnis and Chloe.[59] A worker carrying a child on his shoulders becomes Saint Christophorus,[60] and, of course, Nerval himself occasionally slips into a mythical costume: "Vers deux heures, on me mit au bain, et je me crus servi par les Walkyries, filles d'Odin, qui voulaient m'élever a l'immortalité en dépouillant peu a peu mon corps de ce qu'il avait d'impur."[61]

As far as "souvenir" is concerned, Nerval's focus on "la femme divine" is considerably more narrow than the scope of Rousseau's memories. Yet the range of that ideal archetype extends from real memories of his youth via historical epochs (the Middle Ages, the sixteenth century) into the mythical past which is (by virtue of metempsychosis) also an eternal present. In terms of this temporal range Nerval's "memory" clearly surpasses Rousseau's.

What is behind Nerval's obession to deify real persons or to transfigure them into mythical beings? It is, first of all, his deep desire to overcome nothingness and time.[62] What is eternal in man must be protected "contre le double effort de la mort ... ou du néant...."[63] But it is also his will to cast off the fetters of an all-too material reality in order to reach out for the divine: "Quelle âme généreuse n'a éprouvé quelque chose de cet état de l'esprit humain, qui aspire sans cesse à des révélations divines...."[64] Thus Nerval's ultimate motivation is not political or social but both psychological and religious.

However, Nerval does not always consider the idealization of memories (be they real or mythical) – and here his attitude differs from Rousseau's – as a source of happiness. They can as well produce "douleur" and are sometimes – in moments of sobering self-criticism – reduced to the status of mere "fantômes." In *Sylvie* the sight of the real girl Sylvie whom Nerval has known since his childhood, prompts a critical and almost angry assessment of his quest for the unreal idol of Aurélie: "Tout à coup je pensai a l'image vaine qui m'avait égaré si longtemps."[65] We find a similarly negative statement in Nerval's story *Octavie*. Once again he attempts to free himself from a specter of a woman (an actress) elevated by virtue of his imagination to the loftiest level of the divine: "Je m'arrachai a ce fantôme qui me séduisait et m'effrayait à la fois."[66]

Whenever the skeptical realist in Nerval prevails, he ceases to look for happiness among the idealized mythical fabrications of his fancy. Instead, he slowly and gradually convinces himself that – contrary to Rousseau's beliefs – real happiness can be found only in "la douce réalité." The work of Nerval contains a much stronger degree of self-criticism with regard to "souvenir" and "imagination" than the comparable writings of Rousseau. The age of realism has already begun to undermine the romantic spirit.[67]

NOTES

[1] I am indebted to Karl Maurer of the University of Bochum (West Germany) who first suggested the topic of this paper to me in a slightly different version.
[2] Cf. J. Kneller, 'Nerval and Rousseau,' *PMLA* 68 (1953), 150–69.
[3] Voltaire, *Dictionnaire Philosophique* (Paris: Pourrat Frères, 1838), XVI, 152.
[4] Denis Diderot, *Oeuvres Complètes*, ed. J. Assezat (Paris: Garnier, 1876), XV, 35.
[5] Ibid., p. 36.
[6] Friedrich Wilhelm Joseph Schelling, *System des transzendentalen Idealismus* (1800) (Stuttgart/Augsburg, 1858), pp. 559ff.
[7] Cf. Lester G. Crocker, *Jean-Jacques Rousseau: The Prophetic Voice (1758–1778)* (New York: Macmillan, 1973), vol. 2, esp. chaps. 2, 4.
[8] Albert Béguin, *L'Ame Romantique et le Rêve: Essai sur le Romantisme Allemand et la Poésie Française* (Paris: 1964), p. 335.
[9] *Rousseau Juge de Jean-Jacques: Dialogues*, ed. M. Musset-Pathy (Brussels, 1928), II, 40.
[10] Jean-Jacques Rousseau, *Les Rêveries du Promeneur Solitaire*, ed. Henri Roddier (Paris: Garnier, 1960), p. 14.
[11] *Rousseau Juge*, pp. 212f.
[12] Cf. B. Munteano, *Solitude et Contradictions de Jean-Jacques Rousseau* (Paris: A. G. Nizet, 1975), pp. 187–95.
[13] Jean-Jacques Rousseau, *La Nouvelle Héloïse*, in *Oeuvres Complètes*, vol. 2, ed. Henri Coulet and Bernard Guyon (Paris: Gallimard, 1964), p. 693.
[14] Jean-Jacques Rousseau, *Les Confessions*, ed. A. van Bever (Paris, 1926), p. 231.
[15] Rousseau, *Les Rêveries*, p. 74.
[16] *Rousseau Juge*, p. 216. Cf. Rousseau, *Oeuvres, Correspondance*, vol. 3 (Paris: Lefèvre, 1820), p. 390: "... quand ma cervelle s'échauffe trop, le calmer en analysant quelque mousse ou quelque fougère; enfin me livrer sans gêne a mes fantaisies ..." (1767).
[17] Jean-Jacques Rousseau, *Emile ou de l'Education* (Paris: Garnier, n.d.), p. 59.
[18] *Rousseau Juge*, p. 230. Cf. *Les Rêveries*, p. 114 and *Les Confessions*, IV, 219.
[19] Rousseau, *Correspondance*, III, 457.
[20] Rousseau, *Les Rêveries*, p. 28.
[21] *Rousseau Juge*, pp. 41f.
[22] Rousseau, *Emile*, p. 167.
[23] *Rousseau Juge*, p. 205.
[24] Ibid., p. 203.
[25] Rousseau, *Les Confessions*, p. 135.
[26] Cf. Rousseau, *Correspondance*, III, 392: "Le souvenir de mes amis donne a ma rêverie un charme que le souvenir de mes ennemis ne trouble point."
[27] Rousseau, *Les Confessions*, p. 155.

[28] Rousseau, *Correspondance*, IV, 248.
[29] Rousseau, *Les Confessions*, p. 274.
[30] Ibid., pp. 279f.
[31] Rousseau, *La Nouvelle Héloïse*, p. 289.
[32] Ibid., p. 667.
[33] Ibid., p. 509.
[34] Ibid., p. 675.
[35] Cf. Kurt Schärer, *Thématique de Nerval ou le Monde Récomposé* (Paris: Minard, 1968).
[36] Gérard de Nerval, *Oeuvres Complémentaires*, vol. 1, *La Vie des Lettres*, ed. Jean Richer (Paris: Minard, 1959), p. 13.
[37] Gérard de Nerval, *Oeuvres* (Paris: Gallimard, 1956). I (*Isis*, var. 1845), 1204.
[38] Cf. Francois Constans, 'Nerval et l'Amour Platonique: La Pandora,' *Mercure de France*, 324 (1955), 97–119.
[39] Gérard de Nerval, *Oeuvres*, ed. Henri Lemaitre, 2 vols. (Paris: Garnier, 1958), I, 502 (*A Alexandre Dumas*).
[40] Nerval, *Oeuvres* (Garnier), I, 268 (*Les Confidences de Nicolas*).
[41] Ibid., p. 591 (*Sylvie*).
[42] Ibid., p. 142 (*Les Confidences de Nicolas*).
[43] Nerval, *Oeuvres* (Garnier), II, 289, 309, 312, 314, 315, 342, 511, 524.
[44] Ibid., pp. 309, 310, 323, 330.
[45] Ibid., p. 253.
[46] Nerval, *Oeuvres Complémentaires*, I, 20.
[47] Ibid., p. 20.
[48] Nerval, *Oeuvres* (Garnier), II, 336.
[49] Nerval, *Oeuvres Complémentaires*, I, 21.
[50] Nerval, *Oeuvres* (Garnier), I, 776 (*Aurélia*); italics mine.
[51] Nerval, *Oeuvres* (Garnier), II, 407 (*Druses et Maronites*); italics mine.
[52] Cf. M.-J. Durry, *Gérard de Nerval et le Mythe* (Paris, 1956).
[53] See J.-P. Richard, 'Gérard de Nerval ou la Profondeur Délivrée,' *Critique*, 97 (1955), 483–509.
[54] Nerval, *Oeuvres* (Garnier), I, 819 (*Aurélia*).
[55] Ibid., p. 805.
[56] Ibid., p. 767.
[57] Nerval, *Oeuvres* (Garnier), II, 406 (*Histoire du Calife Hakem*).
[58] Nerval, *Oeuvres* (Garnier), I, 422 (*Les Nuits d'Octobre*).
[59] Ibid., p. 466 (*Promenades et Souvenirs*).
[60] Ibid., p. 801 (*Aurélia*).
[61] Ibid., p. 809 (*Aurélia*).
[62] Jean Gaulmier, *Gérard de Nerval et les Filles du Feu* (Paris, 1956) describes Nerval's desire "d'établir la durée par la puissance du souvenir" (p. 36).
[63] Nerval, *Oeuvres* (Minard), II, 1292 (*Voyage en Orient; Vers l'Orient*, var. 1844).
[64] Nerval, *Oeuvres Complémentaires*, vol. 1, *Préface de la Première Edition du Faust*, p. 6.
[65] Nerval, *Oeuvres* (Garnier), I, 611 (*Sylvie*).
[66] Ibid., p. 643 (*Octavie*).

[67] This is most poignantly expressed in Nerval's letter to Gautier of 7 September 1843: "Toi, tu crois encore a l'ibis, au lotus pourpré, au Nil jaune; tu crois au palmier d'émeraude, au nopal, au chameau peut-être. ... Hélas! l'ibis est un oiseau sauvage, le lotus un oignon vulgaire; le Nil est une eau rousse a reflets d'ardoise, le palmier a l'air d'un plumeau grêle, le nopal n'est qu'un cactus, le chameau n'existe qu'a l'état de dromedaire...."

RICHARD T. WEBSTER

INTUITIONS

I

Tai de skiai aisousi, "they flit as shadows," says Plato, in *The Republic*, about the world of illusions. Yet we have to ask: "what is a shadow?" or "to flit?" It is not possible to say of anything that "it is not" or "it does not exist" *tout court*. One can only say, or imply, that *it*, which must be something, does not exist as one thing, but only as another, not as a body but as a shadow, or a word; not as a physical, but only as a mental phenomenon, for instance, or as one kind of such thing and not another. Thus we should not exactly be asking "what *is*, or really exists?" but "*as what* do the things that exist, exist or not exist?" What is thus called for is some regional ontology, some provisional overall view, at least − a *Jungle Book*, so to say, as distinct from the growling of only one lion. If this is difficult to obtain in a satisfactory way, it is nevertheless worth striving for, and we can but make our suggestions.

To begin with, what is it that does not flit, and is not merely a shadow? Truth, we must say. Although it may not be to all immediately apparent, truth as such is absolute, without *Abschattung*. (1) It is intuitively absolute; and we have not space here to argue with the point of view which confuses so many particular truth-possibilities with the ultimate and necessary condition of their possibility, or which confuses reality or *being*, which is for us always *abgeschattet*, with truth as such, which is not so. (2) Truth can become asymptotically absolute in induction. (3) It is absolute, saving as to discrepancies in the definition of terms, in regard to the past. The difference, then, between Plato and Husserl would appear to be that while they both found truth ultimately on intuition or *gnosis*, Plato requires this to be intellectual intuition − and intellectual intuition of a rather exceptional kind − whereas Husserl allows also, at least within intuition considered as intellectual, for the daticity of experience as such. "Intuition" in a popular or Romantic sense of the word, which may easily boil down to mere freedom of hypothesis, is of course irrelevant here, since what has to be implied is an apodeictic certainty. One would then have to distinguish between different kinds of mere presentation and different kinds of truth-certainty or eidetic reduction. Our concern here, however, is not with all these distinctions, but with the implied contrast between truth as such (granted its experiential foundations) and experience as such in general, not necessarily giving rise only to truth as such. Since, in other words, our concern will be with general values as well as with truth, or with axiological as well as logical experience, the positions adopted will to that extent necessarily be closer to those of Scheler than to those of Husserl.

In experience as such at the extreme limit of mere presentation or knowledge-by-acquaintance, where we have to speak of states of mind as giving in themselves neither truth nor untruth, but merely themselves, intuitively, preexisting constructions of various kinds must always be presupposed. Nevertheless, when we have allowed for all these, there remains over a necessary division and dialectic between truth as such and experience as such. Apart from the distinction between noema and noesis, that is to say, there are states of mind which give rise to what are only problematically noemata, and there is the factor of what will have to be called dumb experience. To this suggestion there will naturally be a resistance in the mere fact that philosophy is taken to be the search after truth as such. Philosophy thus tends to try and analyze away the old antithesis between ratiocination and intuition, or anything of that kind, submerging them in a more general notion of understanding; and if we wish to maintain the antithesis, we shall no doubt be asked for a more precise definition of terms. This, however, as it seems to me, can only abut in a "clarification" whereby we are brought back to the absolute of truth considered as absorbing all significant experience into itself. If, however, the preliminary assumption is that experience necessarily cannot be entirely absorbed, by us, into truth, and yet is not to that extent necessarily only untruth – the false or the meaningless – either, then no further definition of terms for the sake of truth-as-such will serve to remove the division and the dialectic between truth and experience, between "pure truth" and "mere experience." To keep on insisting, as linguistic analysts do, on ever-further definitions of terms, apart from the fact that this can and does (as it were) go on *ad infinitum*, will in effect amount to the rejection of that assumption: the assumption of the absolute importance of experience – and the absolute importance of life, which is experience.

Experience is what has happened, or is now happening, in one's consciousness, and one cannot not be, or have been, conscious of it, although one may in a good many cases remain uncertain as to its further interpretation. Merely on this showing, it would seem that experience as such, where it is not *logical* experience, is the equivalent of *truth about the past* or the nonfuture, of which it is the foundation. If, however, as Husserl suggests,[1] experience as such is absolute in its daticity and without *Abschattung*, and this only means that where it is not logical experience, it is the equivalent of truth about the past or the nonfuture, then, again, outside the absolute of truth as such there can be no real meaning — unless it be that of possibility in the future, of the play between necessity and contingency, of some uncertainty principle. This, however, if what it suggests is correct enough so far as it goes, does not do justice to what is meant by "experience" when we use the word in a certain pregnant way. It is not only a question of the certainty of experience as the general certainty of the past, or of allowing for mere possibility in the future, but of so many particular and characteristic experiences requiring to be "described," or at least, let us say, circumscribed.

Fundamentally, I suggest, one has to make a distinction in principle between an *interpretans* as such and an *interpretandum* as such. A noema is an *interpretans* which justifies itself by way of its clarity and certainty, or clarifiability and ascertainability. While it must not then be misapplied, so far as it goes it requires, not further interpretation, but "description." Any further *interpretans* which would go on to treat the established *interpretans* as in turn an *interpretandum* ought to have the same degree of clarity and certainty. But this is in the logical or formalizable order, and one has to ask what happens outside this sort of order. Sense-intuition is the strongest sort of candidate in the field as a nonlogical or informal certainty, i.e., not merely the certainty of the presence of *some* experience, but a particular, self-revealing certainty: that of seeing a red patch, for instance, when one does see one. This, however, cannot be, saving in a very limited, unevolved sort of way, an *interpretans* in its own right, but is for the most part, in relation to the rest of human experience, an *interpretandum*. The problem is that on more evolved levels, where general values appear, there are states of consciousness which may indeed be in part *interpretanda* for logic or formal procedures, but at the same time lay claim to being themselves the or an *interpretans* in a very wide-ranging way. There is in short the axiological field as the experience of value or good and evil. If this cannot interpret where logic interprets, it nevertheless claims to interpret everything else, although not always with the same clarity or certainty as logic. At this point, however, it will be difficult in philosophy to avoid a certain degree of confusion. In trying to reconstruct the existing construction, to say or repeat what it is that the axiological or similar judgment is "saying," is one really allowing it to remain the *interpretans* or is one reinterpreting it from another point of view, putting it down, as it were, in a condescending sort of way? This is what happens in so many "explanations," whether causal or linguistic, but it is what phenomenology has presumably been trying so far as possible to avoid.

In this way Scheler passed on to the consideration of values or goods as substantive, to value judgments as interpretative in their own right. But even so, the position is not so easy. For one cannot fail to be in some way generalizing; and it is truth, not experience itself, which is speaking, when we try to *say* what our experience has been. What remains over *after* truth has done all its speaking can only be mere existence, the logical or abstract residuum of the *existential*: some undiscernibility or nonessence: names for ignorance or for "nothing." In spite of the difficulties, however, we have to try and maintain the distinction between truth as such and experience as such, and therewith the rights of other kinds of *interpretans* besides logical ones.

If we ask what is the legitimate *interpretans* when it is not of the logical kind, the answer is that it is human life in its wholeness as including both the logical and the axiological sort of interpreting experience: the value judgment in itself. If, then, from a logical point of view human life in its specificity and qualitativeness as human is left over in the last resort as a nonessence or dissident communion with "nothing," the word "nothing" — consequentially enough — gets to be used (also when we say "oh, nothing!" meaning what is all-important) as corresponding to Existence in that sense: the *existentiell* (presumably): that which, if it is necessarily beyond rigorous science, is not on that account insignificant. Even if the word "nothing" had not been used in this way, in any case we have to look at human life as experience-as-such; not only as experience standing for the mere certainty of the past or nonfuture, but as so many characteristic and positively significant experiences; those of each individual in his secret message: personal and yet at the same time more or less universal. What immediately emerges, then, is that fundamental human experience is so largely uncommunicable. We usually have so little idea of the emotive states, problems and perspectives of other people. It is, for heaven's sake, one thing to look at a feeling, a situation or a *Weltanschauung* from the outside, another thing to be having it — to be looking at it from the inside. If this obviously applies to emotive, psychic, and physical experience, it also applies to fundamental cultures and philosophies supposedly expressible in rational terms, as any familiarity with different schools of philosophy with their apparent dogmatisms and mutual incomprehensions will show, not to mention the differences between nationalities, between Eastern and Western points of view, or between the traveled and the untraveled, the invaded and the uninvaded, and so on. This incommunicability applies in the last resort to intellectual experience. "Though I should show you all the nerve-ends on a screen," as T. S. Eliot says, I should not be able to convey it. Something I can always convey, but in the end it is like trying to explain to someone born blind what it is like to be seeing a red patch. The crucial point, however, is that the certainty which extrapolates from discursive reason is not, like the certainty of seeing a red patch, an *interpretandum*, but in the case of fundamental experience the *interpretans*, and the driving force, it may be, of everything that takes place in a given person's life. While it is by no means the case that the *interpretans* is always justified in its interpretation — children can cause accidents, adults can go wrong, and societies can crumble — apart from this we have to consider normal or reasonable interpretative positions.

At this point it may be thought that the problem can be overcome by the distinction between rationalization and *Verstand* or understanding in depth. The point which is then established is that science is one thing and empathy, or rather let us say "en-logy," is another; and there are always enough ingenuous rationalists around to make it desirable for this to be repeated. Yet there is something which is still missed, and that is the endless capacity for development of the asymmetry between the two different sorts of understanding. *Realwissenschaft* and *Geisteswissenchaft* are far from being like two different regiments in the same army. The second grows away from the first, and the asymmetry can be expressed in this way, that for reason misunderstanding is a nonentity, but human life proceeds as much by misunderstanding as by understanding. "If we prosper, it is by miracle," as Lord Burleigh said.

II

And yet, if one wishes to go on from here, one would seem to be trying to take advantage of the mere infinity, "the bad infinity," of relative determinations, justifying obscurity by obscurity, indulging in the mystification which consists in saying "I cannot explain, but I *know*!" What can one ever be said to know if one can never produce a satisfactory "public" (logically public) account of it?

Now if the emphasis is on *episteme*, or on *gnosis* or intuition insofar as the condition for the possibility of *episteme*, one will have to adopt a *geisteswissenschaftlich* point of view for which the unclear, however much appreciated, will always have to be explained in terms of the clear — never vice versa. But implied in the appeal to experience as phenomenology makes it is an allowance *prima facie* for a certain equality of interpretative rights between the clear and the unclear insofar as referable to irreducible experience: an equality of rights as between logic and the *Lebenswelt*. And this, be it remarked, is not all of one piece. There are as many *Lebenswelten* as there are human cultures or subcultures.

Before going further, we have, by the way, to say that metaphysics is still possible. If we put it that we grow *out of* initial experience *into* reason, and then out of reason into further experience, Aristotelian potentiality or the moderate historicism of a Vico will allow philosophy to cope with such a situation. However, that is not exactly our present concern. What has to be affirmed, with or without metaphysics, is the operation, apart from truth as such, of what we shall for convenience call *velle*, meaning the whole range of *voluntas, desiderium*, inclination or disinclination, love or hate, appreciation or laziness, with their correlative types of value or *telos*, action or conation – in short, again, life, and human life in particular. The question then repeats itself: how can one include *velle* in *veritas* without thereby destroying its autonomy or leaving its autonomy as an unilluminating randomness? But if metaphysics admits, as against necessary "emanation," a significant autonomy in *velle*, then it can no longer be purely theoretical, but must be, in part at least, practical – functioning by dint of the autonomous *velle* of which it may be trying to speak in terms of necessary truth. Thus religious philosophy, at least from St. Augustine onward, modified classical Greek philosophy by the introduction of religious principles – religious and therefore *practical* principles – for which there became central the autonomy of the will. One does not on this account have to reject metaphysics, but one has to distinguish between what is purely truth-intentional and what is *velle* over and above that intention. And it must be reiterated that phenomenology has to try and determine, not only "the truth about" *velle*, but what, if anything, *velle* "says" before one has theorized about it, although it may not always be easy to keep the two procedures apart.

III

Velle insofar as it is not capturable in the network of *veritas* – truth as such – splits up into a spectrum of categories which, however characteristic and insistent, remain only vaguely outlined: are always *abgeschattet*. What is an artistic intention? and what is each of its many subspecies? What is esthesis? What is a moral intention, either in a narrow, Kantian sort of sense, or in the broader sense rooted in "way of life," culture, or ethos? And what – a question oddly neglected by modern philosophy – is love? Love, one must say, can be sensitive, existential, or illuminated; but these categories can not only vary in themselves, but conflict or conflate with each other, understand or misunderstand. In any case, there can be no ultimate understanding.

Let us put it that an educated but prosy man may understand the literal words, but may not understand the real meaning of the poet, while the sensitive reader, understanding the poem well enough, may put upon it one of various possible legitimate interpretations. But then it is not only a question of understanding poetry *qua* poetry, as though the aim of life were to be a professor of poetry. The poetry *qua* poetry is embedded in an ethos or culture which transcends it, i.e., includes it *as it is*, but subordinates it to another order of *velle*, that of living in general, just as the poem as a poem has already subordinated its own *literal* meaning; and as, again, in the literal meaning, the empirical is subordinated to the logical. Furthermore, granted all this, the whole situation may once more be transcended in another order of *velle*, yet another type of *interpretans*, the religious, the mystical, or the meta-mystical. Ultimately, there would be required a phenomenology of the state of mind or no-mind of the Buddha under the Bo tree. Suffice for the moment, however, to say that the multiple transcendences in *velle* are not logically ordered, or can only be so in part, and superficially. There can be no guarantee that an apparently more inclusive order does justice to its subordinates. The poet's literal meaning may be false: a travesty of history. The moralist may not do justice to the poet. The religious may not do justice to the moralist, and the mystic may cut himself off too much from the world. In any case, there can be no logical certainty as to what is meant by "too much" or by "not doing justice." One can only attempt to "describe" each *interpretans* as it appears, and something of its relations with the others as these relations appear.

On the whole, the continuum constituted by the categories of *velle* distinguishes itself ever more sharply the higher it goes − "higher" in the order of values − from the material or natural matrix reducible in principle to certain knowledge, although this does not take place in a regular way, so that even "higher order of values," let alone "logically higher order," can be misleading. The relations between ethics, theistic religion, and mysticism are particularly difficult to settle. Of the ethical level in the narrow sense, however, Kant can say that it isolates itself as another absolute in opposition or antinomy over against the absolute of truth. But then − metaphysics apart − *velle*, including the moral *velle*, cannot in any case be separated from reality in the way suggested by Kant in his special preoccupation with the problems of formal truth. If anyone *vult*, then he must be wishing or willing a *quid*: he must be appreciating something (or failing to appreciate something else), and something is not nothing. "Something" must have an essence or nature open in principle to some kind of investigation. What we must say, however, is that the correspondence between the meaningfully autonomous *velle* and the thing willed or wished can never be so hard or perfect as it can become in the case of a truth-intention and its object. It is only the heteronomous *velle* immersed in nature — as in a great many physiological functions — which can ever be perfectly satisfied.

Abschattung is introduced into the voluntaristic spectrum precisely by the concomitance of the exigency for truth, with which it contrasts. Without the exigency for truth as such — as in the case of lower organisms, which have no such exigency — experience and satisfaction would perfectly coincide. It is light which introduces shade, and it is the intellect which makes *Abschattung* possible by way of contrast with itself. To put this in another way, in discovering *being*, the intellect discovers also the infinity of *being*: discovers both the *peras* and the *apeiron*. But what else can we say?

IV

While truth as such is in principle absolute, it would not seem at first sight that there can be any absolute for *velle* apart from that. While truth can be absolute as logical or relative as empirical — but even then is absolute in regard to the past — *velle* can only be relative. As Kant observes, any felicity in this world is necessarily incomplete and unsatisfactory.

The lacuna in the picture, the absolute for *velle*, cannot be properly filled by discursive reasoning, which would take us back to logical relations. Nor can it be filled by any intuition or experience insofar as one can show its finite condition. The lacuna might be filled by metaphysical argument in a way, as purporting to demonstrate the necessary existence of the Absolute of Being, or at any rate some Absolute which would be an Absolute for *velle* — some "Beyond Being," perhaps, depending on what exactly is meant by "Being." Substantially, however, any such arguments must remain incomplete just because they *are* rational arguments. The lacuna can also be filled by Faith or by *kerygma*; but that would be a somewhat different matter, with which we are not just now concerned. Here we must claim — and try to claim as a matter of "description" or circumscription rather than of rational argument — that the lacuna can only be filled by an intuitive absolute: analogous to the law of noncontradiction in being absolute, but differing from it as providing an absolute of happiness or satisfaction for *velle*; and among human states of consciousness this is not, after all, difficult to find, since it appears in mystical states, or, more loosely and polyvalently speaking, in "the sense of the Divine."

This is a different kind, or rather range, of *interpretans* from any other, and arrives at some degree of the ineffable, although not to the same degree in every sort of case. But we can use a rough analogy. If the *interpretans* in question were poetry, poetry in the technical sense — which it is not — one would say that the poem had a prose content which could be understood without understanding the poem. One could 'describe' the outside of the poem without arriving at a description of the inside: the poem as its own *interpretans* — one could see the *vehicle* of its meaning without seeing its meaning. In this way the normally describable contents of mystical states or "the sense of the Divine" seldom provide access to those states of consciousness, which are in any case, quite apart from their outward contents or vehicular form, very varied in themselves. Thus the vehicular form may or may not contain words like "God" or "Divine," which one might expect on the basis of other levels of discourse; and they are likely to contain the unexpected. Furthermore, the mystical interpretative state itself may amount to a God-consciousness in a sense compatible with ordinary uses of the word "God," or it may not: it may be a consciousness of something human, and so only in a manner of speaking "divine" because invested by intuition with an absolute of happiness. The only condition necessary for filling the lacuna as we have framed it is that there should be some kind of intuitive absolute, an absolute for *velle* or absolute of happiness *now*, even if, in terms of time, there may be in one way or another — not always in the same way — a question of limited duration.

Here there is likely to be little agreement, but we would suggest that there can be more agreement than there has been in the past about at least some points.

Evidently, one cannot understand the experience of another when one has had no comparable experience on one's own account. One cannot explain color to those born blind, saving perhaps in a limited way by comparison with heat and cold. One must then add that some experiences are *interpretans* in a much more wide-ranging way than others, and what one will then not understand, if one has not had any similar experiences, is the implied system of interpretation. Either, then, one will interpret the unfamiliar *interpretans* by way of one's own more familiar ways of interpreting, and so largely misinterpret it, or else one will want to say that it can have no meaning. But some open-mindedness is required.

There should be no great difficulty about the fact that one cannot "explain" an intuition which is in some way ultimate. An analogy is provided by the law of noncontradiction. This is intuitive and absolute, but cannot be rationalized. For one cannot rationalize what is itself the foundation of reason. The attempted rationalizations necessarily disguise, distort, or unduly limit the original intuition, either saying something definite which is not the original, or reproducing the original by way of saying something which is in effect nothing. Whereas, however, the law of noncontradiction explicitates itself in rational discourse as such, the mystic intuition does not; and the mystic, we must say, necessarily appears as an idiot, either unable to explain himself, or else, in order to be able to explain himself, having to abandon his original position. But we are not stuck here. For if the mystic intuition cannot explicitate itself in rational discourse — although it may find some correspondence in religion or metaphysics — it explicitates itself in another way — in the way appropriate to the *velle* of which it represents the absolute, namely, in a *practical* way. If it is the property of logic to explain, it is the property of voluntaristic intuition, not in itself to explain, but to *treat* other experience or reality in its own way, i.e., in the way of action or attitude. The object of *velle* is a value or good, to which it relates itself by principles of action or attitude. *Velle* tells me, not what to think, but what to do — in a broad sense — and if the *velle* is absolute, then it will tell me what I ought to do absolutely, subordinating all else to a new fundamental attitude of message and metaphor.

As for the general confusion in mysticism and its consequences, one might say that on the terms of our account of it, this was only to be expected. More exactly, however, we have to say that the Absolute appears, if it ever does appear to *velle*, as incarnated in the contingent, just as does the Absolute for logic; and incarnation is not total identification any more than it is total separation.[2] Thus it is not the complete Absolute that anyone can ever attain, but only, in the absolute for *velle* the "absolute for me" — for the subject in question in his possible experience — of which we have to say that it can only be a symbol or reduced representation of the complete Absolute, and is thus, one must say, a relative absolute. And this, as the absolute for *velle*, but not the total Absolute, cannot share the characteristic of the absolute of *truth as such*, which gives itself as the principle of formal unification, but must be characterized by the possibility of formal confusion — prismatically splitting up the universe into endless autonomous existences: autonomous in their very unification.

V

Further annotations on the absolute for *velle* — let us hope not too controversial ones — can be made as follows.

Since the *velle* of one region is not quite like the *velle* of another, it must be remarked that *velle* in the mystical state is distinct from "will" in the moral or self-conscious sense. The mystical state cannot be consciously sought out or willed (although its consequences can), but is evidently a grace independent of the individual will. If it does not exclude, and no doubt requires, some previous operation of the will — "cooperation with God" or "non-attachment to the world" — it is not the individual will which now counts. How could it, if what is approached is the Absolute of *velle*? "Autonomy" is thus no longer quite the autonomy of the individual as such. Freedom from the heteronomy of rational determination now gives place to the operation of the Divine. "Free will" can be overplayed if what is suggested is freedom from *any* ontology as distinct from the wrong or inappropriate kind of ontology. (If I say that "man is not merely a puppet," everyone will know what I mean, and that is good enough; but to be complete, one should answer more precisely the question: "not a puppet *of what*?" Furthermore, "puppet" for a mystic might have another meaning generally unknown: "the puppet of Being," let us say.)

If one has to distinguish *velle* as such from truth as such, they must not be dichotomized in reality. It should not be a question of trying to escape from the truths of rational determination insofar as they are true. The absolute of truth is included in the absolute of *velle* at least in this sense, that the absolute of *velle* being in its way all-inclusive, truth-as-such will also be its object, included in absolute felicity, not rejecting science, but appreciating it precisely for what it is. While truth and *velle* in themselves are only identical in God or the Absolute (call that metaphysics, if you like), they are intentionally united in one way in the absolute for *velle*, as they are already united, in a different way, in the absolute of truth. In the relative sphere there are various forms of disunity or incompleteness; but the object of *velle* as absolute can be said to be *Being* (subject to many disagreements about the use of the term "Being"), of which truth is one aspect, as goodness is another. By way of excessive attachment to the contingent symbol, by way of ossification, imperfection, or the urge to polemics, the mystical *interpretans* may get a somewhat distorted view of other regions of experience; but the mystical state of consciousness in itself — whatever declines it may suffer — allows them to be, philosophically speaking, "what they are and not another thing."

In other words, intuition here is by no means "intuitionism," deliberately favoring intuition against reason, but tries to see realities as they are. There is thus a certain natural reunion with the *Sachen selbst* of phenomenology. This statement could be misleading, for phenomenology remains, after all, on the theoretical side of things, mysticism on the (in a philosophical sense) practical side of things; but both are concerned in various ways with the absolute of experience-as-such.

No doubt it can be said that for most ordinary purposes the negative component in religious faith, i.e., the confession of ignorance and the immanence of *Endsituation*, is more efficacious as an answer to the material world than the rare positivity of the absolute of intuition. Yet it would seem strange if the negative were never to be balanced by a positive, as though the asymmetry to which we referred at the beginning of the previous section must be ultimate, especially since the kind of experience called for is so abundantly, if imperfectly, recorded. Seeing besides that theoretical metaphysics cut so little ice in societies so peculiarly unstable as those of today, one may wonder whether the gap which it has left should not be occupied — perhaps it will eventually be occupied — by a phenomenology of intuition. In any case, we can no longer afford to carry on the old wars between "the two cultures" — or the twenty cultures.

All this may suggest turning to writers like William James, Rudolf Otto, and Marghanita Laski from the empiricist side, or Augustin Poulain from the point of view of faith and metaphysics. The phenomenological approach here, however, is not fully developed and scarcely radical. More collaboration between East and West would also be desirable. But one can see a way ahead for what Paolo Valori has called "la terza via intermediaria tra Weltanschauung arbitraria e aridità verificativa."[3]

VI

In the aesthetic field the particular case has a significance beyond itself such as is more than, though not exclusive of, the significance of the logical in the empirical or of the organic in the physical. Of esthesis we must also say, at least under modern conditions of specialization, that it is characteristically not action, but contemplation. At the same time, it is a synthesis of a certain range of experience. On the other hand, what characterizes the ethical region, in the narrow sense, is the stress on action, and in modern times, in the West, the absence of synthesis, since it tends to remain in a more or less Kantian dualism. We thus have to say that esthesis under recent conditions has tended to become *synthesis without action* (*engagement*, "happening," etc., have not really done anything to alter this) while moral action has resolved itself, generally speaking, into *appropriate action without synthesis* — moralism, *esprit de sérieux*, or the unconvincing sermon. Spiritual aspirations oscillate uneasily between these two partial views, while what is missing is a persuasive ontology. Perhaps it will not be too speculative to say that in the vague continuum of *velle* esthesis is a sort of halfway house between knowledge and ethics; but an adequate ontology of the absolute tends to be missed, since it is thought that this must depend on "science." As said, there is faith and there is metaphysics, but originally the absolute was read into *velle* also by way of experience, and the mystical or God-conscious state of mind — not to mention any more complete human ethos such as can be found in the older cultures — is both a synthesis or harmonization of experience, and not only of a particular regional range of experience, but of all experience, and at the same time a requirement for moral action in the narrow sense. The mystical thus reassumes in its characteristic absolute, in various ways, the whole gamut of *velle* as autonomous: the purely aesthetic, the purely ethical, and their conflations. Since the mystical state cannot be maintained for long, however long-term its consequences, this state is likely to be substituted by others — sometimes even by undesirable ones — but each state of consciousness has to be considered on its own merits in its intentional significance. It is at this point that phenomenology can act as a peacemaker, allowing to every man his due — to every "man" in the man its due — without condescension.

NOTES

[1] 'Ein Erlebnis schattet sich nicht ab' (Husserl, *Ideen* [The Hague], I, 42).
[2] Cf. the following words of Etienne Gilson, which I have taken from an unpublished letter dated July 1966: "He [Wittgenstein] never asks the only question in which I am interested, and of which the answer has always eluded me: What is, in fact, the relationship of the *material* word to *immaterial* meaning? It seems to me that the answer to that elementary question, should we find it, would reveal to us the secret behind the development (or absence of development) of the history of metaphysical speculation from Plotinus to our own days. Why is what comes after *One*, a *Nous*? I feel more and more inclined to see no difference between the relationship of thought to language and that of Being to beings."
[3] In the article 'Fenomenologia,' in *Enciclopedia Filosofica* (Centro di Gallarate, 1957).

CHRISTOPH EYKMAN

EIDETIC CONCEPTION AND THE ANALYSIS OF MEANING IN LITERATURE

The study of literature has several dimensions each of which, it seems, was once in vogue and consequently elevated above the others to the point of their exclusion or neglect. The biography of the writer, the text taken in its immanence (intrinsic criticism), text plus author as a mere function of the socioeconomic dynamics of a given time, and finally (most recently), the role of the reader as reconstituting mind (*Rezeptionsästhetik* as practiced in Germany) — all these approaches to literature have had a time when they almost eclipsed any other method of literary analysis. Yet, applied in a less absolutistic fashion, i.e., as limited but equally valid elucidations of a literary text, they can all serve what we still consider to be the central purpose of literary studies: understanding a literary work.

We shall focus first on an aspect of the author-oriented approach (somewhat neglected in recent literary studies) and shift our attention later to the problem of textual meaning which in turn brings up to some extent the role of the reader. Right at the outset, we wish to acknowledge that our discussions and analyses owe much to the phenomenological philosophy of Edmund Husserl, Roman Ingarden, and their students and followers.[1]

Numerous authors reflecting upon the creative process state that they have a general and sometimes rather vague idea, a "plan," a seminal vision, or conception of the work they want to write. This conception is never entirely rational, nor is it fully conscious. Therefore, it is more than the mere "message" or "moral" or "thesis" of a text. Although it may shift during the process of writing, it still functions as the spiritual center or nucleus around which the text gradually crystallizes. It functions as a "constitutive prescription" which, in part, is the product of artistic creativity. This notion of a central and guiding conception should not be confused with the notion of "intention" as criticized in W. K. Wimsatt's and M. C. Beardsley's well-known paper 'The Intentional Fallacy.'[2] Their notion of "intention," restricted to poetry and used mainly for the purpose of judging the success of failure of a poem as a literary achievement, is too rational (being equated with the notions of "plan" or "designing intellect").[3] Thus, Wimsatt's and Beardsley's criticism does, in our view, not apply to the notion of creative conception. Few writers start writing a literary text *without* any conception of their project in mind. Even the most "meaningless" dadaist texts or surrealist "automatic" writing do not represent an absolute and totally unguided *ad libitum.*

Elaborating on a brief remark made by Fritz Kaufmann,[4] we claim that the above-mentioned creative conception is one which aims at the essential. It intuits — blending observation with creative imagination — the essence, the eidos of what was, is, or might be.

At this point, a brief recapitulation of the use of the notion of essence (eidos) in Husserl's writings will help to clarify our terminology. The "sphere" of essence, says Husserl, is the phenomenologically purified consciousness, i.e., pure consciousness after the "bracketing" of any positing of reality.[5] On the other hand, essences can also be taken as real, i.e., having ontological status.[6] Therefore, we have to distinguish between essences as an *a priori* "part" of transcendental subjectivity and essences which (though ultimately identical with the former) as empirically-psychologically conceived — and "conceived" means, of course, "based on an encounter with ontic essences." There are immanent essences (as regards acts of consciousness) and transcendent essences (Husserl mentions persons, qualities of a person's character, and physical objects as examples).[7] Every possible object of human experience (*Erfahrungsgegenstand*) has its essence.[8] Essences *qua* "general" essences point to universals (as abstracted from particulars) shared by entities of the same type. They refer to constants, not individualized/concretized variables.[9] The latter are part of the "individual" essence of an entity. General essences are always contained in individual essences.[10]

Husserl's favorite examples by means of which he illustrates his theories are taken from the realm of sensory perception. Yet, in literature, we are mainly dealing with "cultural objects," with meaning-complexes (*Sinngebilde*) derived from the historically conditioned interplay of various phenomena: events, acts, utterances, situations, etc. They show — and this is our own, not Husserl's, view — three dimensions. First, they have "roots" in history, they *became* what they are, they are, so to speak, part of the "syntagmatic" string of historical events. Second, they are what they are since they are always seen in a context of other, contrasting, different phenomena within the total "system" of beings (the "paradigmatic" dimension). Finally, since, according to Husserl, essences are never given in their totality,[11] the attempt to "see" as many different aspects of them as possible needs to be supplemented by an attempt to penetrate ever more deeply into any aspectually limited view of an essence (within the confines of that aspect), into its inner horizon.[12] This latter attempt aims toward the discovery of evermore essential features, toward a closer observation of already discovered features. That is certainly true of an author's eidetic conception (conception focusing on an essence or essences) which he never "possesses" in an exhaustive and definitive fashion.

In *Logical Investigations II*, Husserl distinguishes between linguistic expression (*kundgebender Akt*), ideal meaning, and the expressed *Gegenständlichkeit* ("object" in a rather broad sense of the word) itself.[13] If we apply this to the creative act of writing a literary text, it means that we have to distinguish between the eidetic meaning the text "captured" and expressed and the intended essence itself (no matter whether it is ontically pregiven or created). In Husserl's terminology *das Vermeinte* is more than *das Gemeinte*.[14] Ingarden saw this when he wrote that *Gehalt* (the meaning-substance of a work) is "vor allem eine *intentionale*, d.h. eine auf etwas von ihm selbst Verschiedenes, über sich selbst *hinausweisende* Einheit."[15] Fritz Kaufmann points out that, in the case of paintings, the "noematic intention," which points beyond itself, constitutes what he calls the *Bildfigur* (configuration of the essential formal elements of the image) which "represents" the intention and lets the *Bildwesen* (the eidos of the image) shine through.[16]

One major reason for the just mentioned nonidentity of essence and expressed essence is the fact, recognized by Husserl himself, that semantic meaning has to be *adapted* to the meaning of essences.[17] This problem constitutes one of the central issues in L. S. Vygotsky's *Thought and Language* where he opposes "inner speech" as a distinct plane of verbal thought and as the thinking of pure meanings to external speech. The transition from thought to word leads through meaning (thus, his notion of "thought" is similar to our notion of eidetic conception). In speech, there is always, as a subtext, the hidden thought. "Hidden" then must mean present, yet distant, unexpressed (what Ingarden meant by the expression "über sich hinausweisend"). According to Vygotsky, a direct transition from thought to word is impossible.[18] According to phenomenological theory, not only words but also thoughts often do not grasp an essence adequately. This is also due to the fact that an author "sees" his eidetic conception in a historically conditioned and delimited perspective. The eidos of a cultural *Gegenständlichkeit* is never understood in a definitive and exhaustive fashion independent of the flow of historical time.

Through the medium of the text, the reader has to arrive at the author's conception which shows the above-mentioned limitations. Behind this historically determined and aspectually limited authorial view there looms a richer and more complete eidos. The author's limited conception offers certain clues to the interpreter, so that he may *expand* the author's original vision beyond the level of the conscious explicitness of the text. This does not mean that the interpreter is to go outside the structural and thematic confines of a text or that he is expected to philosophize *ad libitum* about the content of the text. The eidetic conception "behind" the text, no matter whether its author had a full understanding of it at the time of its writing or not, reveals itself in an ever-fuller measure to every new generation of interpreters. For the history of the interpretations of a work (or an author), its *Wirkungsgeschichte*, unfolds as the never-ending and ever-progressing revelation of its full eidetic meaning which was particularized (or only partly represented) when the text was first created.[19]

How is the phenomenological method of variation to be applied to literary texts? First, it proves to be applicable only to the analysis of relatively short texts (for example, poems). Second, as to the meaning of, and the results yielded by, the process of variation, two different routes can be followed. We can analyze the possible semantic meanings of every relevant word in the text. In doing so, we register all lexical nuances of meaning, all intentional ambiguities or (if the context narrows down a meaning) all unequivocal semantic contents. In case of doubt, we shall have to compare to the work under analysis other texts by the same author or even texts by other authors belonging to the same epoch.

If my text contains many images, I can try to substitute similar images for those used in the text. This substitution technique reveals quickly which elements of the text are replaceable by others (without drastically changing its thematic essence) and which are not. What proves to be irreplaceable (invariable) must be congruent with the eidetic conception represented by the text in its entirety. The more complex and interdetermined the imagery of a text representing an *individual* essence, the greater the degree of congruence between text and seminal conception (the larger also the peripheral sphere of variable images). A *general* essence, however, may be represented by a variety of linguistic images which in turn can be semantically polyvalent. "Fire" can represent "love" but also "destruction." If a general essence is represented on a general and abstract linguistic level, the possibilities of variation *on that same linguistic level* decrease as the degree of generality increases. On that level, any change in language might deviate from the original general conception. For example, let us take the following sentence from the closing words spoken by the chorus in Sophocles' *Antigone*: "Wisdom is the supreme part of happiness, and reverence towards the gods must be inviolate."[20] We can say that this sentence represents a general and abstract subeidos of the complex total eidos of this particular play. If I replace the word "wisdom" by "prudence," I can no longer claim to have *varied* the sentence in the sense in which Husserl uses the notion of "variation." I have, in fact, *altered* the meaning (and the represented eidos). The ensuing comparison between the meanings of "wisdom" and "prudence" can, however, aid me in grasping the subeidos and — ultimately — the main and central eidos.

The only comprehensive phenomenological theory of literature written to date is Roman Ingarden's *Das literarische Kunstwerk* (1931).[21] In this work, Ingarden, a student of Husserl's, makes a point which seems to be so obvious that it has often been overlooked and not fully understood as to its theoretical consequences: there is a basic difference between the degree of determinateness of a physical object and the literary (linguistic) presentation of that same object. Looking at a real house, we must admit that its determining features (from its general shape, type of doors, windows, etc. down to the molecular structure of any of its parts) are almost infinite. A *literary* description of the same house, be it ever so detailed, has to remain *schematic*. Its degree of determinateness will always be much lower than that of the real object. It contains "empty spots" (*Leerstellen*) like the white areas on a map which stand for unexplored territory. Needless to say, this ought not to be misconceived as a flaw in the nature of literature. On the contrary, it can function as a highly flexible and ingenious artistic device.

By its own nature, literary presentation, which uses words, expressions, sentences, has to limit itself largely to essential features. Although a word is a type when isolated from other words (when we say "tree," we mean only the general notion, not a particular individual tree), the words in a literary text form a context, i.e., they determine and thereby individualize one another until — at the end of the text — a relatively complex and highly interrelated structure has been constituted that manifests an individualized content and its meaning. A simple experiment suffices to illustrate this point. If one starts reading a novel which he has never read before and stops after four or five lines, he finds in his memory only a very general and vague impression of a few partial outlines which are almost meaningless without any further determination or specification. These outlines call for a certain measure of "filling" with individualized content. Ingarden's comparison between reality and literature focuses on the perceivable physical object. What he points out, however, is equally applicable to nonphysical "objects" (feelings, thoughts) or partly physical and partly spiritual entities like characters.

Looking once more at the transformational process that leads from an eidetic conception in the mind of a writer to the written text, we must consider that inevitably something is "lost" during that process. The written manifestation can only come close to but never fully "cover" the partly prelinguistic conception, a fact which is all too familiar to everyone who writes. One need only be reminded of the "that's it!" experience when one finds the right word or expression for a thought. During the above-mentioned transformational process, both individual and general essence are translated into the written word. An individual and concrete essence will tend to generate a text which will itself bear a concrete and individual character (for example, by means of using a relatively complex and detailed type of imagery), whereas a general essence will require a more general language unless the general essence transforms itself into an individualized exemplification before it is textualized. The number of possible integral texts which manifest a given eidetic conception (or at least part of it) seems to decrease the more concrete the underlying essence is. A highly individualized and concrete eidos will seek an equally unique and individualized textual manifestation, and the choice between possible creative manifestations *qua* total and comprehensive manifestations appears to be limited. At the same time, texts with a highly concrete content allow a high degree of variability of *non*essential textual elements. The fringe, so to speak, of textual material that could be replaced with something similar (be it by the reader or by the writer) grows proportionally to the increase in concreteness and individuality on the part of the eidos. The essentials of the text, however, which are most intimately bound up with its eidetic "deep-structure," remain outside the scope of variability.

In a long and complex text, the directness of the understanding of a general meaning has been replaced by the indirect nature of a composite individualized meaning. To illustrate this point, we might say that a sentence like "Let each son always nourish this great virtue to shine forth through gentleness, a loving spirit, deeds of kindness, and a most fervent piety and devotion to his God!"[22] is directly grasped *in its generality*, whereas the comprehensive meaning of a character in a play or in a novel is distilled with greater difficulty und uncertainty from a given number of textual constituents. In the latter case, which also involves the process of extracting *implicit* meaning, a certain degree of hermeneutical ambiguity will always have to be reckoned with.

We have just explored the three-fold relation — eidetic conception, text, reader — as a process initiated by the conception and moving toward the medium of the text which in turn is aimed at the reader's mind. Let us now start with the reader who has to work his way "back" toward the author's conception to the extent in which this conception has become textual *Gehalt*. But he can go one step farther. He can transgress this *Gehalt* toward an eidos which is not fully textualized but construable from the "clues" offered by the text. If the conceived essence is of an individual rather than general nature, the reader reaches it by means of a double mediating process which *typefies*. First, there is the word as type (see above). Second, the act of cognition is based, according to Husserl, on a function he terms *habitus*. In experiences of any kind, we always fall back on previous cognitive acts and on the patterns which they have established. Our ego contains a large number of cognitive "sediments." These function as molds into which our acts of consciousness cast any new data or experiences. Thus, every act of cognition is grounded in "habit." Nothing radically new or unfamiliar enters our mind.

As the reader reenacts in his own mind the gradual constitution of meaning-complexes by means of type-words (and type-expressions), he carries out three different operations. First, he directly "picks up" a single meaning conveyed by a type-word or type-expression. Second, in the case of a coherent string of such words or expressions, he *abstracts* an all-comprehensive meaning from all meaning-constituents, a meaning inherent in *each* of those constituents. For the latter (be they ever so varied) contain a common core of meaning that can be identified and conceptualized.

However, this process of abstraction plus identification describes only one possible strategy as to understanding the meaning of a multiplicity (a "string") of coherent linguistic units. The reader may also be confronted with a sequence of "carrier"-meanings which do not necessarily have a common element but which *generate in their entirety* a meaning not contained in the individual words or expressions of that sequence. The meaning produced by the entire sequence transcends each of the constituting parts of the sequence. In an oversimplified manner, the aforementioned abstraction plus identification method can be represented in the following symbolic notation (X standing for a variety of words or expressions, the subscripts for elements of the meaning inherent in them):

$$X_{yab} - X_{ycd} - X_{yfg} - X_{yhi} \longrightarrow y \text{ (abstracted meaning)}$$

The generative process could then be described by the following formula:

$$X_{ab} - X_{cd} - X_{ef} - X_{gh} \longrightarrow z \text{ (new generated meaning)}$$

An example for abstracted meaning would be:

> He writes, he discusses, he philosophizes, he does research
> ⟶ he engages in intellectual activities

A simplified example for generated meaning might be:

> He drank only three glasses of water per week. He ate only one slice of bread per day. He slept only three hours per night. Every four hours he prayed and castigated his body.
>
> ⟶ he was an ascetic

The identification of y in the first formula and the equivalence of $a+b+c+d+e+f+g+h$ and z in the second formula depends on the extralinguistic and historically conditioned "repertoire" of meanings present in both the author's and the reader's minds.

A preliminary formal categorization of textual meaning-constituents which build up a comprehensive textual meaning will be a useful tool for literary interpretation. Whether the majority of words or expressions in a given text or textual segment represent emotions, rational thought, descriptions of nonhuman material being, etc. can be highly significant as to the nature of that text. The categories presented here are based on ontological, logical, and aesthetic functional principles. A description of any literary text will register the preponderance or the lack of certain categories of meaning, their distribution across phases of the text (equal distribution versus sporadic accumulation). Also, the *degree* of determinateness or indeterminateness of complexes of meanings (characters, places, action, etc.) has to be observed carefully.

We propose the following scheme:

human (individual-collective)
nonhuman (individual-collective)
(organic-inorganic)

activity	reason	positive
event	emotion	negative
situation	will	neutral
relation (+space/time)	sensory perception	
species/type		
essence		
quality		

intentional (designating)	"real"	
linguistically immanent	imaginary	
extratextual (real)	symbolic	
idiosyncratic (intratextual)	potential	
constituted	conditional	modality
constituting	negating	
	hypothetical	
	interrogative	
	doubtful	

The above categories and their (unsystematic) grouping are of a very tentative nature and require further clarification and explanation. The category "essence" defines meanings pertaining to the true and essential nature of a given referent (for example, a landscape, a character, an event) in contrast to any peripheral meaning. That Hans Castorp in Thomas Mann's *The Magic Mountain* is fascinated by death, is essential to the meaning-complex "Hans Castorp." The "fact" that he studied at the technical institutes of Danzig, Braunschweig, and Karlsruhe does not constitute an indispensable and central feature of his total being.

The category "real" is placed between quotation marks, since it refers to fictitious reality and not to extraliterary historical reality. Within the scope of a literary text there can be a rich and subtle gradation from meanings presented as "real" to others which are explicitly or implicitly characterized as the product of imagination, dream, or vision.

The categories "positive," "negative," and "neutral" apply to evaluative meanings of any kind (ethical, aesthetic, etc.). The opposites "intentional" versus "linguistically immanent" are derived from the following observations. Normally, in literary texts of any genre or any historical epoch, language has an intermediary function. It designates "something" other than itself. The words and expressions act as signs which "stand for" something (be it a fictitious object or an existential state of mind). In that sense, language can be called "intentional." There are, however, literary works (especially certain recent works of poetry) the language of which abandons its medial nature. In such texts, words relate to nothing outside the dimension of language. The linguistic elements still mean something for themselves but they designate nothing. The "immanent" interplay of words and phrases detached from their intentional correlates is totally self-sufficient.

It is a fact well known to historians and theoreticians of literature that literary texts may contain nonfictitious elements taken from historical reality or other extratextual spheres of meaning such as myth or fairy tale.

Elements of this sort are then not part of the poetic fabric invented by the writer. They are, so to speak, "prefabricated" and integrated as such into the texture of a literary work.[23]

The category "idiosyncratic" applies whenever the semantic content of a word or expression (its "semes"[24] or semantic features) deviates from the "ordinary" lexical meaning, i.e., when a context establishes a special — and sometimes genuinely unique — meaning of a word. The meaning of the adjective "blue" in Gottfried Benn's poetry, for example, differs substantially from the meaning of the same word in the poetry of the German postromanticist Joseph Freiherr von Eichendorff. Thus many texts have their own semantic code which can be deciphered only by means of a careful comparison of the contexts in which a given word stands.

The categories "constituted" versus "constitutive" differentiate between meanings which have a subordinate or "carrier"-function vis-à-vis a superior and more complex meaning. In other words, a comprehensive complex meaning in a literary text will always be the composite product of a certain number of constitutive meanings. In a work of fiction, single actions as well as descriptions of space can thus be regarded as constitutive elements. The composite meaning of a character, for example, is recreated in the reader's mind by means of both the abstracting and the generative process (see above). In the case of the abstracting process, there have to be identical semes (basic elements of meaning) in all of the meaning-units from which a common meaning is abstracted. If a character in a play is supposed to be "evil," he must be presented to the audience (reader) as the doer of unethical deeds.

Finally, constituted meaning does not necessarily have to be explicitly formulated in the text itself. Both the abstracting and the generative process of interpretation may have to extract an implicit sense from explicit meaning according to the formula "a means b." First of all, a means whatever meaning-content it stands for, but it also implies the (unexpressed) meaning b. In the abstracting process, significant identical semes of constitutive meanings will be made explicit in interpretation, whereas in the generative process they may remain unexpressed.[25]

NOTES

[1] Cf. Christoph Eykman, *Phänomenologie der Interpretation* (Bern: Francke, 1977).
[2] W. K. Wimsatt, *The Verbal Icon: Studies in the Meaning of Poetry* (Lexington: University of Kentucky Press, 1954), pp. 3–18.

EIDETIC CONCEPTION 453

[3] Ibid., pp. 3f.
[4] Cf. Fritz Kaufmann, 'Das Bildwerk als ästhetisches Phänomen,' in *Das Reich des Schönen: Bausteine zu einer Philosophie der Kunst* (Stuttgart: Kohlhammer, 1960), pp. 20f. Kaufmann defines the nature of art as *Wesensschau*. We claim that this definition is also applicable to literature.
[5] Edmund Husserl, *Ideen zu einer reinen Phänomenologie und phänomenologischen Philosophie*, vol. 1, *Allgemeine Einführung in die reine Phänomenologie*, ed. Walter Biemel (The Hague: Nijhoff, 1950), p. 142.
[6] Edmund Husserl, *Ideen zu einer reinen Phänomenologie und phänomenologischen Philosophie*, vol. 3, *Die Phänomenologie und die Fundamente der Wissenschaften*, ed. M. Biemel (The Hague: Nijhoff, 1952), pp. 85f.
[7] Husserl, *Ideen*, I, p. 143.
[8] Edmund Husserl, *Phänomenologische Psychologie: Vorlesungen Sommersemester 1925*, ed. Walter Biemel (The Hague: Nijhoff, 1962), p. 92.
[9] Cf. A.-T. Tymieniecka, 'Eidos, Idea, and Participation: The Phenomenological Approach,' *Kant-Studien*, 52 (1960–1961), 59–87.
[10] Husserl, *Ideen*, I, p. 31.
[11] Cf. Edmund Husserl, *Cartesianische Meditationen: Eine Einleitung in die Phänomenologie*, ed. Stephan Strasser (The Hague: Nijhoff: 1963), pp. 82, 84.
[12] Edmund Husserl, *Analysen zur passiven Synthesis (1918–1926)*, ed. Margot Fleischer (The Hague: Nijhoff, 1966), p. 103.
[13] Edmund Husserl, *Logische Untersuchungen*, vol. 2, *Untersuchungen zur Phänomenologie und Theorie der Erkenntnis: I. Teil* (Tübingen: Niemeyer, 1968), p. 53.
[14] Husserl, *Cartesianische Meditationen*, p. 105.
[15] Roman Ingarden, *Das literarische Kunstwerk* (Tübingen: Niemeyer, 1965), p. 112.
[16] Fritz Kaufmann, pp. 52f.
[17] Husserl, *Cartesianische Meditationen*, p. 105.
[18] L. S. Vygotsky, *Thought and Language*, ed. and trans. Eugenia Haufmann and Gertrude Vakar (Cambridge.: M.I.T. Press, 1975), pp. 148–50.
[19] Cf. Hans-Georg Gadamer, *Wahrheit und Methode: Grundzüge einer philosophischen Hermeneutik* (Tübingen: Mohr, 1965), pp. 279, 289, and passim.
[20] Sophocles, *Antigone*, in *Greek Drama*, ed. Moses Hadas (New York: Bantam, 1968), p. 110.
[21] See note 15, above.
[22] Gotthold Ephraim Lessing, *Nathan the Wise*, trans. Walter Frank and Charles Ade (Woodbury, N.Y.: Barron's Educational Series, 1972), p. 90.
[23] Cf. Christoph Eykman, 'Erfunden oder vorgefunden? Zur Integration des Ausserfiktionalen in die epische Fiktion,' to appear in *Neophilologus*, Spring 1978.
[24] For the term "seme" cf. Algirdas Julien Greimas, *Sémantique Structurale: Recherche de Méthode* (Paris: Librairie Larousse, 1966). In recent linguistic research (especially in the new field of text semantics), methods are being developed which bear a certain resemblance to phenomenology, for example, the differentiation between deep structure and "surface text" (a "deep," i.e., unexpressed, meaning-structure can "generate" various "surface texts"). Cf. Teun A. van Dijk, *Beiträge zur generativen Poetik*, Grundfragen der Literaturwissenschaft, vol. 6 (Munich: Bayrischer Schulbuchverlag, 1972).
[25] Note that several categories of the scheme are, of course, applicable to one textual meaning at the same time.

ANNEX

Programs of the Seminars and Conferences (1974–1980)
of
THE INTERNATIONAL SOCIETY OF PHENOMENOLOGY AND
LITERATURE
(organ of *The World Institute for Advanced Phenomenological Research and Learning*, Belmont, Mass.)
from the research work of which the studies here published have been selected

THE INTERNATIONAL HUSSERL AND PHENOMENOLOGICAL
RESEARCH SOCIETY
an independent forum for specialized study of phenomenology
SOCIETÉ INTERNATIONALE POUR L'ÉTUDE DE HUSSERL
ET DE LA PHÉNOMÉNOLOGIE INTERNATIONALE
FORSCHUNGSGESELLSCHAFT FÜR PHÄNOMENOLOGIE

ANNOUNCEMENT

of the Foundation of

THE INTERNATIONAL ASSOCIATION OF PHILOSOPHY
AND LITERATURE

Dear Colleague:

We wish to inform you that on May 8, 1974 THE INTERNATIONAL ASSOCIATION OF PHILOSOPHY AND LITERATURE are inaugurated.

This Association was founded in response to the widely felt need for close contact and understanding between the two disciplines. Among those who attempt to make the teaching of philosophy accessible to the contemporary student, or who try to make the understanding of literature relevent to the needs of man, there is a growing recognition that philosophy and literature do, indeed, share basic interests and preoccupations. To investigate and elucidate them constructively in a direct exchange will be the principal aim of the Association.

The major common concerns that the Association proposes to investigate at the outset are:

(1)   The image of man in literature
(2)   The writer and the critic in their cultural world
(3)   Literary analysis and phenomenological aesthetics

The first general meeting of the Association is planned for the winter of 1975 in the United States. Suggestions and proposals concerning the aims and program of the Association are urgently requested.

The Association is affiliated with THE INTERNATIONAL HUSSERL AND PHENOMENOLOGICAL RESEARCH SOCIETY; however, all philosophical perspectives and all literature are welcome.

The Executive Committee is, in its present initial stage, composed of:

ANNEX 457

Jean-Jacques Demorest,	Department of Romance Languages, Harvard University
Claude Levesque,	Department of Philosophy, Université de Montréal
Waldo Ross,	Department of Spanish, Université de Montréal
Anna-Teresa Tymieniecka,	Secretary General of the IHPRS

If you are interested in joining the Association or being entered on the mailing list, please write to Professor Demorest, Department of Romance Languages, Harvard University, Cambridge, Mass. 02138. Any sum you would be willing to contribute to the Association to allow it to meet the expenses of correspondence and of planning its first congress will be gratefully received.

We trust that you share our enthusiasm for the prospects of our Association and we look forward to your active participation. The cause is unimpeachable.

THE INTERNATIONAL ASSOCIATION OF PHILOSOPHY
AND LITERATURE

INAUGURAL MEETING

Harvard Divinity School, 45 Francis Ave., Cambridge

May 14–16, 1976

Conference sponsored by THE WORLD PHENOMENOLOGY INSTITUTE

TOPIC

IMAGES OF THE HUMAN BEING: Pessimism vs. Optimism

PROLOGUE, Friday, May 14, 1976

7:00 p.m. — Welcome by Dean Krister Stendahl
7:15 p.m. — Remarks by Professor J.-J. Demorest
followed by a reception at the Jawett House

OPENING SESSION
Saturday, May 15, 1976

9:00 a.m.

A.-T. Tymieniecka — Pessimism vs. Optimism in the Human Condition and the Creative Spontaneity
The World Phenomenology Institute

A. M. Vazquez-Bigi — Images of the Human Being – A Survey
U. of Tennessee

SESSION II
THE DIVIDED SELF

Hugh Silverman — Beckett, Philosophy and the Self
SUNY at Stonybrook

Martin Dillon — Sartre's Inferno
SUNY at Binghamton

Christoph Eykman	'Souvenir' and 'Imagination' in the Works of Rousseau and G. de Nerval
Robert C. Carroll *University of Maine*	The Dualistic Humanism of Gerard de Nerval

SESSION III
THE UNITY OF MAN WITHIN THE COSMOS AND NATURE

Benjamin I. Schwartz	The Affirmation of the Cosmos and the Chinese Approach to Man
Jeffner Allen *U. of Florida, Gainesville*	Homecoming in Heidegger and Hebel

SESSION IV
THE SEARCH AFTER THE KEY TO THE HUMAN CONDITION

John R. Maier *SUNY at Brockport*	The Epic of Gilgamesh
Solomon Lipp *McGill University*	The 'Little Man' in Paradise
Jean Murray *Rosary College*	Bernanos and the Prophet of Hope
M. Mendoza *Stephen F. Austin State U.*	The Myth of the Future and the Human Being as the Person in Latin American Literature

Dinner together in Rockefeller Hall
followed by "Business Meeting" chaired by J.-J. Demorest

SESSION V
Sunday, May 16, 1976
THE UNITY FROM WITHIN

Beverly Ann Schlack *The New School, New York*	Woman as Hero, or Virginia Woolf on Creativity

James B. Sipple
Millersville College

The Image of Man as Daimon: The Destructive and Creative Power According to D. H. Lawrence

Jorge Escorcia
West Chester State College

Bradomin: Four Images of Man Through Time; The Unifying Factor of Love

Douglas G. Stewart
Brandeis University

Friendship in Plato's View of Socrates

SESSION VI
MAN AS TRANSCENDING BEING

Brita Stendahl
Cambridge, Mass.

Kierkegaard on Repetition – The Novelist and the Interest of Metaphysics

Michel Masson
Harvard University

Repetition in and out of the Western Tradition

Veda Cobb
U. of Lowell

Amor Fati and the Eternal Return in Nietzsche

SESSION VII
THE UNIQUE VERSUS THE UNIVERSAL IN HUMAN EXPERIENCE

T. R. Hartland
SUNY at Albany

Glorification of the Human Condition in Religious Literature

Reinhard Kuhn
Brown University
Providence, R.I.

The Enigmatic Child – or the Privileged Phase of Existence

Peter Salm
Case Reserve U.
Cleveland, Ohio

The Discomfiture in Literature and the 'Eternal Moment'

THE INTERNATIONAL ASSOCIATION FOR PHILOSOPHY AND LITERATURE

SECOND ANNUAL CONVENTION

Sponsored By
THE WORLD INSTITUTE FOR ADVANCED PHENOMENOLOGICAL
RESEARCH AND LEARNING
In collaboration with
The Department of Germanic Studies, Boston College

May 1–4, 1977

Murray Conference Room, McElroy Building, Boston College
Chestnut Hill, Massachusetts

Theme: Nature, Man, and the Literary Work of Art

CONVENING OF THE PARTICIPANTS, SUNDAY, MAY 1, 1977

6:00 p.m. — Welcome by the academic authorities and Professor Christoph Eykman, President of the Conference, and Professor Anna-Teresa Tymieniecka, Program Chairman

OPENING SESSION
Monday, May 2, 1977

Morning Session
9:15 a.m.–12:45 p.m.

Introductory Lecture

Anna-Teresa Tymieniecka Creative Apperception and Human Re-
World Phenomenology Institute ality: *The Phenomenological "Return to Experience Itself"*

Nature Within and Without

Presided by Cécile Cloutier, *University of Toronto*

Margaret Collins Weitz
Harvard University

The Pastoral Paradox

Edwin Cranston
Harvard University

Pastoralism in Early Japanese Poetry

Afternoon Session
2 p.m.–5 p.m.

E. F. Kaelin
Florida State University

The Inversion of Values: The "Unhappy Consciousness" in the Works of Samuel Beckett

Gary Shapiro
University of Kansas

The Owl of Minerva and the Colors of the Night

Contribution to the Debate

Brian T. McDonough
Duquesne University

Values of Life: From Nietzsche to Kazantzakis

SESSION II
Tuesday, May 3, 1977

Philosophical Clarification of the Literary Means and Ends

Morning Session
9:15 a.m.–12:45 p.m.

Presided by Jacques Taminiaux, *Husserl Archives, Louvain*

Christoph Eykman
Boston College

Literary Hermeneutics in the Husserlian Perspective

Contributions to the Debate

Georges Grabowicz
Harvard University

Ingarden's Structural Approach to Literature

Ramona Cormier
Bowling Green University

Sartre on the Reader-Writer Relationship

Afternoon Session
2 p.m.–5 p.m.

Presided by Giancarlo Rota, *Massachusetts Institute of Technology*

Morris Weitz Philosophy and Literature: Implacable
Brandeis University Enemies?

Mary Mothersill Toward a Foundation for Aesthetic
Barnard College Theory: The Beautiful

Contribution to the Debate

Heni Wenkart Santayana on the Beautiful
Harvard University

7:30 p.m.: Reception for all participants offered by The World Institute for Advanced Phenomenological Research and Learning, 348 Payson Road, Belmont, Mass.

SESSION III
Wednesday, May 4, 1977

The Transformation of Experience in the Literary Endeavor

Morning Session
9:15 a.m.–12:45 p.m.

Presided by Richard Stevens, *Boston College*

Jean Bruneau The Real World and the World of
Harvard University Balzac

Wictor Weintraub Jan Kochanowski: Religious Rational-
Harvard University ism in Poetry

Contribution to the Debate

Veda Cobb Reason and Feeling
Lowell University

Afternoon Session
2 p.m.–5 p.m.

Alonzo Lingis　　　　　　　　The Ritual Drama of Balinese Culture
Pennsylvania State University

David Norton　　　　　　　　Nature and Personal Destiny
University of Delaware

INTERNATIONAL ASSOCIATION FOR PHILOSOPHY AND
LITERATURE AFFILIATED WITH THE WORLD PHENOMENOLOGY
INSTITUTE, BELMONT, MASSACHUSETTS

PHILOSOPHY AND LITERATURE RESEARCH SEMINAR

March 16, 17 and 18, 1978

The World Phenomenology Institute, 348 Payson Road, Belmont, Mass. 02178

THEME: PHENOMENOLOGICAL FOUNDATIONS OF THE LITERARY STUDIES

March 16

Morning Session
9:30 a.m. – 12:15 p.m.

Eugene Kaelin, *Univ. of Florida, Tallahassee*	The Dispute between Phenomenological Aesthetics and Structuralism

Afternoon Session
3:00 p.m. – 5:30 p.m.

Anna-Teresa Tymieniecka *The World Phenomenology Institute*	The So-called "Truthfulness" of the Literary Work according to Ingarden, and the Lyrical Subject

March 17

Morning Session
9:30 a.m. – 12:15 p.m.

Robert Magliola *Purdue University*	The Problem of Truth in the Literary Work according to Ingarden and Heidegger

Afternoon Session
3:30 p.m. – 5:30 p.m.

General Debate

March 18

Morning Session
9:30 a.m. – 12:15 p.m.

John Hoberman Psychological Problems in the Literary
Harvard University Analysis: Søren Kierkegaard

PUBLIC INVITED

THE INTERNATIONAL ASSOCIATION FOR PHILOSOPHY AND
LITERATURE
Affiliated with the World Phenomenology Institute

THIRD ANNUAL CONVENTION

Sponsored by
THE WORLD INSTITUTE FOR ADVANCED PHENOMENOLOGICAL
RESEARCH AND LEARNING

March 22–24, 1978

Cronkhite Hall, Harvard University, 6 Ash St., Cambridge, Mass.

Theme

THE HEROIC ELEMENT IN LITERATURE AND THE AUTHOR'S
COMMITMENT

(Program Chairman: Eugene Kaelin)

FIRST SESSION
Wednesday, March 22
9:30 a.m.–12:45 p.m.

Opening Address: Eugene Kaelin

Presided by: Christopher Eykman, Boston College
Anna-Teresa Tymieniecka

Introductory Lecture

Anna-Teresa Tymieniecka	The Literary Paradox and the Writer's
World Phenomenology Institute	Dilemma

Conceptions of Heroism in Literature

Margaret Collins Weitz	The Epic Hero: Neither Sinner nor
Harvard University	Saint

A. M. Vasquez-Bigi
University of Tennessee,
Knoxville

The Heroic Involvement of Cervantes

Afternoon Session
2:00 p.m.–5:00 p.m.

Robert Magliola
Purdue University

Permutation and Meaning: a Heideggerian *Troisième Voie*

James Sipple
Millersville College

Guernica and the Heroic Response

Dan Vaillancourt
Mundelein College
Chicago

The Battle with Alienation and the Twentieth-Century Heroes

SESSION TWO
Thursday, March 23, 1978

Morning Session
9:30 a.m.–12:45 p.m.

The Literary Message and The Social World

Presided by: Betty T. Rahv, *Boston College*

Benjamin Schwartz
Harvard University

The Question of the Autonomy of Literature

William Cloonan
Brandeis University

The Author as a Hero: Celine

Hans Rudnick
Southern Illinois University

The Hero in Contemporary Literature and His Attitude towards Responsibility

Afternoon Session
2:00 p.m.–5:00 p.m.

Personal Search for the Significance of Life and Its Literary expression

Presided by: A. M. Vasquez-Bigi, *U. of Tennessee*

Beverly Ann Schlack *Cooper Union, New York*	Heroism and Creativity in Literature – Some Ethical and Aesthetic Aspects
Mark Sheldon *Indiana University* *Ft. Wayne*	The Search for Wisdom in Philosophy and Literature
Joan Williamson	The Problem of Suicide

7:30 p.m.: Reception for all participants offered by the World Institute for Advanced Phenomenological Research and Learning, 348 Payson Road, Belmont, Mass.

SESSION THREE
Friday, March 24, 1978

The Author and His Work

Morning Session
9:15 a.m. –12:45 p.m.

Presided by: Hans Rudnick, *Southern Illinois University*

Gila Ramras-Rauch *Bar-Ilan University and* *Brandeis*	Protagonist, Reader and the Author's Commitment
John Hoberman *Harvard University*	Heroism in Literature and the Author's Involvement: Søren Kierkegaard
Joan E. Emma *Windham College* *Putney, Vt.*	The David Myth and the Conception of Power in the Judaic Tradition

INTERNATIONAL SOCIETY FOR PHENOMENOLOGY AND
LITERATURE AFFILIATED WITH THE WORLD PHENOMENOLOGY
INSTITUTE, BELMONT, MASSACHUSETTS

PHENOMENOLOGY AND LITERATURE RESEARCH
SEMINAR

April 4, 5, 6, 7 and 8, 1979

The World Phenomenology Institute, 348 Payson Rd., Belmont, MA 02178

Program Chairman: Robert Magliola, *Purdue University*

THEME: FEELING AND FORM

April 4

Afternoon Session
3:00 p.m.–5:30 p.m.

Conducted by Professor A.-T. Tymieniecka, *W. P. I.*

Alan M. Hollingsworth *Michigan State University*	On Roman Ingarden and Louise Rosenblatt

April 5

Morning Session
10:00 a.m.–12:30 p.m.

Conducted by Professor Eugene H. Falk
University of North Carolina, Chapel Hill

Afternoon Session
3:00 p.m.–5:30 p.m.

Hans Rudnick *Southern Illinois University*	On Modern "Systems Analysis" and Ingarden: An Alternative Approach To Feeling and Form

April 6

Morning Session
10:00 a.m.–12:30 p.m.

Morris Weitz　　　　　　　　　　Shakespeare and Idea of Perfection
Brandeis University　　　　　　　(*Othello* and *Cleopatra*)

Afternoon Session
3:00 p.m.–5:30 p.m.

Jacques Garelli　　　　　　　　　Le Temps D'Un Désir
New York University

General Debate

April 7

Morning Session
10:00 a.m.–12:30 p.m.

Robert Magliola　　　　　　　　Heidegger and *La Critique Thématique*:
Purdue University　　　　　　　Can They Accommodate Form and Feeling?

Afternoon Session
3:00 p.m.–5:30 p.m.

General Debate

THE INTERNATIONAL SOCIETY FOR PHENOMENOLOGY AND LITERATURE
affiliated with the World Phenomenology Institute

FOURTH ANNUAL CONVENTION

sponsored by
THE WORLD INSTITUTE FOR ADVANCED PHENOMENOLOGICAL RESEARCH AND LEARNING

April 9–11, 1979

All meetings are to be held in:
Cronkhite Hall, Harvard University, 6 Ash St., Cambridge, Mass.

THEME: NATURE AND FEELING

Program Chairman: Eugene Kaelin
Florida State University

FIRST SESSION
Monday, April 9
9:00 a.m.–1:00 p.m.

Opening Address: Eugene Kaelin
Florida State University

Presided by: Anna-Teresa Tymieniecka
The World Phenomenology Institute

9–10:15	A. M. Vasquez-Bigi *University of Tennessee Knoxville*	The Present Direction of the Phenomenological Search in the Light of Schiller's Departure from Kant
10:30–11:45	Jesse Gellrich *University of Santa Clara*	The Book of Nature, Mythology and Fiction
11:45–1:00	Don Castro *Notre Dame University*	The Phenomenological Approach to the Concept of Genre

ANNEX 473

Afternoon Session

Presided by: John Hoberman
Harvard University

2–3:15 George W. Linden — Metaphor and Metaphysics, East and West
Southern Illinois University Edwardsville, Illinois

3:30–5:00 Peter Stowell — Phenomenology and Literary Impressionism: The Prismatic Sensibility
Florida State University

Evening Session

Presided by: Margaret Collins Weitz
Harvard University

8:30–10:00 David N. Dobrin — Blake, Mozart, Gauss: The Creative Act and the Beginnings of Romanticism
Miami University Oxford, Ohio

SECOND SESSION
Tuesday, April 10

Morning Session

Presided by: Eugene Kaelin

9–10:15 Hans Rudnick — Nature and Feeling: The Constitutive and the Subjective
Southern Illinois University Carbondale, Illinois

Beverly Ann Schlack — Nature and Feeling in Virginia Woolf's "The Waves"
Marymount College

10:30–12:00 Carter Martin — Commitment as a Phenomenological *Vorhabe*: Flannery O'Connor
University of Alabama Huntsville, Alabama

Afternoon Session

Presided by: Veda Cobb
Lowell University

2–3:15	Jeanne Ruppert *Florida State University*	Nature and the Feeling Response in Wallace Stevens
3:30–5:00	Keith Gould *Norwich University*	The Aesthetic Approach: Some Thoughts on Wordsworth and Hartshorne
7 p.m.	Banquet Cronkhite Hall	

THIRD SESSION
Wednesday, April 11

Morning Session

Presided by: Jean Bruneau
Harvard University

9–10:15	Peter McCormick *University of Ottawa*	Literary Truths and Metaphysical Qualities
10:30–12	Anca Vlasoposos *Wayne State University*	The Ruling Passion as Key to the Artist
12–1:15	William S. Saunders *University of Iowa*	"Prometheus"

Afternoon Session

2–3:15	Sura Prasad Rath *Tulane University*	Playing Is Knowing
3:30–5	Robert Shearer *Florida State University*	Donald Barthelme and the Literature of Liberation
Closure:	Anna-Teresa Tymieniecka	

INTERNATIONAL SOCIETY FOR PHENOMENOLOGY AND
LITERATURE AFFILIATED WITH THE WORLD PHENOMENOLOGY
INSTITUTE, BELMONT, MASSACHUSETTS

PHENOMENOLOGY AND LITERATURE RESEARCH
SEMINAR

March 23, 24 and 25, 1980

All meetings are to be held in:
Cronkhite Hall, Harvard University, 6 Ash St., Cambridge, Mass.

Theme

PHENOMENOLOGICAL, STRUCTURAL, AND POST-STRUCTURAL
CRITICISM

Program Chairman: Anna-Teresa Tymieniecka
World Phenomenology Institute

March 23

Morning Session
10:00 a.m.–12:30 p.m.

Anna-Teresa Tymieniecka *World Phenomenology Institute*	Philosophical Issues and Literary Criticism
Hans H. Rudnick *Southern Illinois University* *Carbondale, Illinois*	On Ingarden as a Literary Systematist

Afternoon Session
3:00 p.m.–5:30 p.m.

John Fizer *Rutgers University*	Ingarden's and Mukarovsky's Binonal Definition of the Literary Work of Art: A Comparative View of Their Respective Ontologies

March 24

Morning Session
10:00 a.m.–12:30 p.m.

Marlies Kronegger　　　　　　　Phenomenological Approaches To
Michigan State University　　　Literature

Afternoon Session
3:00 p.m.–5:30 p.m.

Paul Privateer　　　　　　　　　Genevans and Deconstructionist Pluralism Beyond Crisis
University of California
at Davis

March 25

Morning Session
10:00 a.m.–12:30 p.m.

Jesse Gellrich　　　　　　　　　The Deconstruction in the Literary Theory: Ruturning to Mythology
University of Santa Clara

Afternoon Session
3:00 p.m.–5:30 p.m.

General Debate

THE INTERNATIONAL SOCIETY FOR PHENOMENOLOGY AND
LITERATURE
affiliated with The World Phenomenology Institute

FIFTH ANNUAL CONVENTION

sponsored by
THE WORLD INSTITUTE FOR ADVANCED PHENOMENOLOGICAL
RESEARCH AND LEARNING

March 28–30, 1980

All meetings are to be held in:
Gutman Library, Harvard University Graduate School of Education
6 Appian Way, Cambridge, Mass. 02138

Theme

THE EPIC GENRE AND MAN'S HISTORICITY

    Chairman: Eugene Kaelin
                          Florida State University

    Program Director: Anna-Teresa Tymieniecka
                          World Phenomenology Institute

FIRST SESSION
Friday, March 28
9:30 a.m.–1:00 p.m.

Introduced and Presided by: Eugene Kaelin
                                  Florida State University

Anna-Teresa Tymieniecka *World Phenomenology Institute*	Literary Genre and Man's Existential Quest
L. M. Findlay *Univeristy of Saskatchewan*	The Shield and the Horizon: Homeric Ekphrasis and History

                         Afternoon
                    2:30 p.m.–5:30 p.m.

Christopher Macann   Historicity in Martin Heidegger
World Phenomenology Institute

Gila Ramras-Rauch   The Hebraic Epic — The Myth of Man
Ohio State University

                       SECOND SESSION
                       Saturday, March 29
                       9:30 a.m.–1:00 p.m.

Introduced and Presided by: Richard Cobb-Stevens
                            Boston College

Benjamin Schwartz   Epic and History
Harvard University

Christopher Eykman   The Literary Diary as a Witness
Boston College   of Man's Historicity: Heinrich Boll,
                                Gunter Grass, Peter Handke, and Karl
                                Krolow

                         Afternoon
                    2:30 p.m.–5:30 p.m.

Beverly Schlack Randles   Historicity of the Human Existence in
New York   Unfolding Virginia Woolf's Work

Carol Hurd Green   Woman's Autobiography: Search For
Co-Editor Notable Amer. Women Self
Radcliffe College

                       THIRD SESSION
                       Sunday, March 30
                       9:30 a.m.–1:00 p.m.

Introduced and Presided by: Betty Rahv
                            Boston College

Patricia Lawlor Metaphor and The Flux of Experience
Tufts University

Susan K. Harris This Peace, This Deep Contentment:
Queens College of City of NY Images of Temporal Freedom in the
Writings of Mark Twain

Angel Medina Reflection, Time and The Novel
Georgia State Univ.

Afternoon
2:30 p.m.–5:30 p.m.

General Debate Introduced by and Presided by: James Iffland
Boston University

INDEX OF NAMES

Adamov, A 103
Adams, J. L. 242
Aeschylus 161
Alonso, D. 108
Apollinaire, G. 63, 90
Aragon, L. 61–62
Aristophanes 97
Aristotle 7, 18, 48, 50, 54, 69, 119, 176–77, 195, 205, 280, 292, 298, 331, 391, 393–94, 397, 399, 403
Ascanius, B. 280
Auden, W. H. 329, 339, 350–51
Auerbach, E. 84, 108–9, 116–19, 122–24
Austen, J. 162, 164

Balzac, H. de 52, 162, 291–99
Banville, T. de 59, 86
Baroja, P. 102
Barth, J. 103, 137, 150–51
Barthes, R. 159–60
Bataillon, M. 116, 123
Baudelaire, C. 59, 88, 291, 297–99
Beardsley, M. C. 443
Beckett, S. ix, 49, 103–4, 153–60, 163–65, 262
Béguin, A. 416, 426
Binswanger, L. 309
Blake, W. 61, 241, 249
Blanchot, M. 85–87
Blyth, R. H. 368
Boccaccio 281
Boehme, J. 218
Boëthius 281
Bonhoeffer, D. 224
Borges, J. L. 99, 103, 118, 121, 163, 203–11
Brandt, R. 147–48, 151
Brecht, B. ix
Brémond, H. 59–60, 86

Brooks, C. 213
Buber, M. ix
Burckhardt, J. 173
Brunet, J. 401–2, 405
Burns, R. 61
Byron, Lord 61, 335

Callow, P. 215, 241
Calvino, I. 52
Camus, A. ix, 158, 165, 184, 336
Carlyle, T. 62, 90, 331, 334–39
Carter, F. 239
Castro, A. 115, 117–18, 123–24
Cazamian, M. L. 120
Cervantes, M. 98, 105–24, 162, 277–79, 288, 334, 353
Chang Chung-yuan 366–67, 369, 382
Chaucer 247, 262
Chesterton, G. K. 98–99, 103–4, 120–21
Chomsky, N. 86
Coleridge, S. T. x, 5, 61–62, 89
Colletet, G. 280, 288
Conder, C. 99
Conrad, T. 74
Constans, F. 427
Croce, B. 4
Crocker, L. G. 426

Dante 15, 118, 281, 356
Danto, A. 188, 194, 200, 202, 313
Darwin, C. 215, 335
Davidson, J. 99
Defoe, D. 334
DeLaveney, E. 240, 242
Derrida, J. 87, 159, 380
Descartes, R. 153–54, 192
Descourzis, P. 123
Dickens, C. 162, 164, 331, 335
Diderot, D. 415, 418, 426

INDEX OF NAMES

Diogenes 119
Donne, J. 61
Dos Passos, J. 99
Dostoyevsky, F. 241, 339
Dowson, E. 99
Dufrenne, M. 361
Durry, M.-J. 427

Earle, W. 84
Eliot, G. 164, 335
Eliot, T. S. 5, 63, 432
Eluard, P. 61
Empson, W. 285, 288
Erasmus, D. 97, 112
Euripides 161
Eykman, C. 452–53

Feyerabend, P. 207
Fichte, J. G. 416
Fielding, H. 164
Flaubert, G. 55, 161–62
Fleming, W. 27
Føllesdal, D. 306
Fontenelle, C. 278, 288
Ford, G. 213, 218, 241–42
Forster, E. M. 336, 338
Fresnaye, V. de la 279, 288
Freud, S. 190, 200
Friedländer, P. 391, 405
Fromm, E. 216
Frye, N. 27, 50, 84–85, 87, 329, 332, 350

Gadamer, H.-G. 67, 364, 381, 400, 405, 453
Galsworthy, J. 22, 52, 55, 167
Garnett, E. 224
Gautier, T. 291, 297–98, 428
Geiger, M. 74, 92
Gelfant, B. 125, 134
Genet, J. 103
Genette, G. 49, 85
Gert, B. 144–48
Gide, A. 89, 245–46, 262–63
Gilson, E. 441
Goethe, J. W. 23, 50, 54, 85, 92–93, 108, 118, 248, 262–63, 277, 291, 337, 421, 423
Goodheart, E. 239
Grabski, J. 22
Green, O. H. 124
Greg, W. W. 278, 288

Hackforth, R. 401–3, 405
Haeckel, E. 215
Hankin, St. J. 99
Hansen, N. 204, 207
Hardy, T. 101, 103, 249–52, 263, 334
Hartmann, E. von 100–101, 104, 121
Hazard, P. 123
Hazlitt, H. 332
Hebel, J. P. 267–74
Hegel, G. W. F. 100, 134, 159, 165, 218, 272, 274
Heidegger, M. 9, 67, 71–72, 74, 92, 102, 159, 192–95, 199, 201, 204, 207–11, 267, 273–75, 309, 314, 354, 357–70, 373–74, 377–83, 385–89
Heraclitus 4, 198–99
Hesiod 4
Hesse, H. 125, 135
Hintikka, J. 306, 310
Hirsch, E. D. 372–74, 381–82
Hobbes, T. 344
Hoffman, F. J. 159
Hofstatter, A. 88
Hölderlin, F. 50, 61, 85, 120, 208
Homer 4, 161, 297–98, 391
Hope, L. 99
Horace 279–80
Hume, D. 153
Husserl, E. 12, 26, 155, 159, 301–11, 361, 380–81, 429–30, 441, 443–45, 447, 449, 453
Huxley, T. H. 215

Ingarden, R. 66, 68, 70–72, 74, 91–93, 380–82, 443, 445, 447, 453
Ionesco, E. 49, 163–64

Jakobson, R. 87–88, 91, 155, 160
James, H. 165, 246, 262–63, 331–32, 339, 374–75

INDEX OF NAMES

James, W. 440
Jarrett-Kerr, M. 214–15, 218, 241–42
Jaspers, K. 119
Jesperson, O. 155
Johnson, L. 99
Johnson, S. 278, 288
Joyce, J. 163, 165, 338, 344

Kaelin, E. 92
Kafka, F. 103, 118, 163–65, 167–69, 205, 252, 263, 353
Kant, I. 102, 122, 195, 206, 435–36
Kaufmann, F. 444–45, 453
Kaufmann, W. 126, 134
Keats, J. 61
Kierkegaard, S. 30, 84, 119, 156, 173, 177, 190, 205
Klee, P. 308–11
Klein, M. 125, 134
Krasiński, Z. 23
Kuhn, R. 83
Kuhn, T. 207

Lacan, J. 159
Laski, M. 440
Laudun d'Aigeliers, P. de 279, 288
Lawrence, D. H. 213–44
Leavis, F. R. 214
Leibniz, G. W. 102, 192
Leopardi, G. 62, 97
Lessing, G. E. 50, 85, 453
Levin, H. 285
Lewis, R. W. B. 213
Locke, J. 153
Lukes, S. 183

Magliola, R. 379, 385–88
Magnus, B. 201
Mainländer, P. 101
Maldiney, H. 307, 311
Malet, L. 101–2
Mallarmé, S. 59
Mann, T. 22, 52, 89, 122, 124, 256–63, 451
Manzoni 22, 80
Marcel, G. ix
Marquez Villaneuva, F. 124

Marsan, J. 278, 287
Marvell, A. 371–73
Marx, K. 134
Maslow, A. 181
May, R. 216–17, 232, 240–41
Miko, S. 218
Mill, J. S. 173, 188
Miller, H. 74
Millet, K. 238
Milton, J. 277–78, 283, 286, 346
Montaigne, M. de 173
Moore, G. 101
Moore, H. T. 242
Murdoch, I. 32
Murray, M. 92, 238

Nabokov, V. 163
Nerval, G. de 59, 415, 420–28
Niebuhr, R. 226, 229
Nietzsche, F. 4, 73, 97, 102, 104, 118, 160, 173, 177, 184–203, 209–10, 241, 274, 313–18, 353
Nobel, A. 133
Norton, D. L. 184

Oakeshott, M. 175, 181, 184
O'Connor, F. 353–56, 359, 364, 378–79, 383
Olson, E. 84
Olson, R. G. 151
Ortega y Gasset, J. ix, 4, 83, 175, 184, 310
Otto, R. 219–20, 225, 440

Panofsky, E. 285
Paracelsus 218
Parmenides 4, 197–98
Petrarch 62, 281, 283–84
Pindar 177
Plato 119, 176–77, 185, 195, 197, 199, 337, 345, 348, 391–405, 429
Plotinus 441
Poe, E. A. 59
Poggioli, R. 279, 288
Polanyi, M. 207
Poulet, G. 74, 92
Pound, E. 303

INDEX OF NAMES

Predmore, R. L. 122
Proust, M. 42
Pynchon, T. 103

Racine, J. 255–56, 297
Renan, E. 238
Rennert, H. 282, 288
Rey, J.-M. 159–60
Reynier, G. 278, 287
Richard, J.-P. 427
Richards, I. A. 213, 223
Ricoeur, P. 67, 87
Rilke, R. M. 60, 89, 350–51
Rossman, C. 238–39, 242
Rousseau, J.-J. 415–22, 425–27

St. Augustine 434
Saint-Exupery, A. de 263
Saint-Simon, H. de 291
Santayana, G. 179–80, 321–26
Sartre, J.-P. ix, 84, 125, 134–35, 154, 159, 165, 173, 175–76, 181, 184, 214, 252–53, 255, 407–13
Saussure, F. de 155, 160
Scaliger, J.-C. 280
Scaramuzza, G. 92
Schacht, R. 134
Schärer, K. 427
Scheler, M. 430, 432
Schelling, F. W. J. 218, 416, 426
Schopenhauer, A. 99–102, 179–80, 193, 340
Schorer, M. 216
Scott, N. A. 229
Scott, W. 294, 334
Seiling, M. 121
Sextus Empiricus 205
Shakespeare, W. 174, 277, 291, 331, 333, 345–46, 349
Shaw, G. B. 99
Shoemaker, S. S. 154–55, 159
Sienkiewicz, H. 22
Simon, P.-H. 86
Sittler, J. 242
Smith, P. C. 201
Snell, B. 199
Socrates 31, 176–77, 391–404

Solzhenitzyn, A. 125, 135
Sophocles 161, 340, 343, 346, 349, 446, 453
Spengler, O. 97
Spitzer, L. 105, 108, 118, 121–22, 124
Staiger, E. 74, 92
Stambaugh, J. 200, 202
Sterne, L. 162
Stevens, W. 87, 331, 336–37
Stoppard, T. 164
Strindberg, A. 120
Sully, J. 100
Swedenborg, E. 120

Tagore, R. 125, 136
Taine, H. 291, 297–98
Teeter, L. 371
Thackeray, W. M. 162, 164, 335
Thales 199
Tillich, P. 217–18, 220, 242
Todorov, T. 51–52, 85
Tolstoy, L. 80
Toulmin, S. 207
Trilling, L. 221
Trollope, A. 335
Tymieniecka, A.-T. 83, 85, 91, 97–98, 102, 453

Unamuno, M. de ix, 4, 336
Unger, R. M. 175, 184

Valéry, P. 5, 30, 59, 168
Valori, P. 440
Verlaine, P. 60, 88
Vico, G. 434
Virgil 255, 280–81, 285
Voltaire, F. M. A. 415, 426
Vonnegut, K. 103
Vygotsky, L. S. 445, 453

Wagner, R. 102
Wahl, J. 4
Warren, A. 85, 382
Watanabe, J. 88
Wedgwood, C. 278, 288
Wellek, R. 50, 66–67, 74, 85, 91, 371–73, 382

Wharton, E. 338
Whitehead, A. N. 42
Wilde, O. 337–38
Williams, B. 151
Wilson, C. 127, 134
Wimsatt, W. K. 443, 452
Wittgenstein, L. 277, 362, 381

Wojtyła, K. 135
Wolff, C. 102
Woolf, V. 23, 165, 343–51

Yeats, W. B. 350

Zeno 119, 205–6

ANALECTA HUSSERLIANA

The Yearbook of Phenomenological Research

Editor:
ANNA-TERESA TYMIENIECKA
*The World Institute for Advanced Phenomenological Research and Learning
Belmont, Massachusetts*

I. *Analecta Husserliana.* 1971.
II. *The Later Husserl and the Idea of Phenomenology. Idealism-Realism, Historicity and Nature.* 1972.
III. *The Phenomenological Realism of the Possible Worlds. The 'A Priori', Activity and Passivity of Consciousness, Phenomenology and Nature.* 1974.
IV. *Ingardenia. A Spectrum of Specialised Studies Establishing the Field of Research.* 1976.
V. *The Crisis of Culture. Steps to Re-Open the Phenomenological Investigation of Man.* 1976.
VI. *The Self and the Other. The Irreducible Element in Man, Part I: The 'Crisis of Man'.* 1977.
VII. *The Human Being in Action. The Irreducible Element in Man, Part II: Investigations at the Intersection of Philosophy and Psychiatry.* 1978.
VIII. Yoshihiro Nitta and Hirotaka Tatematsu (eds.), *Japanese Phenomenology. Phenomenology as the Trans-Cultural Philosophical Approach.*
Phenomenology as the Trans-Cultural Philosophical Approach. 1979.
IX. *The Teleologies in Husserlian Phenomenology. The Irreducible Element in Man, Part III: 'Telos' as the Pivotal Factor of Contextual Phenomenology.* 1979.
X. Karol Wojtyła, *The Acting Person.* 1979.
XI. Angela Ales Bello (ed.), *The Great Chain of Being and Italian Phenomenology.* 1981.
XII. *The Philosophical Reflection of Man in Literature.* 1982.
XIII. Eugene F. Kaelin, *The Unhappy Consciousness. The Poetic Plight of Samuel Beckett, An Inquiry at the Intersection of Phenomenology and Literature.* 1981.